The Natural History Of Living Mammals

The
Natural History
Of Living Mammals

by William Voelker

Plexus Publishing, Inc.
Medford, NJ

Dedicated to Betsy, Erica, and Christopher

What is man without the beasts? If all the beasts were gone, man would die from great loneliness of spirit, for whatever happens to the beasts also happens to man. All things are connected. Whatever befalls the earth befalls the sons of the earth.

—Suquamish Indian Chief Seathl, 1885.

TABLE OF CONTENTS

INTRODUCTION

Our knowledge of the animal class Mammalia has increased so vastly within the past few decades that it has become virtually impossible for a single book to cover that knowledge completely. Most mammal books are thus specialized, covering single species, families, or orders. Only a few books, like this one, try to touch upon certain aspects of all mammals.

No single person could possibly have witnessed or have had firsthand experience with all the facts and behaviors presented in this book. The book is possible at all only because thousands of individuals have spent long hours of work and hardship discovering the amazing lives of mammals.

Except in unusual cases, this book omits wordy physical descriptions, lists of habitats, gestation periods, and other factual data. Most of these facts, which often make for dull reading, are readily available in other popular works or encyclopedias. This book is in essence a comprehensive review of the popular literature published on mammal behavior. It tries to resolve the discrepancies and disagreements by using reputable sources and attempts to avoid perpetuating errors. Some errors undoubtedly remain, but only until the next edition.

The classification of mammals used here is loosely based upon the system arranged by George Gaylord Simpson in 1945. This volume closely follows the fourth edition of *Walker's Mammals of the World* with regard to scientific names and number of species. A few liberties have been taken with classification to make for smoother reading and arrangement. These will undoubtedly cause some taxonomists to cringe.

There are somewhere between four and five thousand species of mammals alive on Earth. This is a large number, but when compared to the number of all other animals, the class Mammalia represents less than half of one percent of all living creatures. Mammals differ from other animals in four basic characteristics (and in many minor ways): 1) Mammals use mammary glands to feed their young, which are born alive; 2) Mammals have hair on their bodies; 3) Mammals are homiothermic or "warm-blooded"; and 4) Mammals have a four-chambered heart. As we shall shortly see, there are occasional exceptions to even these basic characteristics defining mammals.

Since this book often deals with the unusual and spectacular, one should keep in mind that maximum, often record, figures are quoted and that these quite often do not indicate the animal's usual characteristics. More often, the typical individual of the species lives a much shorter time, eats less, grows to a smaller size, and so on. In generalizing about species, one should also remember that the behavior of many animals varies with geographic location and even in individual animals within the same area. Some of the specific behavioral acts recorded here may not be typical of the species as a whole, though they have been observed in at least one of its members at some time. Herein are many generalizations, and as we all know there are exceptions to every rule. One learns to avoid the words "always" and "never." The reader may soon become aware of and even annoyed at the frequent use of "usually," "sometimes," "occasionally," "rarely," etc. These qualifiers are necessary, however. Just as annoying may be the anthropomorphic and unscientific ascriptions of motives to many behaviors. They ease the writing and they do represent human explanations of behaviors. They may or may not be accurate, but most certainly animals do not reason their actions through as our theories often imply. The behaviors are usually innate or learned through trial and error.

A common problem one encounters in the study of animals is that of nomenclature. For example, the koala bear is also known as the bangaroo, buidelbeer, colo, cullawine, karbor, koolewong, koala wombat, narnagoon, native bear, and New Holland sloth. As one studies more and more, he finds that it is actually unusual for an animal not to have more than one common name. Carolus Linnaeus realized this hundreds of years ago and devised a system to give each animal a "scientific name" composed of two Latinized words. The system has been accepted the world over and is used in this book along with the better known common names. Since the same common name is frequently applied to different species of animals, the scientific name is crucial to positive identification. Thus the koala's positive identification is *Phascolarctos cinereus*. Unfortunately, there are often disputes over the classification of an animal as a true species or to its appropriate scientific name.

Being mammals ourselves, we find in this class of animals our closest relatives. Perhaps this explains the thrill we experience when siting a mammal in the wild—the rarer the animal, the greater the excitement. Even more rewarding are those rare instances when, unobserved, we are privileged to observe animal activities and behaviors in a truly natural state. This book is filled with thousands of such instances as witnessed by dedicated researchers. Although not quite as exciting as observing firsthand, these fascinating accounts of animal behaviors and their postulated motives will surely capture the interest and imagination of anyone pursuing the natural history of living mammals.

1

ORDER MONOTREMATA
(3 SPECIES)

This most primitive of mammalian orders is the Monotremata, composed of the echidnas and the platypus. These animals each possess a cloaca or "monotreme" as in birds and reptiles. A cloaca is a chamber into which the urinary, gastrointestinal, and reproductive tracts all open. The cloaca in turn communicates with the outside via a single orifice. Thus, these animals eliminate all their fluid and solid wastes, have sex, and, in the females, give birth through the same opening.

Unlike most other mammals, the monotremes lack vocal cords in their larynx and are unable to closely regulate their body temperature. Except for hibernating mammals, these strange creatures function with the lowest body temperatures of all mammals, ranging from 72 to 97 degrees Fahrenheit, depending on the species. The monotremes are also the only mammals which lay eggs instead of bringing forth their young alive.

Family Ornithorhynchidae (platypus—1 species)

The duck-billed platypus (*Ornithorhynchus anatinus*) is well adapted to its aquatic lifestyle. Like the beaver, it uses its flat tail as a rudder and to slap the water surface as a warning to others before diving. It

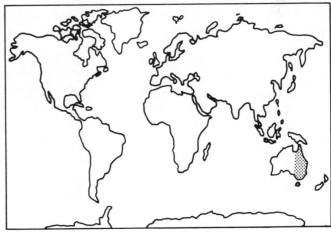

Family Ornithorhynchidae

Platypus (*Ornithorhynchus anatinus*)

walks on its knuckles to prevent damage to the extensive webbing of its feet. The eyes and ear openings are automatically covered by special folds of skin when the platypus dives underwater. Since it is thus deaf and blind while swimming, it must depend upon the finely developed sense of touch found in its soft, flexible bill. At times, platypuses swim with their eyes open underwater. When on land the ears open and a small ridge of skin bulges up to form a temporary external ear to increase sound collection.

A platypus can remain underwater for up to ten minutes. Its heart rate and metabolism are slowed when diving. This particular mechanism has been labeled the "seal reflex" since it was first studied and best known in seals, even though it is now known to take place in almost any animal that ventures into the water (including humans). It is, however, more developed in certain species, as will be discussed later. Like many mammal babies, newly hatched platypuses are more tolerant to hypoxia than adults. Young platypuses have survived being submerged for 3½ hours.

Platypuses lose their unerupted teeth when young but develop horny ridges along their jaws for chewing. These are often aided by the taking in of sand and grit to help pulverize crustaceans and other prey. They have been known to eat as much as their own weight in food during a 24-hour period. If a day's hunting is unusually good, the surplus catch and grit can be stored in special internal cheek pouches for later use.

Platypuses and echidnas differ from all other mammals in that the males do not have a Y sex chromosome. All other normal mammals have two sex chromosomes. (Except some male shrews which have three.) One is inherited from the mother and the other from the father. If there are two X chromosomes the offspring becomes female, whereas if there is one X and one Y it will become a male. The mother always transmits an X chromosome, so that it is the male's sperm that decides the sex of offspring.

In the monotremes, the sperm carries either an X or no sex chromosome at all. In the latter case, the offspring become male and have only a single sex chromosome obtained from their mother. It is also interesting to note that the female's right ovary and oviduct are almost always nonfunctional, and that the male's testes are located internally. In most other mammals the testes are located externally to the body cavity since the internal temperatures are often too high to permit spermiogenesis (the formation of sperm). It may be that the normally low body temperatures of monotremes are responsible for the internal placement of these organs, although this is questionable since cetaceans, elephants, seals, sirens, and sloths all have internal testes and still produce sperm.

Most of the time, platypuses remain underground in burrows. The two sexes generally live together until

Platypus (*Ornithorhynchus anatinus*)

6

the female is about to give birth, when she ousts the male. Most burrow systems have an entrance strategically located underwater. Some experts say that this is because the water level has risen after construction of the entrance. The tunnels are often equipped with air shafts and are narrowed at key points such as entrances and just prior to various chambers. The narrow portions are used to squeeze out excess water from the fur. The same technique is also used by otters and water shrews. The nesting chamber is located as much as 100 feet from the entrance. It is lined with leaves and other materials carried in by the female, held against her abdomen by her tail.

The male begins courtship by swimming circles around the female. Next he grasps her tail, and they swim circles together. Mating takes place in the water with the couple belly to belly but with their heads at opposite ends. Later the female retires to her nest but is careful to blockade the tunnel at various intervals to discourage predators. About 15 days after mating she produces 2 leathery, 0.75-inch eggs that stick together to decrease the chance of one accidentally rolling to a corner of the nesting chamber. The pouchless female incubates them by holding them against her abdomen and then curling her body around them. She keeps them moist by bringing in wet leaves. After about 10 days, the young hatch out with the aid of a temporary eggtooth. Instead of nipples the females bear large pores from which milk oozes when stimulated by the babies, who quickly lap up the nourishing substance.

Man is the playtpus's chief enemy, although the animals are not totally defenseless. On the inner sides of the males' hind legs can be found a 0.75-inch spur. It is hollow and connected to a venom gland. If molested, the platypus jams the spur against the attacker to deliver an immediate dose of toxin capable of causing a severe local reaction, although no human deaths have ever been reported.

The venom glands enlarge and fill with venom during the mating season. There are reports that these spurs are used to sedate uncooperative females to facilitate trouble-free copulation and to fight off other males during the breeding season. In at least one documented case, a male killed another male in just such a manner. Although the spurs have been used to hold the female during mating, it is doubtful if the male injects her. The one time this was observed, the female almost died as a result.

Family Tachyglossidae (echidnas—2 species)

The **echidnas** or **spiny anteaters** (*Tachyglossus aculeatus* and *Zaglossus bruijni*) are the only mammals without teeth or even the vestiges of them. (Most anteaters possess vestigial teeth, and even baleen whales have tooth buds as embryos, although the buds disappear and are replaced by the baleen plates.) *Zaglossus* favors earthworms as food. *Tachyglossus* echidnas feed primarily on ants and

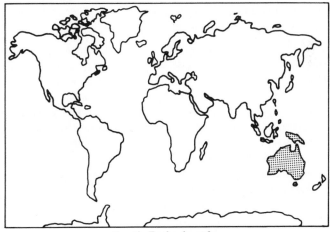

Family Tachyglossidae

termites, gathering them with a fast-flicking tongue coated with a sticky saliva; the tongue can be

Echidna (*Tachyglossus aculeatus*)

protruded up to 7 inches from the mouth. The prey is then "chewed" with special spines located at the back of the tongue and on the hard palate. The spines grind together as the tongue is flicked in and out to trap more ants. The formic acid of the ants is said to be evident when one eats an echidna. The animals also take in pebbles to aid in grinding the food in both the mouth and the stomach. If necessary, echidnas can fast for up to a month.

If a male attempts to mate with an unwilling female, she may attempt to jab him in the genital area with her spines. As in all monotremes, the penis is used to transfer sperm only and never urine. The female's egg is deposited directly from her cloaca into an abdominal pouch. This pouch is only present during the breeding season and disappears at weaning approximately 10 weeks later. Some males are also known to develop a rudimentary pouch every 28 days or so. To transfer the egg the female curls her body into the shape of a ring. The egg is actually glued to the inside of the pouch by a sticky substance accompanying it from the cloaca. Sometimes fecal material is also transmitted into the pouch with the egg. The egg remains in the pouch until the young hatches out with the aid of an eggtooth. The young, the size of a raisin when newly hatched, obtains its nourishment from the nippleless mammary glands through gentle massage and sucking. It continues in the pouch until the growing spines become more than the mother can bear. She then quickly "shows it the door."

Although echidnas are covered with spines, the **curved-beak** or **long-nosed spiny anteater** (*Zaglossus bruijni*) loses its spines with time so that it is often spineless in old age. Echidnas are among the most long-lived mammals, having been known to exceed 50 years. Those inhabiting the southernmost areas of their range do not hibernate during the winter as once thought, but may go into torpor if there is a lack of food. The **short-nosed spiny anteater** (*Tachyglossus aculeatus*) has the lowest normal body temperature of all mammals, 72°F. It may actually succumb to heat stroke at 99°F.

When alarmed, echidnas quickly dig themselves in so that only their protective spines are exposed. If really frightened, they can completely bury themselves in less than 10 minutes; if the ground is exceptionally hard, they may roll into a ball, take shelter in a rock crevice and wedge themselves in with their spines, or even take flight on their hind legs. Male echidnas are equipped with spurs on their hind legs, but most experts feel that the spurs are not venomous despite a few unsubstantiated reports to the contrary.

2

ORDER MARSUPIALIA
(258 SPECIES)

The marsupials are one of the most primitive of mammalian orders. They stand apart from other mammals primarily in the way they reproduce: each female has a pouch or marsupium in which to hold the immature young. In those marsupials that jump or climb trees, the pouch normally opens towards the head of the female; in those that run on all fours or burrow, it usually opens towards the rear. Some marsupials, such as the **short-tailed opossums** (*Monodelphis* spp.) and the **numbat** (*Myrmecobius fasciatus*), have no pouch at all.

Most marsupial young are born after only partial development and must conclude the process while in the pouch. In some species, the vagina is present seasonally so that at birth the young must actually crawl out through the female's tissues. They are aided by special hormones that dissolve a pathway of least resistance. Many marsupial females, including bandicoots, kangaroos, opossums, and wombats, have dual vaginas and uteri while many of the males possess a forked penis. ("Uterus duplex" is also seen in many rodent species.) The duplicate reproductive organs are said to permit rapid replacement of a newborn animal lost for some reason. Except for Notoryctidae, the scrotum is located anterior to the

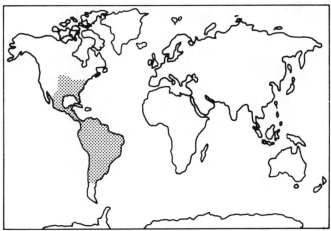

Family Didelphidae

Gray Kangaroo (*Macropus giganteus*)

penis. More specific details in reproduction will be discussed with the appropriate species.

Family Didelphidae (opossums—76 species)

The Didelphidae are one of the two marsupial families with members found outside of Australasia. It also boasts the only aquatic species, the **water opossum** or **yapok** (*Chironectes minimus*). The yapok has water-repellent fur, fully webbed hind feet, and both sexes sport a pouch. The female's is waterproof and is used to hold the young; the male's contains the scrotum. The young go diving with the mother both while enclosed within the protective pouch and, when they've outgrown it, by clinging to her back before she dives. In general, marsupial young can withstand up to twenty times the normal concentration of carbon dioxide without ill effect when "sealed" in the pouch.

Yapoks can fold up their ears to prevent the entry of water. Other marsupials including the **common** or **Virginia opossum** (*Didelphis virginiana*), the **murine** or **mouse opossums** (*Marmosa* spp.), and **brush-tailed possums** (*Trichosurus* spp.) are known to fold up their ears when they sleep. Common opossums are known to lose ear tips and tails as a result of frostbite in the northern areas of their range.

Most opossums are omnivorous. Some, such as the **thick-tailed opossums** (*Lutreolina crassicaudata*), collect fat deposits in their tail areas to tide them over lean times. The **monito del monte** (*Dromiciops australis*) may double its weight in one week in this way. (This species is sometimes placed in a family of its own, Microbiotheriidae.) The Virginia opossum can feed on poisonous toadstools without ill effect.

Mouse Opossum (*Marmosa robinsoni*)

Most of the species in this family lack a pouch (e.g., the mouse opossums) or exhibit only a vestigial pouch (e.g. **woolly opossums** (*Caluromys* spp.). At birth, mouse opossums are the size of a grain of rice and are considered the smallest young of any mammal. They make their way to one of the female's teats and take hold. Occasionally, one of these dangling infants gets knocked off while the mother forages. When older, they cling to her fur so tightly that one can lift the mother off the ground before being able to pluck a youngster loose. The female mouse opossum has been observed collecting her young by tossing them into the air with her snout and then running ahead so that they land on her back. Females occasionally take up residence in abandoned bird nests or hollow cactus. They have prehensile tails.

The most familiar Didelphid is the common or Virginia opossum. The female carries nesting material by compressing it between her belly and

PHOTO: J.F. EISENBERG

tucked-under tail. Tail carrying is also seen in murine opossums, rat kangaroos, and South American opossums. The males have forked penes but, despite legend, do not mate through the female's nostrils.

The common opossum holds the record for the shortest gestation period of any mammal, being able to give birth within 8 days of copulation, although 12–13 days is more typical. When born, the young are smaller than honeybees; as many as 24 can fit into a tablespoon, and 23 equal the weight of a penny. Litters number up to 25, but the common opossum does hold the mammalian record of 56 young in a single birth. Since most females only have 13 teats (range 9–17), only the quickest 13 can survive. (Female opossums are one of the very few mammals to have an odd number of teats.) Some studies have found that only about half the naked, blind, deaf young can complete the one-minute, 3-inch trip from the vagina to the pouch. As in kangaroos and others, the tips of the teats swell after being swallowed and positioned in the infants' stomachs. Thus each of those successful in grasping a teat becomes "welded" to its mother. Special embryonal claws used to climb to the pouch are then shed. The babies undergo a ten-fold increase in weight within their first week in the pouch. The female periodically cleans the pouch of wastes. When older, the young ride on their mother's back while clinging to her fur. They occasionally gain added purchase by wrapping their tails around the base of hers, but the mother's tail is never arched over her back, causing the young to dangle in the air as is often depicted. The tails are prehensile, and the young will occasionally hang by them, but adults rarely do.

Opossums cope with danger with several defense mechanisms. Some, such as the yapok, can secrete a repulsive odor. The best known defense is thanatopsis or "playing dead." (Hardly unique to opossums, this defense can be seen in African ground squirrels, beetles, birds, brown bears, caterpillars, foxes, honey badgers, jackals, moths, mouse deer, palm squirrels, snakes, spiders, striped hyenas, and others.) This may seem a rather foolish defense mechanism since it leaves the animal open to attack, but many predators refuse to eat something they haven't

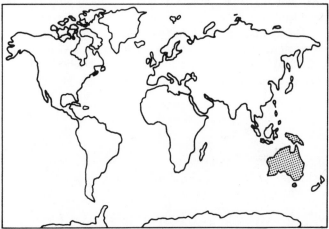

Family Dasyuridae

killed, while others give up the chase because the stimulus (a fleeing animal) which triggers the response (hunt and kill) is no longer present. Recent research has shown that the feigning of death is actually a physiological faint brought on by fear. The opossum, at least, goes into a comatose state. Its respiration and heart rate both decrease. It apparently loses all responsiveness to external stimuli and can be dropped, kicked, poked in the eye, thrown, and even cut without response. It is claimed that, if it is cut while in the faint, the wound will bleed very little if at all. Opossums may remain in such a state for only 8 seconds or as long as 6 hours. Most opossums—over 90 percent—never use this defense at all. They prefer to flee any danger immediately or even to bluff by hissing, salivating, baring their teeth, excreting wastes, or releasing noxious anal gland secretions.

Another defense mechanism is found in the **four-eyed opossum** (*Philander nudicaudatus*). It actually has only two eyes, but each one is surmounted by an eyespot so that it seems to have four eyes. The eyespots serve to direct attacks away from the animal's true eyes in order to preserve its sight should the attack not prove to be fatal.

Family Dasyuridae (native cats and marsupial mice—51 species)

The Dasyuridae include some of the strangest of all marsupials. The family boasts the smallest, the 2 inch, 1/5 ounce **Kimberley planigale** (*Planigale ingrami*). Like others in its genus of **flat-headed planigales**, it has quite a flat head. One species has a skull about 1/8 inch deep, or about as thick as a half dollar. These flat heads enable them to take refuge in small cracks and crevices to elude predators.

The family also includes the largest carnivorous marsupial, the **thylacine** or **Tasmanian wolf** (*Thylacinus cynocephalus*). This animal is currently near (or past) extinction. It is said to be capable of launching vicious attacks, although there are no recorded attacks on humans. Their mouths, as in many marsupials, can be opened almost 180 degrees. If pursued, these dog-like marsupials resort to "kangarooing," i.e., hopping along on their hind legs only. Some place it alone in the family Thylacinidae.

The **Tasmanian devil** (*Sarcophilus harrisii*) is more affectionate and playful than devilish. The name stems from its eerie night-time cries and from earlier reports dealing with vicious caged individuals. When alarmed, their ears turn bright red. They reportedly eat anything remotely edible, dead or alive, including the feathers, fur, and bones of their prey. They even eat deadly tiger snakes.

Tasmanian Devil (*Sarcophilus harrisii*)

The young are kept in a pouch that remains completely closed during most of their stay. It is waterproof to prevent the young from drowning. The males occasionally help care for the young. Tasmanian devils will often escape danger by diving into water and swimming for some ways before surfacing for air. They have also been known to bend metal bars apart to escape cages.

Some **broad-footed phascogales** (*Antechinus* spp.) have grooved soles, sucker-like pads, and long claws, and they can run across the ceiling of sandstone or rock caves. These "mice" are thus able to chase down insects and lizards. The males of some species have a rather drab sex

14

life since they expire soon after losing their virginity. It is thought that copulation produces hormones that destroy their immunity and leads to death from blood parasites.

The **brush-tailed phascogales** (*Phascogale* spp.) are the marsupial equivalent of the pack rat. They are occasionally called the "vampire marsupials" because of their habit of raiding chicken coops and killing more than they can eat. This led early explorers to believe, incorrectly, that the animals merely sucked out the blood of their victims. The **bottle-brushed phascogale** (*Phascogale tapoatafa*) is so named because when active its tail hairs stand on end to resemble a bottle-brush. When alarmed, brush-tails may stamp their forefeet or drum their tails as a warning.

Some other notable members of this family include the **dunnart** or **fat-tailed sminthopsis** (*Sminthopsis crassicaudata*), which stores fat in its short, stumpy tail when food is plentiful. The **mulgara** or **crest-tailed marsupial mouse** (*Dasycercus cristicauda*) obtains all its moisture from the fluids in the bodies of the insects it eats. Like many other desert-dwelling mammals, it has an enlarged auditory bullas allowing it to hear the ground vibrations set up by approaching predators. It is also noted for its expert skinning of prey during consumption.

During courtship, male **tiger cats** (*Dasyurus machlatus*) often bite the females severely. Many of the species found in this family of "cats and mice" have no pouch at all or only one composed of a small flap of skin. The young are often dragged along behind the female attached to her teats, or are carried on her back. In many, such as the **quoll** or **dasyure** (*Dasyurus viverrinus*), the female gives birth to many more young than she has teats. As in the opossum, this unfortunately dooms a number of the young to death from starvation. The function of this excessive birth may be to weed out the weaker offspring, but quoll production of 24 young when there are usually 6 nipples seems to be taking the point to the extreme.

Family Myrmecobiidae (numbat—1 species)

The **numbat** or **banded anteater** (*Myrmecobius fasciatus*) is one of nature's best termite exterminators. One expert estimated that a single numbat consumes about 7 million termites every year, efficiently using a sticky tongue that grows to about half the animal's body length. Much soil is also consumed.

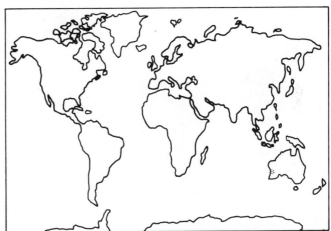
Family Myrmecobiidae

The numbat hides and sleeps in a hollowed log or burrow. This becomes an effective fortress since the head is kept toward the center while the heavily muscled rump blocks the entrance. The body then swells in a process known as phragmoticism to ensure a tight fit and make it virtually impossible to extract the animal by brute force.

The numbat female has no pouch. The young cling to the nipples and are dragged along beneath

15

her when she travels. Numbats are inoffensive animals which have never been known to attack or bite anyone. They "hum" when frightened and may bound away on their hind legs.

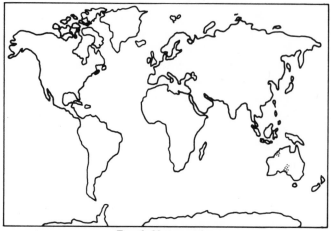

Family Notoryctidae

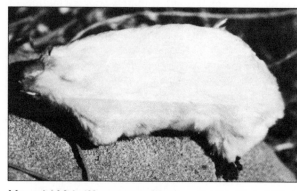

Marsupial Mole (*Notoryctes typhlops*)

Family Notoryctidae (marsupial mole—1 species)

The **marsupial mole** (*Notoryctes typhlops*) is very similar to placental moles in appearance. Its subcutaneous, vestigial eyes contain neither a lens nor a pupil, and the small ear openings can be closed off. A large nose plate covers the snout to protect it when burrowing. Its burrows are not kept open but are closed behind it as it digs out new dirt. Its method of burrowing is sometimes described as "swimming through the dirt," and like a swimmer it "surfaces" often.

The young finish development in a small rearward-facing pouch, but prior to weaning they are too large to fit completely inside. At this stage they are in constant danger of getting knocked off since they merely drag along the side of the mother as she tunnels.

Family Peramelidae (bandicoots—19 species)

Bandicoots (8 genera) resemble large rats. Some, including the **short-nosed bandicoots** (*Isoodon* spp.), are known to dispatch prey by repeatedly trampling the victims with their front feet. The thoroughly pulverized mass is then inspected for edibility. If it qualifies, it is quickly devoured; if not, it is left to rot. When feeding on cockchafer grubs and worms, the bandicoot carefully wipes the meal clean of dirt before consumption. **Rabbit-eared bandicoots** (*Macrotis lagotis*), placed in their own family (Thylacomyidae) by some experts, sleep while propped back on their tails with their heads tucked between their forelegs and their ears folded down over their eyes.

One **short-nosed bandicoot** (*Isoodon macrourus*) is credited with the shortest normal gestation period for a mammal. A normal pregnancy lasts 12 days. In bandicoots the half-inch, 0.01-ounce newborns make their way to the female's pouch with the aid of large claws that are shed once the

16

destination has been reached. As with some other marsupials, the female's pouch faces backwards. **Long-nosed bandicoots** (*Perameles* spp.) and rabbit-eared bandicoots are the only marsupials that possess a true placenta although there are no villi.

Family Caenolestidae (shrew opossums—7 species)

The Caenolestids and Didelphids are the only marsupials found outside Australasia. They are pouchless and resemble shrews. Some are thought to subsist on eggs and small animals, and fat can be stored in their tails. They are very rare, and few have been seen or caught.

Family Phalangeridae (phalangers or possums—46 species)

Probably the most loveable of all marsupials is the teddy-bearlike **koala** (*Phascolarctos cinereus*), sometimes accorded its own family, Phascolarctidae. These small marsupials look good but smell even better, like eucalyptus oil or "mentholated cough drops." They can, however, be quite vicious if handled and are known to produce a harsh, grating noise, one of the loudest produced in Australian mammals (second only to the great glider). When unhappy, koalas are said to weep; when annoyed they produce a ticking sound. A pig-like grunt can be heard when they are foraging at night.

Koala means "no water," but despite their being able to go long periods without it they have been observed to drink on many occasions. They are also good swimmers and will readily take to the water.

Koalas feed almost exclusively on the leaves of about 20 of the hundreds of species of eucalyptus trees. In general they prefer about 5 of these 20. Each adult consumes about 2.5 pounds of leaves each

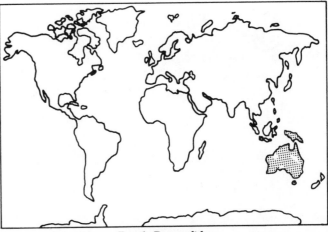
Family Peramelidae

Short-Nosed Bandicoot (*Isoodon* spp.)

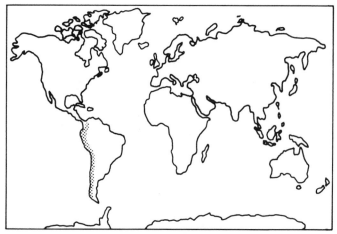

Family Caenolestidae

Rat Opossum (*Caenolestes obscurus*)

day. A few animals have been seen to supplement their eucalyptus diet with other types of plants, such as mistletoe. Koalas carefully avoid eating young eucalyptus leaves since these contain a high concentration of toxic prussic or hydrocyanic acid. Eucalyptus leaves are poisonous to most other mammals. The natural eucalyptus oils, besides giving the koalas a pleasant smell, are also known to keep them free of certain vermin, including lice. In northern areas where the climate is warmer, koalas feed mainly on eucalyptus species that contain the chemical cineol. It lowers blood pressure and metabolic rate. In the much cooler southern areas, they feed from tree species producing phellandren, a chemical that increases metabolism and therefore heat. Koalas occasionally descend their trees to lick the soil and consume some gravel, which may aid in digestion. They also have cheek pouches capable of temporarily storing excess food.

Males mark territories with secretions from large chest glands (as do some other phalangers; male sugar gliders even mark other members of their group with scent). At times, quarrels arise as to ownership of a particular territory. When one combatant has had enough fighting, it surrenders by assuming a submissive posture. To do this, it pretends to be losing its balance and clings to the tree by only one or two of its limbs.

Koalas are generally solitary by nature. During the breeding season, the males gather harems and guard them against other males. The females give birth only every second year. When born, koalas are about as big as an inch-long segment of pencil and are deaf and blind. They remain in the pouch and suckle for their first 6–7 months of life. The female's pouch faces the rear, thus avoiding the hazard of snagging on small branches, and enables the young to engage in a unique feeding mechanism. The female's digestive system produces a special food for her offspring. It's a broth composed of partially

18

digested eucalyptus leaves. It is presented, not mixed with feces, to the young from her anus and is thus within reach from the pouch. This unique process only takes place every two to three days for about a month at the time of weaning. The special meal is highly nutritious and accelerates the young one's growth in addition to supplying it with needed gum-digesting bacteria enabling the youngster to advance to leaf-eating.

The pouch's rim muscles are strong enough to keep the young from falling out on their heads. The females are not possessive with their young and will occasionally accidently switch offspring with other nursing females in the same vicinity. No harm results, but it is strange behavior for a mammal. When a youngster displeases its mother, she is quick to discipline it. At times this is done by turning the offender over her knee and paddling its bottom with the flat of her hand, continuing for several minutes despite its heart-rending pleas to stop. Male koalas don't take to the young, and should the latter attempt to force the issue the adult becomes quite annoyed and either pushes it away or leaves the scene.

The koala is the largest of the phalangers, also known as possums but not to be confused with opossums. The remaining 45 species or so vary in appearance and behavior. The word phalanger means "web" and refers to the webbing between these animals' second and third toes, the claws of which are divided to form natural fur combs. More recent works divide the phalangers into several families including Petauridae, Burramyidae, and Tarsipedidae.

Several species have taken to gliding through the air in the manner of flying squirrels. Included

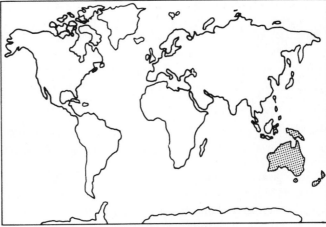

Family Phalangeridae

Koala (*Phascolarctos cinereus*)

in this group are the **feather-tailed phalanger** (*Distoechurus pennatus*), the **sugar gliders** or **flying phalangers** (*Petaurus* spp.), the **great glider** (*Schoinobates volans*), and the smallest of marsupial gliders, the **gliding feather-tail** (*Acrobates pygmaeus*). The sugar gliders carry nesting materials wrapped in their tails, but not while gliding. They have been known to glide for distances exceeding a hundred yards. They will even take to the air to catch flying prey such as insects in mid-flight.

The various species of **cuscus** (*Phalanger* spp.) are unusual in that the sexes of some (including *P. maculatus* and *P. rufoniger*) differ in coloration (sexual dichromatism). This led to the supposition that there were more species than actually existed. To make it even more confusing, the juveniles may be differently colored than either of the adults, and the adults will sometimes change color with a change in season, diet, health, or location. These animals have a prehensile tail, as do many other phalangers, and can hang from a tree limb while leisurely devouring a morsel of food held in the front paws. They also give off an unpleasant musky odor. Some have bald spots on their coats due to constant friction from sitting on that spot.

As a group, the phalangers are omnivorous. The **honey sucker** (*Tarsipes spenserae*) has a long tongue. It is coated with hairs at the end to form a cylindrical brush used to obtain pollen and nectar from flowers. In so doing, it has become an important factor in the pollination of certain Australian flowers. Honey suckers will occasionally leap into the air to obtain a tasty insect, but they usually spend the day sleeping in an abandoned bird's nest. The **brush-tailed possum** (*Trichosurus vulpecula*) prefers the evergreen parasite known as mistletoe, which is poisonous to most other animals. The greater glider prefers eucalyptus trees as food. The **dormouse phalangers** (*Cercartetus* spp.), often found nesting in deserted birds' nests, are primarily insectivorous. They alternate wakefulness with periods of torpor and exhibit a winter sleep, although they are not true hibernators. The **thick-tailed dormouse phalanger** (*Cercartetus nanus*) stores fat in its tail for use during the winter sleep.

Like marsupials in general, the phalangers give birth to underdeveloped young. They continue their growth within the pouch, which faces forward in all phalangers except the koala. The single young of the cuscus spends its pouch time bathing in sticky secretions produced by the female. The purpose of the fluid is to prevent dehydration in the offspring. Should a newborn phalanger be lost, another is quickly produced from the second reproductive tract of the female. This is common to many of the species including the brush-tailed possum. Female feather-tailed phalangers are unique in often having a single medial teat. The young of sugar gliders ride on the female's back during her glides.

Most phalangers depend on either flight or inaccessability to protect them from predators. The cuscus, brush-tailed possums, and **striped phalanger** (*Dactylopsila trivirgata*) give off foul odors. However, nothing works well against the worst of enemies, humans, who have destroyed habitat, loosed introduced

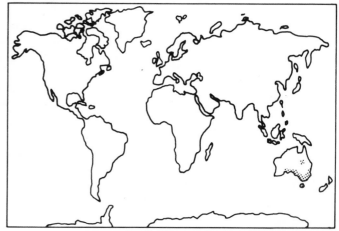

Family Vombatidae

20

predators and competitors, and killed the phalangers for food. In New Guinea, aborigines who happen upon a cuscus and need no food at the moment will break both of its hind legs. This lets the animal continue to feed within a small area until the native gets hungry and returns to collect fresh meat. Natives also exploit the cry of fear emitted by the brush-tailed possum when encountering a predatory goanna (lizard). They imitate the goanna's scratching to induce the tell-tale scream. They then easily locate and dispatch the possum.

Family Vombatidae (wombats—3 species)

There are two species of **soft-furred** or **hairy-nosed wombats** (*Lasiorhinus latifrons* and *L. krefftii*) and only a single species of the more common **coarse-haired** or **naked-nosed wombat** (*Vombatus ursinus*). These animals have the appearance of miniature bears but actually behave more like badgers, a name by which they are often called. They dig underground burrows measuring a hundred feet or more in length and can reportedly outdig a man equipped with a spade. One was observed to dig through six feet of brick-hard earth in an hour's time. Like rodents but unlike all other marsupials, the frontmost incisors grow continually throughout life, making it imperative that they be kept worn down to a

Wombat (*Vombatus ursinus*)

useful size. Actually, all their teeth are rootless and grow continuously. They sometimes feed on certain fungal species known to be harmful to humans without suffering any ill effects. Like the bandicoots, koala, Tasmanian devil, and thylacine, the wombats' pouches open to the rear. This adaptation protects the young from the showers of dirt and debris resulting from the mother's burrowing activities. Their almost armor-plated rump has been used to pin and crush to death burrow invaders such as foxes and dingos.

Family Macropodidae (kangaroos and wallabies—52 species)

The best known of the marsupials are undoubtedly the kangaroos. Much needless confusion has resulted from the use of the terms kangaroo and wallaby. As a rule, the only difference between the two is size. One source states that if the hind feet exceed ten inches the animal is by definition a kangaroo; shorter feet mark a wallaby. In practice, most are referred to by the name they were first given. In both cases, a male is an "old man" or boomer; a female is a flyer, doe, or rover; and a young is a joey.

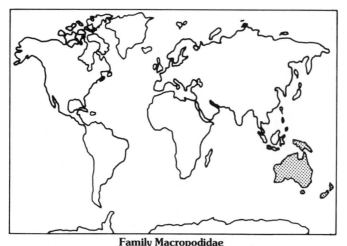

Family Macropodidae

Tree Kangaroo (*Dendrolagus* spp.)

Collectively, groups are referred to as mobs.

The **red kangaroo** (*Macropus rufus*) is the largest of all marsupials, reaching a height of 7 feet on its toes and a weight of 200 pounds. Usually the male is more predominantly red, particularly during the breeding season when skin glands located on his chest secrete a reddish substance. The males dutifully smear it over their head and body. The females are grayish-blue, although in some areas the sexes are opposite in color. Both the red and the **great gray kangaroo** (*Macropus giganteus*) have leaped distances of 40 feet and cleared heights of 10 feet. They can travel short distances at speeds up to 40 mph. Interestingly, at speeds in excess of 10 mph the kangaroo exerts less energy in hopping due to the mechanical arrangement of its leg tendons, which allows a recycling of energy. Almost all kangaroos are good swimmers.

As odd as it sounds, several species of kangaroo have taken to trees and are appropriately known as **tree kangaroos** (*Dendrolagus* spp.). They are actually only fair climbers and ascend trees primarily to feed on the leaves or to escape predators. Once up, they are agile and sometimes are seen making flying leaps from limb to limb with the aid of special foot pads that increase traction and are not unlike the rough soles of the **rock wallabies** (*Petrogale* spp.) which use theirs to scale almost vertical cliffs on their hind feet. Tree kangaroos will occasionally make quick exits to the ground by jumping as much as 50–60 feet without incurring any physical damage. Their hind feet can move independently, unlike many of the terrestrial species whose hind feet must move in unison. Tree kangaroos often rest with their head lowered onto their chest. The hair on their nape grows in a reverse direction so that rainwater is easily shed while sitting out a rainstorm in a tree in just such a position.

The tails in most kangaroo species are large, powerful organs used in balance and support and as an additional thrust when the animal makes a long or high jump. The great gray and others use it when walking about slowly by placing their weight alternately on the tail and hind legs. Rock wallabies have tufted tail tips thought to improve their function in balance. In the **potoroo** (*Potorous tridactylus*), the prehensile tail is the only part of the body with sweat glands, although it does account for 10 percent of its body surface area. When hot, the tail is wagged to increase evaporation and therefore cooling. **Rat kangaroos** (subfamily Potorinae) have prehensile tails used to carry nesting materials. The **musky rat kangaroo** (*Hypsiprymnodon moschatus*) is the smallest species of kangaroo, weighing about 1 pound; it is also the only species to have a naked tail. The **nail-tailed wallabies'** (*Onychogalea* spp.) fifth appendage is tipped with a small spine whose function is unknown. They do have the curious habit of swinging their arms in circles as they bound along giving them the alternate name of "organ grinders."

As a rule most species in this family are browsers or grazers. The musky rat kangaroo is an insectivorous exception. Its musk glands give it its name, and like other rat kangaroos it hops about on all fours rather than on its back legs. It also differs in that it usually bears twins rather than a single offspring.

It is said of many kangaroo species and particularly the **pademelons** (*Thylogale* spp.) that they "chew the cud." Actually, these wallabies and others are able to regurgitate fibrous foods and submit them to a second rechewing and digestion. In true cud chewing, food is quickly swallowed, stored in a special stomach pouch, and then brought back up at a more convenient time for its first mastication. It is then swallowed and transported into a new chamber. Thus no kangaroos truly chew their cud, although behaviorally the processes are similar.

The **short-nosed rat kangaroo** (*Bettongia lesueur*), the only species that lives underground, stores food inside its large burrows. At times, it will move into a rabbit warren and live side by side with this introduced species.

Kangaroos inhabiting some of the more arid regions of Australia, such as the gray kangaroo, can go without water for up to several months. Their intake is usually small even when water is readily available. This is probably due to the fact that the more water they drink, the greater is their loss of nitrogen, an undesirable event in a species that feeds on protein-poor desert vegetation. During dry spells, they are known to dig out their own water holes with their powerful hind legs. These are essential to other animals in the vicinity since they are incapable of digging their own. Some rat kangaroos are known to dig into dried creek beds to obtain water during droughts. The **tammar** (*Macropus eugenii*) can drink seawater without problem, and the **hare wallaby** (*Lagorchestes conspicillatus*) does not drink any free water.

The **quokka** (*Setonix brachyurus*) lives in swamps and brush lands where it forms runways and tunnels through the thick vegetation. Unlike most members of this family, it requires a continual supply of water to prevent death from dehydration. Like many other kangaroos, it rids itself of excess heat by panting and licking its forearms and chest to make use of evaporation. Quokkas can live on meager food supplies, however, and are actually capable of recycling nitrogenous waste through their stomachs where it is remanufactured into new protein by special bacteria.

Dominance in males is often determined through actual sparring contests. In fact, the boxing skills of some individuals have been so developed that they can successfully take on human challengers. Serious fighting by kangaroos is done with the powerful hind legs. Social structure ranges from solitary to gregarious. Dominant males father most of the offspring. In some species the females may only be receptive to the males for a few hours' time.

Due to the dual reproductive anatomy discussed earlier, a female kangaroo is usually pregnant, even while one youngster is in the pouch. The implantation of the second blastocyst is delayed for approximately 200 days (a process known as embryonic diapause) unless the developing joey in the pouch should be lost for any reason except maternal death. It is cessation of teat feeding that triggers immediate implantation of the blastocyst so that a "replacement" offspring can arrive within a very short time, about 4 weeks. Should the first joey progress normally the second blastocyst will not begin development until about a month before his older sibling is ready to leave the pouch for good. In the quokka, the dormant embryo degenerates if the first does well. During severe droughts, female kangaroos become anestrous. With the exception of the musky rat kangaroo, which usually produces twins, most kangaroo females give birth to a single young, although twins and triplets do occur. In one case of twinning, the mother was observed to throw one of the offspring out of the pouch. Others have successfully raised twins. Occasionally male kangaroos will pull very young joeys from the pouch and throw them on the ground to die. They may do this to hasten estrous in the female.

At birth, a kangaroo is about 0.75 inch long and tips the scales at about 0.03 ounces. This is about 1/30,000 its adult weight, the greatest discrepancy between adult and newborn size in the mammal world. It was first thought that the mother would lick or scratch out a pathway from the birth canal to the pouch with her tongue and nails, but this was disproved by later researchers who found that she was doing no more than cleaning the pouch in anticipation. Since the newborn is both deaf and blind, it is thought that it finds its way to the pouch by smell. The naked neonate takes about 3 minutes to make the 6 inch journey using its front legs only. If it falls the female offers no help. Once in the pouch it grasps and swallows one of the teats. The tip swells when inside the stomach so that the infant cannot fall off accidentally. It was once thought that this small morsel of life was unable to suck on its own and that the teat automatically contracted to force-feed milk into the stomach. This is now known to be incorrect, for the newborn is capable of sucking. The teat does not pose a problem to the newborn's breathing since the epiglottis is unusually shaped at this age to allow breathing around it.

The mammary glands are capable of producing two types of milk. The first is very concentrated and is secreted for newborns, which require a great amount of nourishment but due to their size can only take in a small volume. The second, more dilute form is for the larger joeys that come and go from the pouch and are near weaning. Since they can take in much more, the dilution takes some of the strain of production away from the mother's system.

The pouch is lined with a rim of muscle the female can voluntarily contract. When the joey gets in, it is held firmly in place by this built-in "seatbelt" when the female takes flight. She can also contract the muscle to prevent older joeys from hopping into the pouch. The pouch must be regularly cleaned of feces and urine by the mother and actually grows with the joey. When older, the joeys make headlong dives into the pouch and then perform a somersault placing them head-up facing the mother's abdomen. In order to look out, it must twist its head around 180 degrees. In some instances, two joeys will jump into the pouch. Usually only one is her offspring, the second is a playmate belonging to another female and only hops in for a visit. When the young step out of line, they are quickly and

24

literally boxed into shape by the older boomers. Females occasionally adopt motherless joeys.

Some species, such as the pademelons and rock wallabies, warn others of approaching danger by thumping the ground with their feet. In most instances, kangaroos tend to flee rather than fight. If losing the chase while carrying her joey, a flyer will sometimes throw her youngster into the cover of a grassy thicket and proceed to lead the pursuer away. Others feel that the female actually panics and abandons her joey to save herself. There are reports of their returning to seek out such an abandoned offspring. The **desert rat kangaroo** (*Caloprymnus campestris*) has great endurance when fleeing and has been known to outdistance 2 horses over a 12 mile run. Should a kangaroo stand its ground or be cornered, it can put up a good fight. To defend itself, it will rear back on its tail and lash out with its powerful, clawed feet. In large species, these are capable of disemboweling most oncomers. The larger kangaroos will also defend themselves from dingos by heading for water where they stand chest deep facing the dogs. Any dingo foolish enough to follow is grabbed and held under until it drowns.

As a rule, kangaroos not only take flight readily but often can easily be panicked. This occasionally leads to their watching where they've been rather than where they're headed. This has resulted in a number of documented fatalities as a result of head-on collisions with large rocks or trees. Other animals that have been known to kill themselves via such "reckless driving" include humans, rodents, birds, and—surprisingly—rabbits. Kangaroos have been known to backtrack towards a pursuer and actually clear it with a single bound just prior to collision.

3

ORDER INSECTIVORA
(373 SPECIES)

The Insectivora are a hodgepodge of primitive placental mammals. Most are small and nocturnal and feed extensively on insects and other invertebrates.

Family Solenodontidae (solenodons—2 species)

The **Haitian solenodon** (*Solenodon paradoxus*) and the **Cuban solenodon** (*S. cubanus*) are two of the very few mammals whose bite is venomous. Solenodon means "grooved tooth," and indeed, the second lower incisors are grooved to allow the venomous saliva to trickle easily into the wounds inflicted by the teeth. Solenodons are apparently not immune to their own venom.

It is said that when solenodons are in a hurry they run only in zigzags and trip over their own feet as often as not. When truly alarmed, they stick their heads into the ground while leaving the remainder of their bodies completely exposed, possibly relying on the scent glands located in their groins and armpits to deter predators. They probably use a form of echolocation when inside the tunnel systems of their burrows. When sleeping, they

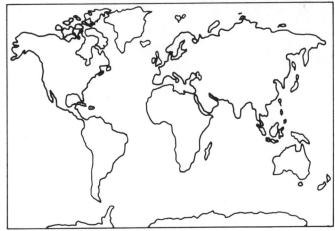

Family Solenodontidae

Hedgehog (*Erinaceus europaeus*)

27

are known to pile on top of each other. The female's teats are located close to her buttocks, and along with elephant shrews and shrews, they are the only insectivores to exhibit teat transport of the young. Both species are nearing extinction, first because they reproduce slowly (the female generally has only one or two offspring), and second because they are being quickly wiped out by cats, dogs, and introduced mongooses.

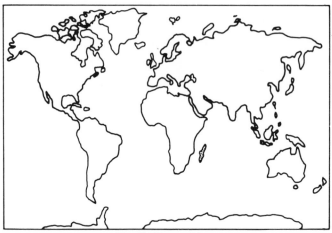

Family Tenrecidae

Family Tenrecidae (tenrecs—28 species)

This strange family of insectivores is found only in Madagascar. The **long-tailed tenrec** (*Microgale longicaudata*) possesses the longest tail in proportion to overall body size of all mammals. It exceeds twice the body length and contains as many as 42–47 caudal vertebrate; its end is prehensile.

There are many forms of tenrecs. Some resemble mice, shrews, skunks, and solenodons. The **hedgehog tenrec** (*Setifer setosus*) has the appearance of a miniature hedgehog and also uses its spines for protection. When sleeping or frightened, it curls into a defensive ball much as does its namesake. The **water tenrec** (*Limnogale mergulus*) has webbed feet and flattened tail to enhance its swimming ability. **Rice tenrecs** (*Oryzorictes* spp.) look like moles and have a similar lifestyle. Some **shrew-like tenrecs** (*Microgale* spp.) store excess fat in

Long-tailed Tenrec (*Microgale* spp.)

their tails. The **common tenrec** (*Tenrec ecaudatus*) and hedgehog tenrec go into a state of torpor during winter, and many can eat their weight in earthworms daily.

Hedgehog tenrecs have an unusual courtship. The males become stimulated by the females' odor. Their harderian glands (eye glands that keep the nictitating membrane moist) then secrete an excess amount of a whitish substance that covers the eyes and often drips down the cheeks and out the nostrils. During this time, the male runs about thrusting his nose into the female's body. She responds with flight or resistance. He overcomes her physically, bites her neck, and scratches her back with his legs. She then lifts her tail to allow mating to take place.

Streaked tenrecs (*Hemicentetes semispinosus*) communicate the presence of food and "all is well" signals from mother to offspring through ultrasonic sounds produced by vibrating their spines. The streaked tenrec is also said to be the "fastest developer" of all mammals. The young are weaned at 5 days, and the females can breed at 3–5 weeks of age.

One of the most notable members of this group is the common tenrec. As a rule, it produces the greatest number of offspring per litter of all mammals. The prolific females bear up to 32 young at a time, although the average is 15–20. The females have up to 29 teats, the most of any mammal (multimammate rats can have up to 24 teats). Female tenrecs usually eat their afterbirth and their youngs' feces.

Streaked and common tenrecs make use of a primitive echolocation. The tongue is used to produce repeated clicks whose echoes the animal analyzes to find its way in the dark. Another unusual behavior in the common tenrec is its "yawn," in which the posterior part of the palate bulges from the mouth, giving the impression that the animal is blowing up a pink balloon. The purpose of this action is currently unknown.

Family Potamogalidae (otter shrews—3 species)

The otter shrews, sometimes placed in the tenrec family, do resemble otters. The **giant otter shrew** (*Potamogale velox*) is one of the largest insectivores alive today. It has a long, flattened tail for propulsion and valved nostrils to prevent the entry of water. The **dwarf otter shrews** (*Micropotamogale* spp.) are much smaller and use their legs instead of their tails for propulsion.

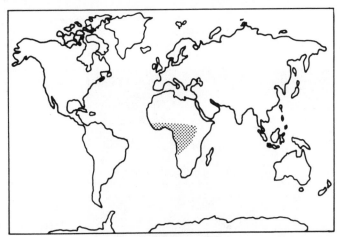

Family Potamogalidae

Family Chrysochloridae (golden moles—18 species)

The golden moles differ from true moles in a number of ways. Their forelegs are directed inwards instead of outwards, they dig with their specially padded snouts rather than the forelegs, and their fur is stiff and set at an angle which prevents them from backing up in tight situations. Instead, they are forced to turn a somersault in order to reverse their direction of travel. In desert areas, they are sometimes observed as a small wave of sand which at times does not even stop when prey is snatched from the surface and consumed while on the move or at a distant site. One **desert golden mole** (*Eremitalpa granti*) can produce up to 3 miles of surface tunnels in a single night.

The small eyes of golden moles are covered with hairy skin. Their ears are concealed by their fur, which in most is iridescent and may appear red, yellow, green, or blue. Iridescent fur can only be found in golden moles; in several species of true moles including **shrew moles** (*Urotrichus* spp. and

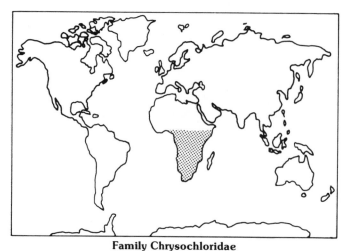

Family Chrysochloridae

Rough-haired Golden Mole (*Chrysospalax villosus*)

Neurotrichus spp.) and the **Pyrenean desman** (*Calemys pyrenaicus*); in a few rodents including the **groove-toothed creek rats** (*Pelomys* spp.); in a few shrews such as the **dwarf shrew** (*Sorex minutus*); and in the marsupial mole. A few other mammals show mild iridescence when their fur is wet.

Golden moles, like many true moles, have a high metabolic rate. In some the body temperature must be so consistently maintained that even during sleep their muscles display a continual twitching to produce needed heat. These small animals are quite muscular and are some of the most powerful mammals for their size. The 1.75 ounce **African golden mole** (*Amblysomus hottentotus*) has been known to lift objects weighing 20 pounds, more than 160 times its own weight. The **DeWinton's golden mole** (*Cryptochloris wintoni*) and **Van Zyl's golden mole** (*C. zyli*) will sham death if touched.

Family Erinaceidae (hedgehogs—15 species)

Hedgehogs come in both quilled and quill-less forms. The largest of all insectivores, weighing up to 4 pounds, is a quill-less type known as the **gymnure** or **moon rat** (*Echinosorex gymnurus*), which is at least partially aquatic. Most hedgehogs have short spines which, unlike porcupine quills, are barbless. The **European hedgehog** (*Erinaceus europaeus*) carries approximately 16,000 spines on its back. The spines are sometimes used to cushion jumps or falls.

Hedgehogs feed on a large variety of small animals including bees, wasps, blister beetles, scorpions and venomous snakes, to whose stings and venom they are at least partially immune. The

30

amount of cantharidin (the toxic substance of blister beetles) necessary to kill a hedgehog is fatal to 25 humans. They are 40 times more resistant to adder venom than guinea pigs of the same weight, and they can tolerate 7,000 times as much tetanus toxin as can a human. Their procedure for taking snakes is actually quite safe. They sneak up on the snake, grasp its tail in their mouth, and then curl into a ball. The snake repeatedly strikes its attacker, but it only winds up impaling itself on the spines, which are considerably longer than its fangs. Meanwhile, the hedgehog slowly dispatches it by chewing down into its spine. After finishing it off, the hedgehog uncurls and proceeds to consume its prey tail first. Some species of hedgehog, such as the **long-eared desert hedgehog** (*Hemiechinus auritus*), can survive for 10 weeks without food or water, but most feed quite frequently.

Several hedgehog behaviors once claimed to be no more than myth or imagination are now widely accepted as fact. For instance, one old tale insisted that hedgehogs have a craving for milk that is satisfied by literally taking milk from the udder of a cow. We now know that hedgehogs can and do suck milk from cows. They can easily get at the udders when the cow is lying down. One consequence is that there have been many documented teat injuries as a result of bites sustained in the process.

Another dispute centers on the idea that hedgehogs use their spines

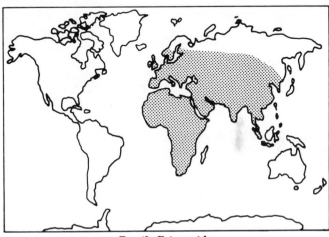

Family Erinaceidae

to transport fruit, rolling over the fruit and thus skewering it. Some claim that this has been observed too many times to be denied and point out that the story was perpetuated by both Pliny and Darwin. Photographic evidence has even been submitted to support the claim, but such "proof" is too easily faked for many to believe. Yet the behavior is possible, and if nothing else it probably takes place accidentally since hedgehogs roll about often and are thus often seen to be cluttered with leaves and other debris. There is, however, no evidence of purposeful thought or known subsequent use of such objects.

When hedgehogs (including *Erinaceus, Hemiechinus*, and *Paraechinus*) meet some astringent substance for the first time, they will often lick or chew it repeatedly to produce copious amounts of saliva. They carefully lick foamed saliva onto their spines. The reason for this behavior has not yet been satisfactorily explained, but the coating of their bodies with saliva mixed with substances such as the poisonous secretions of toad glands may help to repel predators or rid them of some skin parasites.

During the mating season, the male's accessory reproductive glands swell so greatly that he appears to be pregnant. Males court the females by twirling and dancing around them while at the same time rattling their quills. Often the pair will engage in mutual butting for hours on end prior to mating. The females accomodate the males in the actual mating process by flattening down their quills. When born, a baby hedgehog's spines are soft, short, and pliable. As a further safeguard to

31

Hedgehog (*Erinaceus europaeus*)

prevent injuring the mother, the quills can be pushed down into the baby's skin with only slight pressure because the skin is swollen with a large amount of fluid. This excess fluid is absorbed for the most part within 24 hours. The spines gradually harden and become effective for defense at about 3 weeks of age.

When first alarmed, a hedgehog will generally jump immediately into the air, thus jabbing any intruder that has ventured too close. This is often followed by curling into the well-known ball form. This is quite sufficient to ward off most attackers, but they have been observed to take to the water when threatened by predators such as foxes or badgers that can overcome their spiny coat. Quill-less varieties such as the moon rat defend themselves from predators by giving off an offensive onion-like odor from special skin glands. Such glands, although usually much less developed, can be found in one form or another in most insectivores. The **desert hedgehogs** (*Paraechinus* spp. and *Hemiechinus* spp.) protect themselves from the desert heat by becoming torpid during the hottest times of the year, while the **spiny hedgehogs** (*Erinaceus* spp.) and some desert hedgehogs (*Hemiechinus* spp.) hibernate through winter. They are probably the only insectivores to exhibit true hibernation, although tenrecs and others exhibit hibernation-like torpor in both winter and summer.

Family Macroscelididae (elephant shrews—15 species)

Although **elephant shrews** are larger than true shrews, they were not named for their size but for their elongated snout. The long nose is used to probe through leaf litter in search of prey. **Forest elephant shrews** (*Petrodromus tetradactylus*) sometimes tap the ground with their feet when foraging. Some ants respond with warning noises, thus revealing their presence and offering an

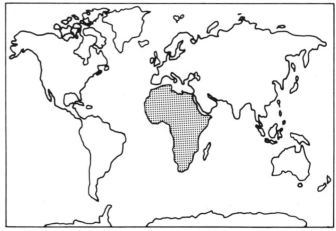
Family Macroscelididae

easy meal. Like true shrews, they are short-lived, with a lifespan of up to 18 months in the wild. Unlike true shrews, however, the young are born at an advanced state; in some species, the young can run almost immediately after birth. As a rule, the more advanced an animal's young are at birth the greater is the chance of their survival. In elephant shrews, this is compensated for by smaller litters; only 1 or 2 of the 80 eggs produced by the female at ovulation are fertilized and mature after she is impregnated by the male's 3-pronged penis. Some pairs may mate for life.

Long-eared Elephant Shrew (*Elephantulus* spp.)

When alarmed, many species warn others by thumping the ground with their tails or hind legs. Their reaction to overhead movement is to jump quickly to one side, an effective maneuver when one considers that birds of prey are their main enemies. When danger threatens a female with young offspring, she momentarily waits to give them a chance to grasp hold of her teats before she bounds to safety kangaroo-style, on her hind legs. The teats are located above the shoulder blades to facilitate this method of transport; clinging to their mother's back, the young run less chance of being knocked off.

Family Soricidae (shrews—265 species)

The shrews truly "live fast and die young." Their average metabolic rate is quite high, compared to other mammals, for metabolic rate decreases with increase in size due to a lowering of the surface-to-volume ratio. If a steer had the metabolic rate of a 1-inch shrew, its surface temperature would have to exceed 212°F (the boiling point of water at sea level) in order to rid itself of internal heat. If the shrew, on the other hand, had the steer's metabolic rate it would require an insulating fur coat 8 inches thick in order to keep warm at room temperature. Though pulse rates vary with species, that of the **cinereous shrew** (*Sorex cinereus*) has been measured as high as 1200 beats per minute during excited states. For comparison, an excited human's heart rate will rarely exceed 150 beats per minute. One desert species, the **eared** or **desert shrew** (*Notiosorex crawfordi*), goes into a daily state of torpor during the dry season. Its body temperature decreases up to 36°F from normal to conserve energy and water.

Clearly, shrews must consume a lot of food to keep their feverish bodies going. It is now generally agreed that they eat about three quarters of their body weight in food daily, and they can eat as much as 3 times their body weight daily, at least when allowed to eat all they want rather than the amount they require for good health. Some species and nursing mothers probably do require more than their own weight in food daily. Like many other small mammals, shrews operate on an activity cycle alternating about 3 hours of activity with 3 hours of rest.

One of the consequences of their high metabolic rate is the ease with which shrews can be thrown into shock. During observation, they have been known to die after sudden, loud noises or even after confinement in unfamiliar surroundings. More recently, this has been attributed to old age or

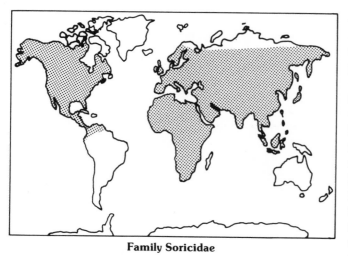

Family Soricidae

debilitation as a result of starvation. Fasting can lead to death from starvation in as little as 3–12 hours (depending upon the source consulted). Some of the more hardy species have been known to last up to 36 hours without food. It has been shown, however, that a shrew deprived of food for several hours can be killed by almost any form of trauma or harassment. They have been known to expire from the sound of gunshot or thunder. Since a few species have average lifespans lasting only a few weeks they are considered the shortest-lived mammals. Some have been known to reach the ripe old age of two in the wild and an astonishing four years in captivity.

Another factor contributing to their fragile nature is their small size. After all, the ribs of many species are about as big around as toothbrush bristles. The smallest of the lot is **Etruscan** or **Savi's shrew** (*Suncus etruscus*). It grows to about two inches in length and at 0.071 ounces weighs less than a dime. This minute beast is so small that it can travel through the tunnels left by large earthworms. It was formerly thought to be the smallest mammal in the world, but that title has since been taken over by the **bumblebee bat** (*Craseonycteris thonglongyai*) whose average adult weight is 0.066 ounce. It would take more than 48 million etruscan shews to equal the average weight of a full-grown female blue whale, the largest of all mammals. It is interesting to note that in mammals smallness is limited by surface area and therefore heat loss, while terrestial largeness is limited by the ability of bone structure to support weight. In marine mammals, largeness is limited by the surface area-related functions of vital organs such as lungs, kidneys, and intestines. A point is reached where such organs cannot keep up with the metabolism of the tissue mass.

Not all shrews are fragile, and in fact some are anatomically built to be ridiculously strong. A case in point is the **armored** or **hero shrew** (*Scutisorex somereni*). It has a unique, intricately-woven, and structured backbone enabling this small, 8 inch mammal (half of which is tail) to support the weight of a 160 pound man without resulting in injury or damage to the shrew itself.

One of many examples of folklore later proved correct is that of the venomous shrew bite. In several species, the saliva is very toxic and is introduced into the wounds inflicted by the teeth. The submaxillary salivary glands are the venom-producing structures. In the **North American short-tailed shrew** (*Blarina brevicauda*), the species with the most toxic venom, there is enough toxin to kill 200 mice. The bite of this shrew is relatively harmless to humans but may cause pain for several days. Other venomous shrews include two species of **water shrews** (*Neomys fodiens* and *Sorex cinereus*). It is thought that the venom is used primarily to paralyze but not kill certain prey such as snails and beetles so that they can be stored and eaten later without spoiling. Other shrews paralyze prey such as insects by crushing their head and/or biting off their legs so that they can be stored alive and used as

34

fresh food at a later time. **Dusky shrews** (*Sorex monticolus*) occasionally store large numbers of incapacitated termites for later feeding. Most shrews shed their first set of teeth as embryos. As a result of their prodigious appetites the permanent teeth can completely wear out in 12–18 months.

The water shrews have adapted well to an aquatic way of life. The **web-footed water shrew** (*Nectogale elegans*) has webbed feet and "suction pads" on its soles for better traction on wet rocks and a better grip on slippery prey. The **Asiatic water shrews** (*Chimarrogale* spp.) have somewhat waterproof fur and can close over their ear openings when diving. **Old World water shrews** (*Neomys* spp.) can regulate their density so that they are capable of floating at the surface, suspending themselves in midwater, or walking along the bottom as if they were on land. This probably is related to the amount of air they trap in their fur which occasionally is of sufficient quantity to give them a silvery appearance when below the surface. Several species can even walk on the water's surface (a behavior known as hydrodromus). Two different mechanisms are used, but light weight (less than 0.5 ounce) is also a common factor. The **Pacific water shrew** (*Sorex bendirii*) and **Northern water shrew** (*Sorex palustris*) have specialized hair-fringed feet capable of trapping small air bubbles, while *Neomys fodiens* has specialized hairs on its feet and tail that swell when wet to increase their surface area for both water walking and increased swimming efficiency. All three have been known to be snatched from the surface by large fish such as trout. The northern water shrew can dive for up to 45 seconds. Strangely enough, most water shrews do not have waterproof coats. After taking a swim, they must force the excess water from their fur, often with the aid of narrow passages in their tunnels. They also shake their fur to shed excess water. The northern water shrew occasionally nests inside beaver lodges.

The **mole shrew** (*Anourosorex squamipes*) spends most of its life in underground tunnels. It both looks somewhat like the true moles and can travel as quickly in reverse as it can forward using the same mechanism as do moles and gophers (discussed later). Some shrews, such as the short-tailed shrews, are able to climb trees. The **forest shrew** (*Myosorex longicaudatus*) is aided in climbing about by its long, prehensile tail.

Water Shrew (*Neomys fodiens*)

Shrews can produce a variety of squeaks and chirps. The **money** or **house shrew** (*Suncus murinus*) emits a sound that resembles the clinking of change. It is believed that some shrews use an echolocation system for navigation and food foraging. Some, like the water shrews and various other animals, are known to memorize certain pathways within their territory. They are thus quite adept at taking quick flight when on their home ground; if they must flee a predator in an unknown area, they blindly run about, colliding with stationary objects. Should a rock be experimentally placed in the middle of a memorized pathway, the shrew will undoubtedly hit it head on. After wandering about a bit, the shrew may pick up his pathway proximal to the obstruction and repeat the stunt a second time. There are even reported incidents in which shrews have broken their necks by diving headlong into a familiar but recently drained pool. Some shrews can run up to 2 mph, which in strict geometric equivalence is comparable to a human sprinting in excess of 100 mph.

Since the bulk of a shrew's food is small animals such as insects, shrews can be very beneficial to humans. Some use ambush techniques in hunting, as when water shrews drift along with the current until they have approached close enough to potential prey to make an easy kill. Many shrews eat their own feces (the process of coprophagy) or the feces of other animals on occasion to obtain vitamins B and K. The **North American least shrew** (*Cryptotis parva*) sometimes enters beehives to feed on the larvae. Occasionally, it will nest there as well. The **vagrant shrew** (*Sorex vagrans*) and others supplement their diet with a fungus (*Endogone* spp.), while **Trowbridge's shrew** (*Sorex trowbridgii*) often consumes seeds and seems to be fond of those from the Douglas fir. The desert shrew obtains all the moisture it needs from the fluids in the insects it eats.

Although the North American least shrew lives in groups, most shrews are solitary, coming together only for mating purposes. Should two males meet underground, they will probably engage in a "shouting" match, squeaking at each other until one gives up and leaves. Sometimes they come to blows until one surrenders by exhibiting a submissive posture, rolling over onto his back. If caged, they will fight to the death, and the victor will as often as not eat his vanquished foe.

One can readily tell when it is the breeding season for the **smoky shrew** (*Sorex fumeus*) since the tail swells during this time. Every so often, the **musk shrew** (*Crocidura russula*) exhibits a population explosion similar to that seen in the infamous lemmings. Musk shrews, however, increase in both numbers and size, some becoming ten times as heavy as normal. The genus *Crocidura* contains more species (143) than any other mammalian genus.

Male shrews usually bite the females during the courting and mating process. This is thought to stimulate ovulation in the females. Many females will again mate the day after giving birth as they can lactate while being pregnant. Mothers generally transport their young while the latter hang from the teats. Several species, including the house shrew, **common shrew** (*Sorex araneus*), and **white-toothed shrew** (*Crocidura leucodon*), form family caravans made up of the mother and her trailing offspring all connected together by mouth-to-rump holds. Thus linked, they march in single file or occasionally in double rows. Presumably this is done so that the poor-sighted young do not become separated from their mother as she leads them foraging and back to the nest. Should the chain break, the mother patiently awaits for the young to hook back up. If the female is lifted up, the remainder of the chain stays intact as the babies dangle in the air. It usually breaks up for good at weaning, but at times a dominant sibling may continue to lead it around for a while longer.

Many shrews, such as the **pygmy shrew** (*Microsorex hoyi*), secrete an offensive odor from special scent glands. This serves to protect them from at least some predators, such as hawks and owls. The musk shrews are reported to be the most offensive of the lot. The desert shrew finds safety when it moves into a beehive, a practice made possible by its small size. Some shrews, when cornered, respond by squeaking while lying on their back with all four legs kicking in the air.

Family Talpidae (moles—27 species)

Moles are best known for and adapted to their burrowing habits. They dig primarily with their large, posteriorly-directed forepaws. Molehills are composed of subsoil thrown out with the front legs, not the snout as commonly supposed. One **Townsend mole** (*Scapanus townsendii*) produced 302 molehills in 77 days, and other individuals have been credited with 500 molehills during a single rainy season.

Even though almost all moles depend on water for survival and are good swimmers, it is surprising to many that some species are aquatic. These include the **musky desman** (*Desmana moschata*) and **Pyrenean desman** (*Galemys pyrenaicus*) which are well adapted with webbed feet and flattened tails. In addition, the tails are covered with ridges of hair to enhance speed and swimming ability. The ears and nostrils are valved. The desman's long snout is used as a snorkel when in the water. On land, the snout can be used in the same manner as an elephant's trunk to place food within the mouth, although desmans rarely stay on land to do more than sunbathe; even the entrances to their burrows are underwater. Another reported use of the long snout is to strike and stun small prey. The Pyrenean desman probably uses echolocation.

Generally, a mole's fur is erect, not set at an angle. This facilitates motion in any direction, allowing moles to run just as easily backwards as forwards. Mole hair is remarkably resistant to dirt clinging to it. As previously mentioned, some have iridescent fur. Fur color in the others is usually brown or black, but in the **hairy-tailed mole** (*Parascalops breweri*) the snout, feet, and tail turn white with age. **Old World moles** (*Talpa* spp.) molt their fur four times per year.

Moles are quite nearly blind and can as a rule only sense the difference between light and dark with their small, atrophied eyes. It has been found that in many the eyes are actually covered with skin (e.g.,

the **Roman mole** (*Talpa romana*), the **Eastern mole** (*Scalopus aquaticus*), and others). Except for **Asiatic shrew moles** (*Uropsilus soricipes*), moles have no external ears.

Moles are said to have more sensitive touch organs per square inch of their body than any other mammals. Many of these organs have special names, such as the Eimer's organs located on the snout and the Pinker's plates found on the belly. These sensitive receptors are used as the mole's eyes and ears. In some species, they are thought to be

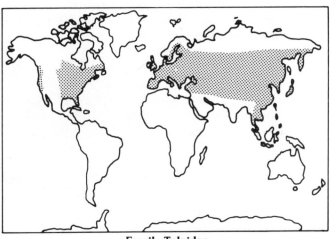

Family Talpidae

sensitive enough to detect the underground movements of nearby worms. In the **star-nosed mole** (*Condylura cristata*) there are 22 "feelers" radiating from the tip of the snout to give a bizarre appearance. These tentacles can be retracted during feeding to avoid being bitten. They are also used to cover the nostrils when swimming and digging to prevent entry of water and dirt.

Digging is definitely the mole's forte. These small insectivores can dig at a rate of 12 feet per hour or about 100 yards per day; some, such as the star-nosed mole, can better these figures by plowing up to 50 yards per hour in very loose soil. Individuals have been observed to dig themselves out of sight in as little as 11 seconds. Many, including the **common American garden moles** (*Scapanus* spp. and *Scalopus* spp.), build two sets of tunnels, one deep and one shallow. The moles take up residence in the former and hunt in the latter. Some tunnel systems have measured more than a mile in total length. The **coast mole** (*Scapanus orarius*) will in addition tunnel down to six feet to obtain water during the dry season. In solitary species, two moles may occasionally share a common tunnel to a water source, but as a rule moles will not tolerate the presence of another except at mating time.

Like shrews and other small mammals, and for the same reasons, many of the moles have a high metabolic rate. Some species can starve in as little as 10–12 hours. Most moles must consume about a fifth of their weight in food daily. Some may need to consume their own weight or more each day in the wild, and captive hairy-tailed moles have eaten three times their weight in 24 hours. Some, such as the **European mole** (*Talpa europaea*), are known to cache stores of up to thousands of earthworms when the latter are plentiful. The moles decapitate the worm, tie the remaining body into a knot, and then bury it in a cavity along with others treated similarly. Since the wounds are not fatal, the worms remain alive and therefore fresh for when the mole returns. The method does possess one drawback in that if the worms are not reclaimed within a short while, the heads regenerate and the worms dig their way to freedom. When eating worms, many moles start with the head end. This way much of the dirt inside the worm's intestines is forced out the anus and is not eaten by the mole. In addition to worms, they are known to eat thousands of harmful insects and have successfully taken on field mice, birds, and even snakes. **Shrew moles** (*Neurotrichus gibbsii* and *Urotrichus* spp.) are the only moles to climb bushes to hunt insects. They also eat some vegetation.

Although the star-nosed moles will sometimes form colonies or pair up for the winter, shrew moles occasionally travel in groups, and *Talpa* species will occasionally construct community tunnels housing as many as 40 moles when food is plentiful, the majority of species are solitary except during the breeding season. The sex organs are often suspended in an inactive state until this time. In fact, female hairy-tailed and European moles have a sealed vagina at all times of the year outside the mating season. The courtship of European moles is rather unusual as well. At mating time, a male will tunnel

Coast Mole (*Scapanus orarius*)

into the female's burrow system (or vice versa). He then immediately takes her captive. Should a second male intrude, the first one will seal the female into a side chamber. An "arena" is set up, and the two moles fight to the death. The victor naturally gets the female but often consumes the remains of his rival first.

Moles have few enemies since they usually remain underground, a way of life known as anachoresis. The musky desman, which does not spend much time below ground, has powerful scent glands located at the base of its tail. These make it unpalatable to most predators. They are occasionally taken by large fish that then seem to acquire the foul odor for quite some time, much to the distaste of fishermen and other predators. Many moles taste bad as well, which explains why pet cats often kill them but don't consume the carcasses. It is said that moles can be deterred by placing garlic cloves in their burrows or by setting empty bottles at an angle so they whistle when the wind blows.

4

ORDER DERMOPTERA
(2 SPECIES)

Family Cynocephalidae (colugos—2 species)

The order Dermoptera consists of only one family and two species of **flying lemurs** or **colugos** (*Cynocephalus* spp.). Colugos is more apt, since they don't fly and they aren't true lemurs. Although virtually helpless on the ground, they are efficient gliders and can travel in excess of 400 feet in the air with as little as 45 feet drop in elevation. The lower front incisors are slotted like combs and are used as such to groom their fur and to strain their food. They feed primarily on fruits and blossoms. Their fur is silky brown but splashed with silver-white spots, causing them to resemble lichen-laden bark, an effective camouflage for animals that spend much of their lives clinging to the trunks and large branches of trees. They often sleep while hanging upside down. During defecation the tail is curled upwards, pulling the flying membrane with it to keep it from becoming soiled. The single young frequently glides along with its mother by clinging to her belly or teats.

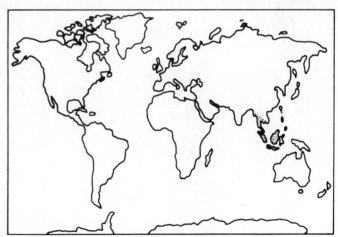

Family Cynocephalidae

Colugo or Flying Lemur (*Cynocephalus volans*)

5

ORDER CHIROPTERA
(942 SPECIES)

Chiroptera is the order of bats, the only mammals capable of true flight. Quite variable in facial appearance, they make up the second largest order of mammals, outnumbered in species only by the rodents. To prevent needless repetition as well as to allow for a smoother presentation, we will consider them as a whole rather than as family groups. To most readers, there are not enough physical differences among bats to divide them into their 19 families.

Bats range in size from the 0.06 ounce (less than a penny) **bumblebee** or **Kitti's hog-nosed bat** (*Craseonycteris thonglongyai*), measuring one inch in length with up to a six inch wingspan, to the almost two pound **flying fox** (*Pteropus neohibernicus*), which grows to 16 inches in length with a wingspan exceeding 5.5 feet. The bumblebee bat recently displaced the Etruscan shrew as the smallest living mammal in the world. This order also contains the most geographically widespread genus, *Myotis*, of all mammals excluding humans and their tag-alongs, domesticated mammals and some rodents.

The flight of bats is not like that of birds. The easiest way to describe it is to say that bats swim through the air using a figure-eight wing stroke, reaching out with their membrane-covered hands to grasp pockets of air.

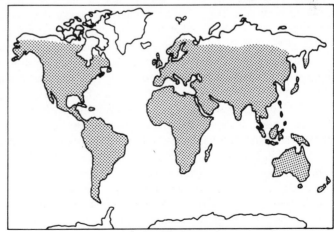

Order Chiroptera

Fruit Bat (*Pteropus* spp.)

43

This is accompanied by a coordinated leg kicking. The resulting flight is often quite remarkable; the fastest bat is the **Mexican free-tailed** or **guano bat** (*Tadarida brasiliensis*), which can reach speeds of 40 mph. In dives, it is reported to reach speeds of 80 mph or more. It is also known to fly up to 10,000 feet to ride air streams at speeds up to 60 mph. Some of the large fruit bats (including *Pteropus* spp.) will occasionally ride thermals in the same manner as vultures. The **lesser horseshoe bat** (*Rhinolophus hipposideros*) can produce close to 20 wingbeats per second. The **pipistrelles** or **flittermice** (*Pipistrellus* spp.) are famous for their aerial ballets. Many species, including the **long-tongued bat** (*Glossophaga soricina*) and the **long-eared bats** (*Plecotus* spp.), can hover, the latter while picking moths and other insects from walls and foliage, the former and others while feeding on the pollen of flowers.

The anatomical structuring of a few bats, including the **giant African leaf-nosed bat** (*Hipposideros commersoni*), makes it impossible for them to regain flight or move about after being grounded. When so stranded, they usually die from starvation or at the hands of a predator. Most species can at least regain flight by jumping into the air, and some, such as the **slit-faced bats** (*Nycteris* spp.) and the **pallid bat** (*Antrozous pallidus*), actually land before giving chase to their prey, primarily scorpions and terrestrial insects located by their sounds. The **greater horseshoe bat** (*Rhinolophus ferrumequinum*) and other species perform an unusual pre-landing maneuver when descending to flat surfaces. It consists of a decelerating somersault that allows the animal to make a two-point landing. Many social bats will land on a conspecific and then jostle for a perch hold. Probably the most agile of all while on the ground is the **vampire bat** (*Desmodus rotundus*). It not only walks about with ease but is capable of running and even jumping in any direction, an ability that can save it from a sudden and potentially fatal swat from its victim's hand or tail. The **New Zealand short-tailed bat** (*Mystacina tuberculata*), which rivals the vampire for the "most agile" title, is aided by long furrows on its soles which give it added traction when running. At such times, its wings are folded into special sheaths. It is known to catch most of its prey while running on the ground or climbing trees rather than in flight. Some bats use thumbhooks to swing from branch to branch.

The thin-membraned wings or patagiums of bats give these animals the highest surface to volume ratio of all vertebrates. The wings increase surface area up to tenfold. The membranes are only 1/5th as thick as surgical gloves but are three times as resistant to puncture. They also exhibit good healing properties in that a hole two centimeters square will completely fill in within less than a month. Some **African big brown bats** (*Eptesicus* spp.) have translucent wing membranes.

Bat wings serve many secondary functions in addition to the primary one of flight. They are important factors in heat regulation. In fact, the wing vessels actually pulsate to ensure an even blood flow. They are used as cloaks or blankets when cold, as raincoats during bad weather, and even as "baseball mitts" to one-hand insects in mid-flight. When the bat is over-heated, it can fan its wings to dissipate heat. If this is not enough, some bats then coat themselves with saliva to make use of evaporation. In addition, some will mass fan themselves when roosting in groups.

Bats fall into two sub-orders. The "insect-eating" or "small" bats belong to the sub-order Microchiroptera. The "fruit-eating" or "large" bats are assigned to the Megachiroptera. Both categories are inexact since many of the insect-eaters feed on fruit and some of the fruit-eaters supplement their diet with insects. Also, some of the small bats are larger than the big bats, and some of the big bats are smaller than the small bats.

44

Probably the best-known feature of bats is their use of sonar or echolocation. It is often incorrectly referred to as radar. Radar is the emission of radio waves, which travel at the speed of light (186,000 miles/sec) and analysis of their reflections; sonar uses sound waves (about 0.2 miles/sec) in the same manner. Sonar is best developed and generally found only in the insect-eating bats; most fruit-eating species rely on sight and smell to get around. As usual, there are exceptions; one genus of fruit-eaters, the **dog-faced bats** (*Rousettus* spp.), uses echolocation although the sounds are tongue clicks, not the vocal squeaks seen in most insect-eaters. As used here, the terms fruit-eating (ers) and insecting-eating (ers) refer to the two suborders and not to the actual diet. Echolocation can be found in all bat families.

It is astonishing that some bats can produce squeaks whose energy output reaches 145 decibels. Such energy is comparable to the noise output of a modern jet during take-off or to that of a pneumatic jack-hammer. Humans incidently experience pain and serious ear damage at sound levels equal to or greater than 140 decibels. Fortunately, bat cries are for the most part in the ultrasonic range (15,000 to 160,000 Hz) and are thus largely inaudible to the human ear which is sensitive only in the range of 20–20,000 Hz. The chirps and clicks of only a few species, including the **Western mastiff** (*Eumops perotis*) and **spotted bat** (*Euderma maculatum*), can be detected by humans. Bats require high-frequency, short-wavelength sonar because prey will only reflect sound waves smaller than their bodies. An **African trident-nosed bat** (*Cloeotis percivali*) can produce calls up to 210,000 hz, the highest pure tone ever recorded in bats.

Bats' ears are quite sensitive to these higher ranges. They often exhibit a special flap or fold known as the tragus which serves to increase their acuity and aids in the localization of sound in some. Vespertilionidae bats require both ears to locate echoes, while Rhinolophidae bats need only one. The former depend on time differences and the latter on pitch changes in order to form their "sound picture." To protect their ears from damage, bats possess small muscles which in effect dampen their acuity during the process of vocalization but relax to give full hearing afterwards to perceive the rebounding echo. When a bat closes in for the kill, the squeaks are produced at such a fast rate that these ear muscles remain constantly contracted; at such close range, the increased acuity is not needed, although protection of the ears is. The inner ears of most species are isolated from the skull by fat and blood-filled sinuses to reduce bone conduction of sound. The **sucker-footed bat's** (*Myzopoda aurita*) ears are unique in that the canals are sealed by a mushroom-shaped valve.

Many bats noses are adorned with numerous ugly folds and skin flaps known collectively as noseleaves. These species beam sound through their noses, although the noseleaves may also serve a function in interpreting reflected sound waves. The shape of the noseleaf flaps can be altered to change the width of the sound beam to accomodate the various distances from their targets and maximize efficiency. Long distances require a wider beam; the beam gradually narrows as the distance shortens. Bats without noseleaves generally emit their echolocation sounds from their oral cavity.

The number of sound pulses per second ranges from 5 to 500. The **Indian false vampire** (*Megaderma lyra*) emits up to 300 squeaks per second. Since most fly by sound, each animal must be able to distinguish its "voice" from those of its companions, which may number in the millions. Some experts feel that each bat emits a slightly different wavelength or depends on some subtle difference to enable it to detect its own echoes and allow harmonious, safe flying. Actually, bats do occasionally collide when leaving roosts in large numbers. **Moustache bats** (*Pteronotus personatus*) and horseshoe bats are known to alter emission frequencies to compensate correctly for Doppler shifts when flying.

Overall, the echolocation system is quite accurate. It is sensitive enough in some to detect flying insects at distances of 45 feet, although in the vast majority of species it is effective for less than 3–10 feet. It has been shown that the Indian false vampire and greater horseshoe bats can detect wires that are 0.004 inches in diameter or about the same thickness as a human hair. It has also been found that some bats can analyze reflected waves despite the presence of background noise that is 2,000 times louder than the echoes. Sonar is also specific enough so that many bats are even able to differentiate small insects during a heavy rainfall where the sizes of raindrops and prey are the same. They are probably able to differentiate insect species as well.

One final note about bat sonar is that once a bat has developed a mental picture of its surroundings it will often fly through the area "blind," that is without using echolocation. This is clearly demonstrated in experiments where bats continue to dodge obstacles that have been removed and will collide with new ones, sometimes fatally, that have been placed in a previously cleared area.

Information on intraspecific communication in bats is sketchy at best. Recently, it has been thought that some bats communicate contentment to each other through body vibrations rather than voice, although the latter is certainly used between young and their mother and is extensively developed in the pallid bat. When on a collison course bats will sometimes alert each other by lowering the frequency of their calls to prevent crashing.

The great majority of bats are covered with a brownish or black fur, but a few sport colorful or unusual coats. One of the African pipistrelle bats has at least three different coat colors dependent upon the region it inhabits. In Europe, this species is dark brown; in the Sahara, it is light brown, in the Algerian hills, it is found with a somewhat greenish adornment. Geographical color variation, also present in many other animals including pocket mice, desert larks, butterflies, etc., is dictated by the environment through natural selection. The **ghost bats** (*Diclidurus* spp.) are named for their white fur, while the **yellow-winged bat** (*Lavia frons*) has wing membranes splashed with various shades of yellow, orange, and green. The spotted bat is pied with black and white blotches, and the **painted bat** (*Kerivoula picta*) is wholly covered with a reddish-orange fur which camouflages it during the day, when it spends its time sleeping among piles of fallen leaves. Probably the most gaudy of the clan, however, is a species of **long-tongued fruit bat** (*Macroglossus* spp.). It is covered with an iridescent orange fur which also serves as camouflage since it gives it the appearance of a clump of colorful fungi while it sleeps in trees. Some of the brightly colored species, and the darker ones as well, are known to become blondish or a drab gray at times through the bleaching effect of the ammonia given off by their guano (wastes). The **hoary bat** (*Lasiurus cinereus*) roosts in trees and uses behavioral camouflage in that it increases its leafy appearance by swaying with the wind. A tightly clumped group of **painted bats** (*Kerivoula argentata*) resembles a mud wasps' nest, while horseshoe bats sleep with their wings and tail membranes wrapped around their bodies giving them the appearance of fruit pods. In the vast majority of bats the sexes are indistinguishable, but in a few, such as the **straw-colored bat** (*Eidolon helvum*) and the **American red bat** (*Lasiurus borealis*), the males are generally larger and even a little more colorful. Male **hammer-headed bats** (*Hypsignathus monstrosus*) are nearly twice as heavy as females, exhibiting the greatest size difference between sexes. In **Rafinesque's big-eared bat** (*Plecotus rafinesquii*), the female is often heavier than the male. In some species, females have

46

larger wings as an aid to carrying young, while some males may have longer tufts of hair in the nape and shoulder area, often in conjunction with scent glands.

Two of the strangest bat species are actually hairless except for small areas on the head, neck, and tail. They are appropriately known as the **naked bats** (*Cheiromeles* spp.). When at rest, their wings are folded into special, unique pouches on each side of their body. Recently it has been found that a small insect related to earwigs lives inside the wing pockets of these bats. When frightened the bats can emit a foul odor. On their first (big) toes, they have special bristles with which they brush off their skin.

Within the order Chiroptera can be found many unusual anatomical adaptations. The **wrinkle-faced bat** or **lattice-winged bat** (*Centurio senex*) has a large flap of skin emanating from its chin. This can be drawn over its face to form a protective hood, usually while roosting. The flap has transparent areas overlying the eyes so that the bat can see gross light changes even with the "hood up."

Mouse-tailed bats (*Rhinopoma* spp.) and slit-faced bats have valved nostrils like camels. These can be closed to prevent the entrance of wind-blown sand and dust. The slit-faced bats' tails have a T-shaped tip, unique to the mammal world, while mouse-tailed bats have extra long tails (nearly equal to the head and body length) used to feel their way as they back into narrow crevices. Many bats, including the **harpy fruit bat** (*Harpyionycteris whiteheadi*), have no tail at all.

The larynx in male hammer-headed bats is so large that it comprises a third of the body cavity and displaces both the heart and lungs. At times, these males will gather in trees to take part in a wing-flapping "singalong," to attract receptive females. They are thus one of the rare species of mammal to exhibit such behavior (lek display).

The **flat-headed bat** (*Platymops setiger*), as its name indicates, has a flattened skull better enabling it to live and hide in crevices and under rocks. The **disc-winged bats** (*Thyroptera* spp.), sucker-footed bats, and a few of the fruit and vespertilionid bats have special "suckers" on their wrists, ankles, and feet. These work so well that many can use them to crawl up vertical, smooth surfaces. In **Spix's disc-winged bat** (*Thyroptera tricolor*) the discs are used to hold the bat while roosting, which is done head-up instead of head-down as in most other bats. The discs not only have muscles to produce suction but produce sticky secretions as well. A single disc can support the bat's whole weight. The **little bamboo bats** (*Tylonycteris* spp.) are named for their habit of taking up residence inside bamboo stems, entering through holes cut by **chrysomelid beetles** (*Lasiochila goryi*). They have both a flattened skull and adhesive discs on the thumbs and feet for clinging. Instead of suckers, the New Zealand short-tailed bats have dual claws, one above the other, for increased clinging ability. They use their teeth to cut out tree cavities for nesting.

Bats often roost in humid areas such as caves. This helps prevent their wings from dessicating. In many insect-eaters, this is complemented by the application of oily secretions obtained from facial glands. Most bats roost in an upside-down position; their femurs are too weak in most cases to support them in an upright position. They do not fall while asleep because when their legs are placed in an extended position by the pull of gravity on their body, the tendons attached to the toes are pulled taut so that the toes are held in the flexed grasping position. In fact, should they die while in this position, they will remain hanging by their feet. In cold weather, bats will often hang by one foot and fold the other leg up under the blanket-forming wing. Mouse-tailed bats often roost rightside up hanging by their thumbhooks, and a few bats, such as the **velvety free-tailed bats** (*Molossus* spp.), usually rest horizontally rather than vertically.

Almost all bats are crepusuclar (twilight-active) or nocturnal, but the fruit bats and certain species of the **sheath-tailed bats** (Family Emballonuridae) are diurnal and can be seen flying about in fruit trees or chasing after insects in broad daylight. Most bats spend the daytime sleeping in the shelter of trees, caves, rocks, loose bark, brush, man-made structures, or crevices. They emerge at intervals set by internal clocks since they often are unaware of light changes outside their roosts. Some hang out in unusual places. The **silver-haired bat** (*Lasionycteris noctivagans*) is frequently found in bird nests, and naked bats often take up residence in ground holes. The **woolly bats** (*Kerivoula* spp.) often are found in quite strange places. *K. lanosa* sleeps in abandoned bird nests, *K. picta* prefers weaverbird nests and flowers, and *K. harrisoni* has been known to spend its day sleeping in spider webs. Slit-faced bats are reported to inhabit mammal burrows and human latrine pits in some areas, while several species of **round-eared bats** (*Tonatia* spp.) and a **spear-nosed bat** (*Phyllostromus hastatus*) have been found roosting inside termite nests. The **Mexican fishing bat** (*Myotis vivesi*) is sometimes found sleeping under rocks or in old turtle shells, and **leaf-nosed bats** (*Hipposideros fulvus*) have been found in **porcupine** (*Hystrix* spp.) burrows. The **proboscis bat** (*Rhynchonycteris naso*) will occasionally sleep in groups lined up in straight rows along tree trunks floating on lakes. These bats resemble lichens and strengthen this appearance behaviorally by curving their heads and noses backwards when threatened to further resemble the curled edge of lichens.

A few bats actually construct primitive shelters. The "tent-building" bats such as the **little fruit bats** (*Artibeus* spp.), the **white bat** (*Ectophylla alba*), and the **yellow-eared bat** (*Uroderma bilobatum*) cut through the ribs of wild plaintain leaves or palm leaves and then fold them over to produce crude tents. **Short-nosed fruit bats** (*Cynopterus* spp.) are the only Old World bats to construct shelters. They do so with folded leaves as well as by cutting out the center of a fruit cluster and then hanging inside the empty space remaining. The **banana bat** (*Pipistrellus nanus*), disc-winged bats, and ghost bat are frequently found nesting inside rolled leaves with some species changing to a new leaf each night because they require a certain size as the leaves unroll.

During their daytime sleep many insectivorous bats reduce their energy requirements by automatically lowering their body temperature, blood pressure, and respiration rate, going into a state of torpor or diurnation. Upon awakening, they must warm themselves through shivering and wing-flapping before they are able to take part in normal activity.

In temperate areas, insectivorous bats adapt to the cold either by migrating, sometimes up to 1500 miles, or, much more commonly, by hibernating. Fruit bats, on the other hand, reside in the tropics and do neither. Sometimes a single species of insect-eater such as the American red bat will do both; that is, some of the group will migrate while the remainder hibernate. **Gray bats** (*Myotis grisescens*) and guano bats are known to migrate north in winter to their hibernating caves which provide a more humid environment. It has been calculated that in some small insect-eating bats that hibernate, 92 percent of their lifetime is spent in sleep or rest, which is probably a mammalian record.

Respiratory rates during hibernation number about 10 per hour, quite slow compared to the normal resting rate of 200 per minute and a flying rate as high as 600 or more per minute. While flying, there is sometimes a synchronization between breaths and wingbeats, both functioning at 10 per second. Hibernating heart rates fall to 25 beats per minute compared to 400 beats per minute at rest and 1000 beats per minute during flight. Body temperatures also drop from around 104°F into the low 30s.

48

Two of the hardiest hibernators are the big brown bat and the American red bat, which are occasionally found covered with snow or icicles and can survive temporary body temperatures of less than 30°F. The **little brown bat** (*Myotis lucifugus*) reportedly has survived a body temperature of 20°F without harm. It has also been found at temperatures of 130°F, making it one of the most temperature-tolerant species of mammal in the world. Another species resistant to cold is the **Northern bat** (*Eptesicus nilssoni*), which can even be found in the Arctic.

Hibernating bats generally awaken at intervals, sometimes to do nothing more than urinate or lick condensation from their fur before nodding back off; they are also known to move to another area or even to do some winter feeding. Natural arousal at hibernation's end is thought to be triggered primarily by loss of fat, internal clocks, and an increase in ambient temperature. It takes them 30–60 minutes to awaken fully.

Some bats require high humidity levels during hibernation. The lesser horseshoe bat requires about 95 percent relative humidity and can be killed if it drops to 80 percent or lower, probably from dehydration secondary to water loss from the wings. Others, such as the Mexican free-tailed bats, can tolerate humidity of 60 percent without problem.

Bats lose from a quarter to a third of their body weight in fat during the hibernation period. In preparation, mouse-tailed bats develop large abdominal pads of fat that may equal the weight of the remainder of their body. In some species, juveniles who are unable to put on sufficient fat reserves before winter will die during their first hibernation from exposure and/or starvation. To help conserve heat and/or water, many bats group together in large clusters. Bats are stated to be the most thigmotactic (contact-seeking) of all mammals. Some can be found in densities as high as 300 per square foot with multiple layers of bat hanging on bat. Should cave explorers (spelunkers) disturb the hibernating bats, it could cost them added energy loss that might lead to death from exposure.

Due to their small size and therefore large surface-to-volume ratio, bats are required to eat large quantities of food. This often causes the fruit-eaters to be a menace and the insect-eaters to be beneficial. For example, one **short-nosed fruit bat** (*Cynopterus sphinx*) can eat twice its weight in fruit within a mere three hours, and the **Western pipistrelle** (*Pipistrellus hesperus*) eats about one third its weight in insects nightly. The little brown bat can make up to 1,200 individual insect captures within a single hour, averaging one catch every three seconds. Individual gray bats may consume more than 3,000 insects during one night's feeding. It has been estimated that Mexican free-tailed bats in the state of Texas alone consume anywhere from 7,000 to 20,000 tons of insects each year qualifying them as one of the best insecticides available. Time from ingestion to excretion can be as short as 15 minutes.

Frugivorous bats are not all bad since they are sometimes the principle and occasionally only pollinators of certain tropical and subtropical plants, including about 20 genera of plants, and at least partial pollinators to an additional 110 genera, a process referred to as chiropterophily. Bats are particularly important to those species that blossom only at night, such as the saguaro cactus (*Carnegia gigantea*) and the midnight horror (*Oroxylon* spp.). The latter's flowers not only open at night but emit a terrible stench that attracts the pollinating bats, as does the African sausage tree (*Kigelia* spp.). The night-blossoming calabash tree (*Crescentia* spp.) is pollinated solely by the long-tongued bat, a species classified among the insect-eaters. Other plants known to be pollinated by bats at least in part are the silk cotton trees (*Bombax* spp. and *Ceiba* spp.) and many fruit trees including dates, figs, peaches, mangoes, avocado, banana, and guava. Occasionally plants are injured and gardens marred because some of the fruit bats eat the blossoms' petals. Some species feed only on plant nectar and on the pollen that sticks to their fur.

Another plus in these bats' favor is that they do aid in the dispersal of the plants they feed on. In fact, some seeds exhibit a higher germination rate after passing through a bat's gastrointestinal tract. This method of dispersal is known as chiropterochory. The pits or seeds are often dropped or eliminated unscathed to new fertile areas. This results from several behaviors. First, most of the fruit-eaters do not actually consume more than the fruit's juices and merely discard the remainder after extracting the liquid. They often will fly away from the original tree to eat the fruit in a quiet place. In addition, fruit bats rarely roost in caves where a tree obviously would have no chance to grow. There are exceptions, such as the **Egyptian fruit bat** (*Rousettus aegyptiacus*), which dwells in caves and old graves where it uses echolocation.

Many fruit-eaters rely on the fruit for their Vitamin C since like humans, they themselves are unable to synthesize it. In seeking their food some, such as the short-nosed fruit bat, are known to fly up to 70 miles in a single night to obtain fruit. One flying fox blown off course was found flying 200 miles out at sea. Some dog-faced bats are known to steal fermenting palm juices from the collecting pots of natives. The bats subsequently become drunk, stuporous, and either fall to the ground or fly off with a definite "stagger."

The fruit bats are well adapted to their calling. To gain the fruit juices they need, they either suck out the juice directly or crush the fruit against special ridges on the tongue and/or hard palate, after which they reject the pulp and seeds. Many have a well-developed sense of smell enabling them to track down ripe fruit. The vision of some flying foxes is ten times more sensitive to light than human vision, which also aids in finding fruit. No bat is truly blind. Many are also capable of using their hands and dexterous thumbs to hold and manipulate the fruit. **Epauleted fruit bats** (*Epomophorus* spp.), named for their shoulder glands, feed on soft, ripe fruit while hanging upside down. In order to continue breathing while feeding, these bats possess a large, inflatable air sac that can be filled and then closed off from the throat while remaining open to the airways to supply fresh air. They are usually very messy and wasteful in their feeding habits and often take only a nibble of each fruit, leaving the rest to rot. Short-nosed fruit bats have a facial groove to channel juices between the nostrils while hanging upside down. This arangement prevents the inhalation and subsequent aspiration of acidic fruit juices. The **tube-nosed fruit bats** (*Nyctimene* spp.) have tube-shaped nostrils to serve the same purpose. The wrinkle-faced bat has papillae between its lips and gums to strain out coarse pieces of fruit that will not fit through its half-inch throat. In most fruit-eating bats, the fruit juice rapidly passes through the digestive tract (average time is about twenty minutes) and often gives their roosts a fruity odor. Nectar-eating bats have rougher textured hair to better entrap protein-rich pollen needed for a balanced diet. Such pollen is consumed when the bat grooms its fur. Unfortunately, pollen is difficult to digest. Remarkably, some of these bats, such as **Sanborn's long-nosed bat** (*Leptonycteris sanborni*), drink their own urine to increase the efficiency of pollen digestion.

The insect-eaters have also adapted well to their way of life and comprise about 70 percent of all bat species. The long-eared bats have huge ears measuring almost as long as the head and body together. They are sensitive to the almost silent flight of the moths upon which they primarily feed. While sleeping, the ears are conveniently rolled up to decrease water/heat loss and avoid injury. They

50

are then tucked into their armpits beneath the wings. Some bat species have cheek pouches to store prey for later eating. The little brown bats crush prey at rates of 7 chews per second.

It is also interesting to note how prey have adapted in response to bats. For example, some **noctuid moths** (Family Noctuidae) have special organs to detect bat sonar and can take immediate evasive action such as flying away or in erratic patterns. If the bat is quite close, the moth will immediately drop to the ground. These organs are occasionally parasitized by mites, a condition that may prove fatal to the moth since infected "ears" don't function. It is possible that some moths can actually "jam" bat sonar with their own special emissions. Recent research, however, seems to indicate that these moths (Family Arctiidae) produce special sounds audible to the bat indicating that they are unpalatable in the manner of warning coloration seen in brightly colored insects. It has even been found that some palatable species mimic the sounds of foul-tasting ones and therefore are not eaten. Some frogs avoid bats by only calling on moonlit nights when approaching bats can be seen. They are also known to call from thorny bushes to deter attacking bats. In these instances, it has been shown that the bats home in on the frog calls rather than using their sonar.

Many of the insect-eating bats, such as the **vesper bats** (Family Vespertilionidae), use their tail membranes or uropatagiums as nets for catching insects as well as for temporarily storing them while on the wing. Prey can be put into the caudal reticule or can be taken out and eaten while in flight. This is one reason for some of the short, sudden drops observed during flight. Some investigators feel that the tail membrane is used only as an aid to gain a better hold on the prey which they say is caught in the mouth only, although photographic evidence exists to indicate otherwise.

Prey can also be caught with the wings. This allows the bat to capture the prey despite its last moment attempts at evasive maneuver. Many bats do catch most of their prey with their mouths, and some such as the **leaf-chinned bats** and **moustached bats** (Family Mormoopidae) have bristles forming a "funnel" into the mouth cavity to facilitate this behavior.

Many insect-eaters actually specialize in other prey. The Indian false vampire preys on small vertebrates such as mice, baby birds, and frogs. Its sonar is specialized to pick up the presence of prey against a solid background. It skins mice and plucks birds before eating them and often consumes no more than the head and trunk. The **giant false vampire** (*Macroderma gigas*) feeds mainly on wild house mice but also eats birds, reptiles, insects, and other bats. The African trident-nosed bat feeds almost wholly on moths. **Linnaeus's false vampire** (*Vampyrum spectrum*) feeds not only on birds and rodents, but on other bats as well, including the spear-nosed bat. One **hollow-faced bat** (*Nycteris thebaica*) feeds primarily on scorpions. The **mouse-eared bat** (*Myotis myotis*) locates the beetles upon which it feeds by smelling them out as it hops along the ground. Insectivorous bats that specialize in fruit include many of the **American leaf-nosed bats** (Family Phyllostomidae).

Bats that feed on fish include the **fishing bat** (*Myotis vivesi*), the **fishing bulldog bat** (*Noctilio leporinus*), Indian false vampire bat, some **mouse-eared bats** (*Myotis macrotarsus*), and a few others. The fish are located by sonar detection of ripples or of body parts that protrude above the water's surface (bat sonar does not appreciably penetrate water). These bats do not catch fish with their tail membrane as has been claimed. At times, the bats skim the surface with their claws trailing in the water until they hook into a fish. Several runs are made over the same area, and it is thought that the white bellies of the fishing bat may attract the fish to the surface so that subsequent runs are much more successful than blind chance. The fishing bulldog bat is known to catch insects in flight with its enlarged fish-catching feet. Sometimes these fish-eating bats fall or dive into the water, but they are capable swimmers and can easily take off from the water's surface. Their fur is greasy to retard water absorption but helps little to protect them from carnivorous fish which are known to snatch them from the surface.

Most bats require a regular intake of water. This is usually obtained by skimming the surface with open mouth. If knocked into the water most can hop or swim to shore, but they may die anyway from exposure. In times of drought the **straw-colored bats** (*Eidolon helvum*) will bore into tree trunks and drink the exuding sap. Some supplement their drinking by licking dew from their fur. Some **flying foxes** (*Pteropus* spp.) supplement their diet with seawater to obtain needed minerals. A few desert species can go without free water, obtaining what they need from fluids in their insect prey.

By far the most feared of bats is the dreaded **vampire bat** (*Desmodus rotundus*). Vampires, the only parasitic mammals on Earth, do not induce hypnotic trances by fanning victims with their wings or spraying them with a "calming soporific scent" as claimed. These stories probably stem from the fact that vampires frequently circle over their victims for up to five minutes before landing. They have soft skin pads on their wrists and feet to help keep them from waking their prey. The vampire takes blood from various mammals while the **hairy-legged vampire bat** (*Diphylla ecaudata*) and the **white-winged vampire bat** (*Diaemus youngi*) prefer bird blood. Vampires detect vascular areas with the aid of heat-sensitive pits on their faces. Experimentally, they have been shown to prefer dark skin over light and young animals to old; they even have anatomical preferences such as in birds the feet, in pigs the nipples (some pigs actually counter by learning to sleep on their bellies), and in humans the cheeks, nose, and digits.

Up to seven bats may feed from the same wound, which they inflict with their razor-sharp teeth (vampires possess only 20 teeth total, the fewest of any bat species; The greatest number, 38, is seen in **funnel-eared bats** [*Natalus* spp.], disc-winged bats, sucker-footed bats, and some vespertilionids). It is the sharpness of these teeth that causes the bite to be painless and not the presence of an anesthetizing agent in the saliva as once postulated. The resulting wounds bleed freely for as long as eight hours due to the presence of an anticoagulating agent known as desmokinase. The blood of victims is not sucked from the body but is instead lapped up with the tongue which can move in and out up to five times per second. Small, coarse projections at the tip of the tongue continuously abrade the wound, adding to the effect of the anticoagulant in keeping the blood flowing. Often the blood is pulled along tongue grooves by capillary action so that little tongue movement is required. When fed out of a dish, the tongue can be rolled up straw-like to suck up the blood. These little beasts will sometimes engorge themselves drinking their own weight or more in blood, occasionally to the point where flight is very difficult if not impossible. (Mouse-tailed bats will sometimes eat so much in summer that they become too fat to fly.) Vampires are not grounded for long, however, since they quickly eliminate about 40 percent of their meal through urination which often begins seconds after the bat begins feeding, a process lasting anywhere from ten to forty minutes. It should be remembered, too, that rabies can be transmitted through the urine as well as the saliva.

A single vampire bat drinks about an ounce per meal and two gallons of blood per year, but it has been estimated that another three gallons is lost to continued seepage so that each bat is responsible for the loss of five gallons of blood per year. A single vampire drinks about twenty-five gallons of blood during its lifetime. People and cattle are sometimes killed by vampires, but not because of blood loss

52

(although birds and small mammals have died of blood loss). Mammals that fall prey to vampires are done in by secondary infections resulting from Chagas disease or, worse yet, rabies. It was once thought that vampires were immune carriers, but it is now known that they succumb to rabies like all other mammals, although some individual vampires may become unafflicted carriers. Both of these illnesses, as well as others such as histoplasmosis, can be transmitted by other bats. What makes the vampires such important vectors in transmission is the fact that they feed solely on blood. This increases their chance of exposure, although bats can pick it up from each other's urine.

As a rule, bats are scrupulously clean, but most have scent glands causing them to give off a musky odor. **Naked bats** exude a very foul odor when frightened while others such as the **fishing bulldog bat** gives off such a strong odor that it can normally be detected up to 100 yards away. The **Neotropical fruit bat** (*Artibeus jamaicensis*) actually has a body odor resembling perfumed soap. Scent glands are sometimes used to hold territory. For example, male **whitelined bats** (*Saccopteryx* spp.) participate in a ritual known as "gland shaking" when they fight for territory. The gland is located in the wing membrane. In shaking, the bat will push his wing forward, open the gland, and shake out some of the secretions. This behavior usually serves to drive away the intruding bat. Some sheath-tailed bats mark their living space with gland secretions, and some male **free-tailed bats** (Family Molossidae) mark their females with chest gland secretions.

Many bats live in groups known as camps or colonies. Some camps become incredibly large. In the recent past, a group of Mexican free-tailed bats in Texas was estimated to contain 20–30 million bats at one time, and another group in Arizona was estimated at 25–50 million, forming the single largest gathering of a mammalian species. When they took to the air at dusk, it appeared as though billows of smoke were pouring from the roosting area. In fact, such camps were so large that it may have been physically impossible for all the bats to leave each night within the appropriate time period for successful feeding. There are reports that the bats rotated turns, going out an average of only once every third night so that only a third of the population would have to come and go daily. They would each consume enough forage to hold them over for a few days. In some species that roost in trees, the numbers have on occasion become so great that large tree limbs snap under the resulting weight. The sexes often roost separately, particularly during the birthing time when females come together to form large "maternity roosts." This may serve to increase ambient temperatures and speed growth in the young. Some, such as the silver-haired bats, are solitary for most of the year but come together to form large congregations that migrate south. If some of the bats arrive too late, they go into hibernation.

Most bats breed only once per year, but some, like the slit-faced bats and the **black tomb bat** (*Taphozous peli*), breed twice a year. In most species, the testes are intra-abdominal and descend into the scrotum only at mating time. Most mate in the usual mammalian manner while hanging from their daytime roosts, but with no premating rituals or courtship. In a few, such as the hoary bat, the male grabs the female in mid-flight whereupon both flutter to the ground where mating occurs. There are even reports of bats coupling in the air. Mating usually takes place just prior to winter. Bats' penes can thrust independently (as in elephants). Sometimes they are reported to mate with sleeping females or even to hibernate through winter locked in copulation. In horseshoe bats, **noctule bats** (*Nyctalus* spp.), and others, the females' vaginas become plugged after mating to prevent sperm loss and/or further mating. Male sac-winged bats will claim choice feeding areas as their territory. They then allow only the females that mate with them to feed there. Male **greater spear-nosed bats** (*Phyllostomus* *hastatus*) take over and defend harems of females. The average harem numbers 18. Painted bats, false vampires, and others form monogamous family groups.

Some bat species use delayed fertilization (unique to bats among mammals), and in a few delayed implantation (also seen in deer, mustelids, rodents, seals, and others) so that the young will be born at the optimum time for survival. The sperm in the former case remain viable for up to 7 months. Most mammals' sperm degenerates within a few weeks' time at most, but in these bats the female's uterus actually secretes a nutritive substance to "feed" the sperm. Sperm storage also seems to depend upon the maintenance of cold temperatures.

Actual gestation ranges from six weeks to eight months. Most females give birth to a single young although routine twinning is common in a few species, including many of the vespertilionids. A few bat species have been known to have up to four young at a time, but this is unusual and only **hairy-tailed bats** (*Lasiurus* spp.) regularly produce three or four young at once. This genus also has the only species in which the females have four functional teats, all others having only two. Amazingly, there are reports that a female can carry her brood about when flying. This is no mean feat since their combined weight at times exceeds her own.

Many female bats, when giving birth, hang by their thumbhooks and form a "cup" with the wings and tail membrane to catch the young. Babies are reported to be born headfirst in some and tailfirst in others. The vast majority of all mammals are born headfirst, and species that regularly give birth tailfirst are very rare. The females often eat the placenta. Insectivorous bats are usually naked at birth, whereas fruit bats already have some fur. Baby bats instinctively climb upwards, and since female bats do not construct nests for their young, the babies must cling to their bodies, the ceiling, or other baby bats. Should one fall to the ground, it will probably die of starvation, predators such as **guano beetles** (*Dermestes* spp.), which do feed on guano as well, or aphyxiation from inhaling the high concentration of ammonia emanating from the guano abundant below large roosting areas. Big brown bat females and some others are known to retrieve young that have fallen to the ground. In fact, the young of this species seem to fall often.

The babies of many are frequently carried upon the mother's back, but in some, such as the horseshoe bats, **leaf-nosed bats** (*Hipposideros* spp.), and others, the females possess false teats on their abdomen on which the young bats take hold. They also may wrap their wings around her body for additional support. Others sometimes use the real teats in like manner, but most merely take hold of the female's fur. After they have grown somewhat, the young are often forced to roost on their own to await their mother's return. It has been found that while in the womb some baby bats become imprinted upon the "voice" of their mother. After birth they are able to recognize her approach with food and can squeak out to her so that she can easily find them. She identifies them through either sound or smell. The young can generally see quite well and are capable of consuming more than their own weight in insects each night, a chore that keeps the mother very busy. In some, such as big-eared bats and **free-tailed bats** (*Tadarida* spp.), the females are known to nurse any hungry baby regardless of whose it may be, an unusual behavior for a mammal. In a very few species, the males may help raise the young. A male Linnaeus's false vampire was frequently observed to wrap his wings around the female and offspring. On initial flights which take place at 20–90 days of age depending on species, the young are led about by their mother who emits orientation sounds the young can follow.

The average lifespan for most bat species is probably in the 5–10 year range. Fruit bats can live for more than 20 years and occasionally past thirty; one **Indian flying fox** (*Pteropus giganteus*) lived in excess of 31 years in a zoo. Insectivorous species generally have a maximum longevity of 17–20 years, although one little brown bat was known to have lived 30 years in the wild.

Bats have very few natural enemies but are preyed upon by various species of bird including the **bat hawk** (*Machaerhamphus alcinus*). It specializes in feeding on bats but has had its numbers drastically reduced indirectly in many areas as a result of insecticides and directly by human destruction of habitat and the elimination of its food. In some studies, bats were found to be more susceptible to DDT than any other mammal tested, a clear case of an inferior insecticide destroying a superior one. Some bats, particularly young red bats, are killed in accidents as attested by bat bodies found impaled on thistles, cactus, and barbed wire. A few species, such as the mouse-tailed bat, will warn others of approaching danger by waving their long tails.

Bat caves are usually marked by the presence of guano, at times in layers as thick as 100 feet. It is interesting to note that some bat species have special respiratory and blood system adaptations allowing them to tolerate high concentrations of ammonia (up to 5,000 parts per million) without ill effect. This same concentration is sufficient to kill a mouse in 10 to 15 minutes. Some dung heaps produced by bats actually form the only niche for unique animals such as a wingless earwig (*Arixenia* spp.) and an unusual gecko, as well as commoner animals such as insects, beetles, cockroaches, crabs, millipedes, mites, and pseudoscorpions. Bat dung was once used extensively as fertilizer, and during the U.S. Civil War was used for its nitrate content to make saltpeter for the manufacture of gunpowder.

Bats have even played a part in warfare. Free-tailed bats were used (at least in preliminary tests) to carry incendiary bombs during World War II. Usually the bats would land, chew off the harness, and fly away. The small bombs produced a 22-inch flame that burned for eight minutes. Known results include the destruction of at least one test village and one accidental burning of an auxiliary air base near Carlsbad, New Mexico.

Bats can on rare occasion be dangerous to humans, particularly spelunkers. They do not, as popularly believed, get tangled in human hair, although they will swoop down close to humans and other animals to capture insects that may be swarming there. The greatest real danger comes from rabies. Tests show that rabies can be transmitted through the mere inhalation of urine excreted by infected bats. To make things worse, bats almost always urinate and defecate upon being awakened or disturbed. This allows for quicker flight due to lightening of the load. It should also be kept in mind that rabies is easily transmitted via a bite, and that a healthy bat is not easily caught. Thus one should never handle a "friendly" or seemingly "injured" bat (or any other mammal, for that matter) even though only eleven cases of bat-transmitted rabies have been confirmed in the U.S. and Canada as of 1983, and only two deaths have been documented in the rest of the world (excepting South America).

6

ORDER PRIMATES
(203 SPECIES)

Most primates are large-brained, tree-dwelling animals inhabiting the tropical and subtropical regions of the world. This order contains not only the simians—monkeys, apes, and man—but also the more primitive prosimians. The prosimians are represented by the first six families below.

Family Tupaiidae (tree shrews—16 species)

Little is known about these squirrel-like animals which should probably be placed in an order of their own, Tupaioidea or Scadentia. They are the only primates in which the male's scrotum is located in front of the penis, and the only primates to have claws on all their fingers and toes instead of nails. Claw-like nails on all but first digits are found in some primates including the aye-aye and some bush babies.

The smallest of all primates is the less than two ounce **pen-tailed tree shrew** (*Ptilocercus lowii*). It has a naked tail with a white-tufted end used to signal others of its kind at night. During sleep, the tufted end is sometimes used to cover the face. It is the only nocturnal tree shrew species, and the only one whose primary

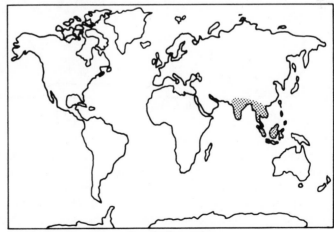

Family Tupaiidae

Red Howler (*Alouatta* spp.)

means of locomotion is hopping.

Tree shrews are omnivorous. When eating, they generally hold their food in the front paws like squirrels. After finishing a meal, they often dip their paws in water and wash their faces. They drink and bathe frequently.

Most tree shrews are solitary, although a few live in family groups. The males in some species have been reported to fight to the death over disputed territory.

In the **common tree shrew** (*Tupaia glis*), the male alone constructs the leaf nest for the offspring, a unique behavior in mammals. The females then enter the nest to give birth, clean the young, tidy the nest, and feed each of the newborn about a third its weight in milk. She then leaves the nest and limits further mothering to ten-minute feeding sessions only once every 48 hours. The males take no part in parental care. Female **terrestial tree shrews** (*Tupaia tana*) and **arboreal pygmy tree shrews** (*Tupaia minor*) also exhibit 48-hour suckling behavior. The fat content in female tree shrews' milk is 25 percent, the highest of all primates. Delayed implantation probably takes place in the common tree shrew.

Except in the common tree shrew, it is the female's duty to make the nest. In some species, two females will work together to make a single shared nest. While one gives birth, the other drives away trespassing males.

When alarmed, tree shrews are known to whistle antiphonally. That is, when one stops, another takes up. This is thought to confuse lurking predators and keep them from isolating a victim.

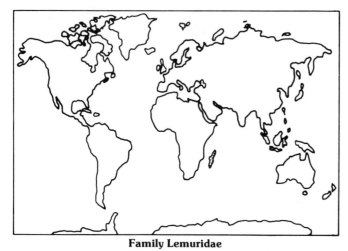

Family Lemuridae

Family Lemuridae (lemurs—23 species)

The lemurs are sometimes split among three separate families, the Lemuridae, Cheirogaleidae, and Lepilemuridae. If not counting the tree shrews as primates, the smallest primate in the world is the **lesser mouse lemur** (*Microcebus murinus*). It tips the scales at 2–3 ounces when full grown. It, the **dwarf lemur** (*Cheirogaleus major*), and the **fat-tailed lemur** (*Cheirogaleus medius*) avoid summer heat by slipping into a dormant condition similar to hibernation but known as estivation. To support this practice, they store large quantities of fat near the base of their tail as well as in their subcutaneous tissues, particularly in the hind legs. The dwarf lemur spends this time buried in rotting tree trunks. Its temperature drops as much as 36°F. The fat-tailed lemur may "sleep" in this manner for as long as 8 months.

As a group, lemurs are scrupulously clean. They spend long hours grooming themselves with their lower incisors, which are notched to resemble combs. Both they and the lorises also have a special structure used to clean away the hair and debris that clog these combs. This horny "toothpick" is found beneath the tongue and is appropriately known as the sublingua ("below the tongue").

Lemurs are primary vegetarians. The **ring-tailed lemurs** (*Lemur catta*) obtain water by lapping up dew from vegetation in addition to that obtained in the fruit they eat. **Coquerel's mouse lemur** (*Microcebus coquereli*) milks flattid bugs for their sweet exudate as some ants do to aphids. The **sportive lemur** (*Lepilemur leucopus*) is coprophagic, the only primate to exhibit this lagomorph habit.

The larger lemurs, such as the ring-tails and the **black lemurs** (*Lemur macaco*), are gregarious, roaming about in small groups of 5–20. The females are almost always dominant to the males. Females usually outnumber the males in any particular group, and it is the females that often fight to defend the group's territory.

In most species, the sexes are similar in appearance. In the sexually dimorphic black lemurs, the males are all black while

Ring-tailed Lemur (*Lemur catta*)

the females are reddish-brown with white bellies. This led them to be classified originally as two separate species.

Male lemurs mark their territories with urine and feces, although a few, including the ring-tails and the **gentle lemurs** (*Hapalemur griseus*), use secretions from special underarm scent glands for the job. Both also have special spurs on their forearms to mark branches, and ring-tails have glands on their forefeet that are used to clean the tail.

As the breeding season approaches, male ring-tailed lemurs become aggressive and engage in what has been called "stink fights." During such confrontations, the smelly secretions from their underarms and forearms are wiped onto the animals' long tails or onto sticks which are then flicked toward other males. In the process, some of the scent is wafted into the air towards the rival and helps to establish a hierarchy among them.

It is interesting to note that in some species such as the dwarf lemurs and mouse lemurs, the female's vagina is sealed except during estrous and birth.

Lemur nests are generally constructed in safe, inaccessible areas. Some, including the **true lemurs** (*Lemur* spp.), do not build nests; instead the young ride the mother by clinging to the fur on her back. **Ruffed lemur** (*Varecia variegata*) females leave their young in a nest lined with some of their own fur.

A few of the smaller species exhibit unusual nesting behaviors. The **reed** or **broad-nosed lemur** (*Hapalemur simus*) builds a nest composed of floating reeds on the water's surface in large marshy areas. The **fork-crowned lemur** (*Phaner furcifer*) frequently takes up residence in hollow trees already occupied by bee swarms. The bees are active by day and keep predators away, while the lemur is most active at night.

Most lemurs are diurnal. Those that are nocturnal include the fork-crowned lemur, the ruffed lemur, "dwarf lemurs," and the **weasel lemurs** (*Lepilemur* spp.). All lemurs, including the nocturnal species, will participate in sunbathing, often referred to as "sun-worshipping" since they assume a prayer-like posture at these times.

Females usually give birth to a single offspring although twins are not uncommon and even make up the majority of births in the ruffed lemur. In at least one instance, a female ring-tail was seen to eat one of her twins but successfully raise the other. In general, female lemurs eat the placenta after birth and then lick their newborn while holding it in the front paws. They also eat the young's wastes. The females transport their young in the mouth or on the back, depending upon the species. Both sexes aid in rearing the young, as do other adults in the larger social lemurs.

At night some, such as the ring-tailed and black lemurs, keep in touch with each other by grunting in unison. This often takes the form of a crescendo and can be repeated many times and as often as every 2–5 seconds throughout the night, an annoying occurrence to anyone attempting to sleep within ear range.

When harassed or chased, lemurs will generally flee. Some take to evasive maneuvers such as jumping from trees and making off through the underbrush, while others, such as the black lemur and fork-crowned lemur, have been observed making flying leaps up to thirty feet through the trees.

Family Indriidae (indri and sifakas—4 species)

The **indri** (*Indri indri*), largest of the prosimians, has secretive habits. Natives believe that it is the reincarnation of their dead and that parent indris teach their young to make long jumps and not to fear heights by throwing them back and forth among the treetops. Whether this is true or not, the indri is an excellent leaper. A vegetarian, it is thought to be monogamous, with the females dominant to the males although the latter defend the territory. Their presence is usually betrayed only by their loud, mournful wailing audible for distances greater than a mile. Indris often descend to ground level to consume dirt, possibly for its mineral content.

The **sifakas** (*Propithecus* spp.) are also great leapers capable of

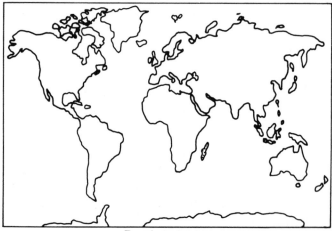

Family Indriidae

60

jumping distances of forty feet in single bounds through the trees, including thorned varieties (*Euphorbia* spp.). Such feats are due in part to the presence of skin flaps, known as patagiums, located on their sides. Sifakas, being primarily arboreal, will sit in trees and sunbathe by the hour, turning frequently so that no spots are missed. While in the sitting position, the tail is rolled spirally and rests between the animal's legs. They are strict vegetarians and feed primarily on fruit, which they peel with their teeth. During those infrequent times when they descend to the ground, they hold their arms up over their heads to maintain balance. Ground locomotion often consists of hopping kangaroo fashion with jumps known to exceed ten feet. They present a somewhat humorous picture with their arms up for steadying when they pick fruit up from the ground with their teeth. Each time a sifaka vocalizes, its head is snapped back suddenly with great speed. All members of this family mark territory with urine and glandular secretions. Troops of sifakas frequently invade the territory of neighboring troops. Observers report that at these times each animal continues to face its rival's original position while hopping forward, sideways, or backward. On occasion, this has led to two rivals ending up back to back.

Family Daubentoniidae (aye-aye—1 species)

The **aye-aye** (*Daubentonia madagascariensis*) uses its extended, middle fingers for scratching, combing its fur, fanning water into its mouth when drinking, picking its teeth, and feeding. The base of this finger is connected to the hand with a ball and socket joint, greatly increasing its mobility. This small primate with large ears has such acute hearing that it is capable of detecting wood-chewing grubs beneath bark. With its continuously growing in-

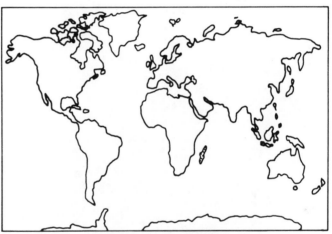

Family Daubentoniidae

cisors, it quickly chews away the bark and proceeds to extract the invertebrate with its elongated finger, a feature and behavior it shares with **striped possums** (*Dactylopsila* spp.). It is reported that should the grubs remain silent, the aye-aye will tap along the tree surface and dig out any hollow sounding areas, a procedure said by some to yield a fair catch of edible bugs. Some zoologists claim that this tapping behavior does not take place, but that instead the animals smell out the larvae with their keen noses. Ayes-ayes also feed on coconuts and other fruit which they open with their large, rootless incisors. They have the peculiar habit of grunting as they eat. Aye-ayes generally live singly or in pairs and are known to spend much time grooming while hanging by their hind legs.

Family Lorisidae

Bush Baby (*Galago* spp.)

Family Lorisidae (lorises, pottos, and bush-babies—10 species)

Both the **slender loris** (*Loris tardigradus*) and the **slow loris** (*Nycticebus coucang*) are reminiscent of sloths since they go about their business of living at a very slow pace. They are capable of moving quickly, however, since once they have crept up on their prey, mainly birds and insects, they make a final pounce with lightning-fast speed. Defenseless prey is often not even killed before eating begins and noxious insects are eaten with impunity. Lorises have comb teeth and a toilet claw like their relatives the lemurs. The forearms of lorises and pottos possess a complex network of blood vessels known as the rete mirabile. This system serves as a blood storage area allowing them to clamp onto a branch and maintain their hold for long periods of time. Their second fingers and toes are stub-like, allowing them to exert a greater pinching force. Lorises obtain moisture by sucking wetted fingers. The slender loris is said to be unable to swim. The slow loris can emit foul odors if threatened.

The **potto** (*Perodicticus potto*), whose movements are similar to those of the lorises, will attempt to bump predators from trees by lunging at them with special spines on vertebrate C7–T2. The elongated spinous processes are covered only with a thin layer of skin and some erectile hairs. Some experts feel that the spines serve a sensory function, enabling the potto to know when to make its lunge since it places its head between its forelegs when threatened. Whatever the case, as it snaps its head upward, the closing predator receives a sharp blow which often displaces it from the tree or at least gives it a sharp rap in the nose or eyes. The potto is one of the few animals reported to mate frontally. The sexes mark each other with genital gland secretions.

The **angwantibo** (*Arctocebus calabarensis*) spends most of its life hanging upside down. Its index fingers are even more stub-like than those of the potto or loris. Its grip is thought to be at least partially involuntary, being mechanically similar to the grip of bats' feet. The angwantibo's hold is so persistent that should one die while hanging onto a limb, which is reportedly not a rare event, the body will not fall but remain in place. Angwantibos can maneuver their way upright by passing their body up through their hind legs from which they are hanging. Their diet contains large numbers of irritant caterpillars. They carefully groom themselves with their teeth after meals. Areas that are inaccessible to the mouth are groomed with the hands after coating them with saliva. When threatened, they too place their head between their forelegs but then distract the predator's attention with the stubby tail which has a special patch of erectile hairs. As the enemy moves in, the angwantibo quickly snaps out and bites it. This often scares it away or causes it to lose its balance and fall.

The **bush-babies** or **galagos** (*Galago* spp.) have suction-padded feet that allow them to gain a firm foothold in trees. This is important since these small animals, in contrast to their lethargic relatives, are quick runners and jumpers known to make leaps of up to 25 feet from branch to branch and vertical jumps of up to 9 feet from a standstill. Their name is derived from the sounds they produce. These are frequently described as being similar to the cry of a human infant. Their necks are quite mobile and can be turned through 180 degrees in either direction, allowing them to look directly to the rear. Bush-babies can move their ears independently, allowing them to locate easily the insects upon which they feed. Some, such as the **dwarf bush-baby** (*Galago demidovii*) and **Allen's bush-baby** (*Galago alleni*), can fold their ears by curling them from the tip or by folding them at the base. At the slightest sound, they quickly unravel and perk up. They are usually closed only when sleeping to shut out excessive noise.

Besides being the smallest, the dwarf bush-baby is also one of the most colorful. It has green fur on its back with yellow on its sides, although the colors usually fade in captivity. In the wild it is known to take up residence in abandoned squirrel nests.

When sleeping, bush-babies huddle together in groups of two to seven to form a comfortable, warm ball of bodies. Some species, such as the **thick-tailed bush-baby** (*Galago crassicaudatus*), are known to supplement their insect diet with flowers. At times they pollinate them in the process, as in the case of the baobab tree. This species of bush-baby is also known to get drunk on palm wine, and many are known to feed on tree gum. Bush-babies often snatch insects out of the air with a single hand.

Lorises, the potto, tarsiers, and bush-babies "wash" their hands by urinating on them. This is done to mark out their territory as they travel about. All species in this family are nocturnal and arboreal. Bush-baby mothers keep their nests clean by eating their offsprings' waste products. Some bush-babies discourage predators by nesting in thickly thorned "wait-awhile" trees (*Acacia tortilis*).

Family Tarsiidae (tarsiers—3 species)

Like the bush-babies, the **tarsiers** (*Tarsius* spp.) have suction-padded feet. They can rotate their

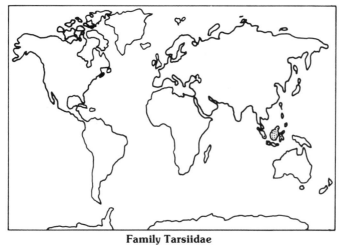

Family Tarsiidae

heads through almost 360 degrees, and they are the only mammals known that can turn one ear forward and the other backward simultaneously. The eyes on these small animals are truly gigantic. Each eye of the **Western tarsier** (*T. bancanus*) outweighs the brain. They are said to be the largest in proportion to head size of all mammals. One authority estimated that should humans have eyes proportionately as large on a weight for weight basis, they would have to be nearly a foot in diameter.

Tarsiers are some of the very few primates that are entirely carnivorous and are even known to feed on venomous snakes and scorpions. Locomotion consists almost wholly of hopping and jumping in the manner of frogs, and tarsiers can jump six feet in a single bound on the ground and up to 20 feet between trees. They usually live singly or in pairs and do not wash their faces except by rubbing them on small branches or leaves. Occasionally females carry their young in their mouths.

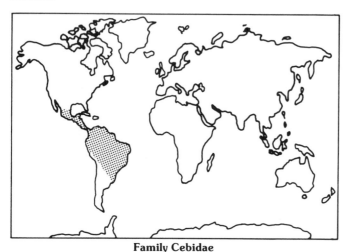

Family Cebidae

Family Cebidae (New World monkeys— 32 species)

The Cebidae and the Callitrichidae contain those primates generally referred to as the New World monkeys. All but one of these animals are diurnal or active by day. The lone exception is appropriately known as the **night monkey, owl monkey,** or **douroucouli** (*Aotus trivirgatus*).

The New World monkeys differ from the Old World monkeys in a number of ways. The New World species have nostrils facing sideways instead of downwards, possess three premolars in each side of both jaws rather than two, and are generally unable to oppose their thumbs to the other fingers although there are exceptions, including the night monkey, **squirrel monkeys** (*Saimiri* spp.), and the **capuchins** (*Cebus* spp.). None of the New World monkeys are ground dwellers, have cheek pouches,

64

or sport calloused pads on their buttocks as is often the case in the Old World varieties, which are also more commonly sexually dimorphic. Exceptions include the **white-faced saki** (*Pithecia pithecia*), whose males are black with white faces while the females are brown with black faces, and the **black howler** (*Alouatta caraya*) whose males are black while females are colored gray or cream.

None of the New World monkeys are tailless and some can use their tail as a "fifth hand". That is, the tail is prehensile; this is never the case with adult Old World monkeys, although very young guenons do have "handy" tails. Prehensile-tailed monkeys include the **spider monkeys** (*Ateles* spp. and *Brachyteles* spp.), who use theirs to obtain fruit or pick up nuts out of arm's reach; the **woolly** or **pot-bellied monkeys** (*Lagothrix* spp.), who rest hammock-style by hanging from their tail and arms or can use the tail for support when standing on the ground; and the **howler monkeys** (*Alouatta* spp.), who shoo away bothersome insects and may at times remain hanging from their tail even after being shot dead.

The many New World monkeys whose tails are not prehensile include the squirrel monkeys, **titis** or **widow monkeys** (*Callicebus* spp.), and the **uakaris** (*Cacajo* spp.). These monkeys' tails do, however, function as rudders and brakes and help maintain balance. Uakaris are the only monkeys in this family whose tails are shorter than the combined head and body length. When titis sit near each other, they often intertwine their tails.

The **bald uakari** (*Cacajao calvus*) is also known as the **blushing monkey** since its face glows a brighter red when it becomes excited. The degree of redness in its normal state is influenced by the amount of sunlight the animal is exposed to. The young possess thick head hair but lose it with maturity.

Monkeys are very sociable animals and generally associate in certain types of group patterns. These will be outlined in the section on Old World monkeys.

Squirrel monkeys huddle together and wrap their tails over their shoulders for added warmth on cool nights. It is interesting to note that, like people, these monkeys become more restless and even agitated when the moon is full. They are extremely agile and are known to perform somersaults in mid-air, seemingly for the mere fun of it. They are also very curious and will frequently be found stalking people who intrude into their territory.

Most monkeys are omnivorous. Squirrel and night monkeys can catch insects on the wing with their quick hands and will even make daring leaps to do so. Night monkeys also lick pollen from flowers to supplement their diet. The capuchins, sometimes called **organ-grinder monkeys**,

Capuchin (*Cebus* spp.)

PHOTO: LOS ANGELES ZOO

65

will crack open hard-shelled nuts by whacking them with bones or rocks. They leisurely feed from fruit held in their tail as well. They occasionally prey on other monkeys including titis and squirrel monkeys. White-faced sakis occasionally raid hollow trees to obtain bats, which they skin before eating.

To drink, most monkeys simply immerse their snouts and take in their fill. The **sakis** (*Chiropotes* spp. and *Pithecia* spp.) obtain their water by wetting the thick fur on the back of their hands and then sucking out the entrapped water. These unusual primates often walk upright in the manner of gibbons.

Much behavior observed in social animals serves to suppress hostility and aggressiveness. In monkeys this often takes the form of mutual grooming. When people observe this grooming of others or self, they almost always assume that the animals are picking out vermin of some sort. Actually, monkeys are usually quite free of fleas and the like and are usually picking out small bits of salt exudate and flakes of skin which they promptly eat. They also pick out stickers, thorns, and scabs, and if by luck they should find a juicy bug in the process they surely will not turn their nose up at it.

Monkey troops generally stake out temporary territories which they guard jealously against neighboring troops until moving on to new areas or establishing new boundaries. The hoots and howls of the howler monkeys and others are actually verbal "no trespassing" warnings to nearby monkeys; most New World monkeys reinforce their cries with physical markings consisting of urine, feces, or special secretions.

The champion vocalizers of the New World monkeys are the howlers and the night monkey. The howlers are among the loudest mammals in the world. They have a modified hyoid bone enlarged to form a resonating chamber known as the corniculum. Only the males produce the booming call, which in the **red howler** (*Alouatta seniculus*) can be heard up to three miles away. The night monkey has an inflatable pouch connected to its windpipe to increase volume and can produce more than fifty different calls. Their vocalizations are variable, ranging from sounds resembling a roaring jaguar or a barking dog to those suggestive of a mewing kitten and even gong-like noises. Needless to say, these communicate much more information than just territorial warnings. They also serve to signal the approach of predators, finding food, and courtship functions, among others.

As previously mentioned, many monkeys mark their territory physically as well. Night monkeys do so by urinating on their hands and feet so that a tell-tale scent is left wherever they go. They also frequently embrace others in their group by hugging, kissing, and entwining their tails to impart a common scent among themselves. Capuchins coat their fur with cologne - fruit, onions, urine, and other strong-smelling substances they find. Squirrel monkeys cover themselves with smelly glandular secretions and copious amounts of their own urine, both to mark out their trails through the treetops and to repulse would-be attackers.

Social hierarchies are the rule for monkeys. In the **common squirrel monkey** (*Saimiri sciureus*), the males threaten other males or exhibit their dominance via erections. Most others use facial expressions or body posture for the purpose.

In most monkey groups, the dominant male has first rights to the female of his choice. Night monkeys are one of the few species thought to be monogamous. Titis also live in pairs, giving the appearance of stability. They mutually defend a territory through which they wander together. When the mating season arrives, however, the pair bond loses much of its strength and the male will mate with the partner of any less dominant male. Afterwards, the couples will reunite and take up life where they left off. In howler monkeys, the female informs the male of her readiness to mate by sticking her tongue out at him. The male acknowledges with the same sign. In spider monkeys, the courtship takes a form known as "grappling," in which the pair sit facing each other on a branch. While so doing they aggressively push, pull, belt, slap, and bite each other. Soon it ends with the male chasing the female about until they "make up," after which the female will usually return to sit on his lap while they manipulate each other's genitalia with hands, feet, and mouths. Paradoxically, the **black spider monkey** (*Ateles paniscus*) females are distinguished from the males by their longer sex organ. This stems from the fact that the females sport a very long clitoris while the male's penis is quite short.

The mothering instinct in monkeys is well developed. Most New World monkey mothers carry their young on their backs, although the sakis and woolly monkeys carry theirs on the belly. Some monkeys will continue to fondle and carry their young even after death, and occasionally after decay has begun to set in. The females usually do most of the caring for the offspring, although in titis the male does most of the work except for feeding. When a baby capuchin first begins to wander from its mother's side, she will grasp its tail to keep it from slipping as well as to ensure that it does not wander away. Baby howlers purr when content, and it is known that the parents will sometimes form a "bridge" with their bodies across a wide gap in the foliage so that the young can easily cross over with the rest of the troop. Spider monkeys are said to help each other across such gaps by holding vines or even rarely by grasping each others' tails to form a monkey "chain" that easily swings over small streams and wide gaps.

New World monkeys usually avoid predators by staying in the trees; some species never touch ground. When intruders roam about below, howlers, and spider monkeys, and others will shower fruit, vegetation, or excrement upon them. Spider monkeys have dropped branches weighing up to eleven pounds towards intruders.

Family Callitrichidae (marmosets and tamarins— 15 species)

The **marmosets** and **tamarins** (many genera) include some of the most adorable of all primates. For instance, the **Emperor tamarin** (*Saguinus imperator*) sports a striking white moustache up to two feet long while the **common** or **white eared-tufted marmoset** (*Callithrix jacchus*) has bushy tufts of fur on the sides of its head. The **golden marmoset** (*Leontopithecus rosalia*) has a golden mane reminiscent of a

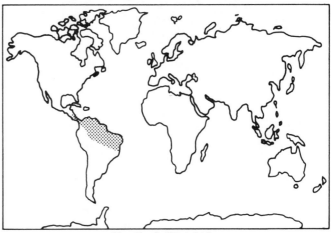

Family Callitrichidae

lion's. Some marmosets can make leaps of up to sixteen feet, and many are known to twitter like birds, particularly after feeding. This is probably most marked in the **crested bare-faced tamarins**

(*Oedipomidas* spp.) sometimes referred to as "**nightingale monkeys.**" Like birds, their singing is thought to "mark" their territory. The calls of some species, such as the **pied tamarin** (*Saguinus bicolor*), range in the ultrasonic frequencies, 10,000–50,000 hz. The long tails present in this family are not prehensile.

Emperor Marmoset (*Saguinus imperator*)

Callitrichids are omnivorous. The **pygmy marmoset** (*Cebuella pygmaea*) chisels holes into trees to feed on the sap and gums. In fact, a square yard of tree bark may contain hundreds of such holes.

The sexes in this family are generally codominant. Female marmosets usually do the territory guarding and are the ones that mark out the area with urine. The common marmoset has the largest brain to body weight percentage of all mammals. Its brain makes up 5.5 percent of its body weight. In comparison a human's brain is about two percent of its weight.

During courtship the males flick their tongues in and out and loudly smack their lips as part of their display. Sometimes the pair will come together to touch tongues. The males also run about pressing their scent gland-laden scrotums against various objects. **Cottontop** (*Saguinus oedipus*) females solicit males during the breeding season through the use of "perfume," actually a combination of her urine and glandular secretions mixed together and carefully rubbed into the tail fur. In many species only the dominant female mates and produces offspring.

Female marmosets and tamarins most often give birth to twins. In pygmy marmosets, the smallest species of true monkey, the newborn are about the size of a bean. Their small fingers can only be seen well with the aid of a powerful magnifying glass. In most species, the father "delivers" the young and then cleans them off. As a rule, the male takes possession of the offspring. In some species, he does so immediately after birth, while in others he may not do so for one to three weeks. They are surrendered to the female for fifteen minutes or so every two to three hours so she can feed them. (This same behavior is also seen in the **titis** (*Callicebus* spp.) of the preceding family.) The males frequently carry the young wrapped around their necks like scarves. When the young begin to take solid food, the father aids them by mashing it up in his fingers or even pre-chewing it in his mouth. The fingertips have sickle-shaped rather than flat nails as in other true monkeys. Both sexes use these claw-like nails in the manner of hooks to support their weight when they are resting in the treetops.

When angry or alarmed, many of the tamarins will raise their furry manes. Cottontops are known to erect the white head plumes for which they are named.

68

Family Cercopithecidae (Old World monkeys—85 species)

The Old World monkeys are divided into two groups, those with cheek pouches for temporary storage of food (Subfamily Cercopithecinae), including the baboons, drills, guenons, macaques, mangabeys, and moors; and those without pouches (Subfamily Colobinae), such as the guerezas (or colobus), langurs, and leaf-eating, proboscis, and snub-nosed monkeys.

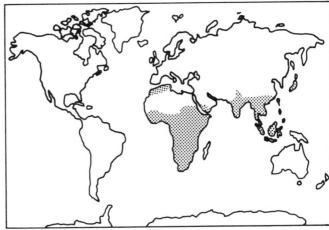

Family Cercopithecidae

Although color vision is often taken for granted by humans, few other mammals possess it. Those that do include the tree shrews, alone among the prosimians; most monkeys, except for the night monkey, a few felids; some mustelids including the mink, stoat, and weasel; a few rodents such as the golden hamster and some tree squirrels; sloths; a few viverrids including the suricate; and finally a few ungulates including the horse, giraffe, and red deer. Color vision is widely present in nonmammals, including birds, reptiles, amphibians, many fish, and a few invertebrates such as the octopus and various insects.

When more than one species of monkey dwells in the same locality, the various species frequently occupy different vegetation levels in order to avoid competition and promote peaceful relations, although there does tend to be a lot of overlap. This behavior has come to be known as arboreal stratification. As a rule, in tropical areas there are four layers of vegetation, the ground layer (0–20 feet), the lower canopy (20–40 feet), the middle canopy (40–75 feet), and the upper canopy (75–90 feet). The **guenons** (*Cercopithecus* spp.) are often used to illustrate this principle. For example, in one forest the **DeBrazza's monkey** (*C. neglectus*) lives at the ground level, the **red-tailed monkey** (*C. ascanius*) sleeps in the middle canopy but spends the day on the ground, the **blue guenon** (*C. mitis*) lives in the upper canopy but forages in the middle, and the **Diana monkey** (*C. diana*) spends almost all its time solely in the upper canopy.

The **macaques** (*Macaca* spp.) are the greatest in number and are the most widespread of all Old World monkeys. The most northern monkey is the **Japanese macaque** (*M. fuscata*). Some, such as the **pig-tailed monkey** (*M. nemestrina*), are found to be right- or left-handed, eyed, and footed as in humans. The **stump-tailed macaque** (*M. arctoides*), like many human males, becomes bald in old age. Its naked face turns red with excitement or heat and bluish when the macaque is cold. One of the best known is the Japanese macaque. It gathers into the largest troops of all monkeys (up to 1,000 individuals in some troops as a result of artificial feeding by humans).

Most monkeys inhabit tropical regions. The few exceptions which have taken to cold areas include the **snub-nosed monkey** (*Pygathrix roxellanae*). It inhabits regions up to 9,000 feet and is protected by fur that can grow up to a foot thick. These monkeys will embrace for hours to keep warm. The

69

Japanese macaque or **snow monkey** has also taken up residence in snowy areas, often strategically located near hot springs from which the monkeys do not wander far when young are present in the troop. These monkeys spend hours on end submerged in these hot baths during winter. Upon exiting, their fur often becomes coated with icicles since the ambient air can be as much as one hundred degrees colder than the water. After a heavy snowfall, these monkeys will construct new "roadways" and will also roll large "snowballs" to be used as stools. During very cold spells, they will huddle together to keep warm.

Most monkeys are capable of swimming and many, such as the **proboscis monkey** (*Nasalis larvatus*), take to it readily for what appears to be the fun of it. The proboscis dives into rivers from the trees and can swim beneath the surface for distances of forty-five feet. Whole groups have made dives from fifty feet high. The **crab-eating** or **dog-faced macaque** (*Macaca fascicularis*) swims underwater while obtaining a portion of its food in the form of aquatic animals, and the Japanese macaques that live on the island of Koshima will actually "skin dive" to depths of five feet in the ocean in order to feed on strands of nourishing seaweed. They are also known to play and bathe daily in the sea.

Old World monkeys are primarily omnivorous. Those with cheek pouches, such as the **baboons** (*Papio* spp.), will feed while on the move since the cache can be temporarily stored in these pouches. Their diet ranges from scorpions to milk. When eating the former, the baboons are careful to pinch off the stinger. They gain the latter from the stomachs of goats and calves they have killed. **Savannah baboons** (*Papio cynocephalus*) are known to chase down and capture hares for eating. During the dry season, grass may comprise 90 percent of many baboons' diet. Many colobines eat dirt, termite mud, sand, and dead wood.

Geleda Baboon (*Theropithecus gelada*)

The crab-eating macaque occasionally "washes" its fruit and seeds by rubbing them down with moist leaves, while Japanese macaques rinse the sand from their sweet potatoes and separate grain from sand by placing both in the water. Some also dip their sweet potatoes in seawater before each bite. This is possibly the only example in the animal kingdom of a nonhuman animal "seasoning" its food.

Talapoin monkeys (*Miopithecus talapoin*), the smallest of Old World monkeys, transform poisonous manioc roots into edible morsels by allowing them to lie out and ferment before eating them. **Guerezas** (*Colobus* spp.) spend almost their entire lives in the trees eating leaves but will on rare occasion descend to the ground in order to obtain needed minerals such as salt. Male proboscis monkeys sometimes require manual displacement of their large noses in order to get their food into their mouths. The large nose straightens out on its own only when the monkey gives its alarm call. During such times the nose acts as a resonator.

Most Old World monkeys readily drink water. A few, such as the guenons, rarely do more than lap rainwater and dew found on tree leaves. In times of drought, baboons and patas monkeys will dig down into dried-up river beds to obtain water.

Baboons, the most terrestial of Old World monkeys, are said to be afraid of the dark. As evening approaches, they take to the trees to sleep until dawn's light.

Many monkey behaviors are either signals of communication or outward expressions of emotions, again a form of communication. Fondness is expressed via mutual grooming, although this is tied to dominance in many species such as baboons, in which the females regularly groom the males while the reverse is seldom if ever done. Male **mandrills** (*Papio sphinx*), largest of true monkeys, show their desire to be groomed by shaking their heads and shoulders. Baboons' lip-smacking is interpreted as "I mean you no harm," and it has been used successfully by humans to avoid attack by hostile baboons. Subordinate baboons often pacify attacking group members by displaying their rumps in a pseudosexual behavior. The dominant attacker then mounts (regardless of sex) as if to say "all is forgiven." The males threaten by raising eyebrows, lowering the lids, and "yawning" to expose razor-sharp canines. "Yawning" is actually a widespread symbol of aggression in primates, although the guenons use it as a tension releaser when frightened. The guerezas release tension through closed-mouth chewing motions, which are sometimes used as greetings as well. The macaques and **mangabeys** (*Cercocebus* spp.) threaten by raising their scalp and eyebrows. In the mangabeys, the upper eyelids are covered with white fur which is thus suddenly exposed to enhance the threat display. Some think that these white eyelids are also used to flash signals of warning since the monkeys blink their eyes frequently when discovered. The crab-eating macaque displays dominance through tail positioning, subordinates letting theirs droop while dominant animals hold their tails up over their backs. Baboons mark their areas with visual cues, especially erections. At these times, an erection has no sexual significance. In the mandrills, friendliness is shown by exposing the teeth, a signal of threat in most other primates.

It is interesting that many monkeys, such as the baboons and macaques, are born into their hierarchy position. That is, the young of dominant monkeys are born to dominance while the young of subordinates become subordinate. The latter can "work" their way up, often by first becoming the "servants" to the dominant animals. The hierarchy is often changed via physical overthrow of dominants by subordinates. At times, as in some baboons, two or three subordinate males will group together to become dominant over a more powerful male who could easily overpower any one singly. These "club" members quickly come to the aid of each other to maintain their dominance. When a male leader of a **langur** (*Presbytis* spp.), red howler, or other troop is beaten by a member of a bachelor group, he is driven out with the other males of his troop. The victor then takes over the females. Sometimes he will even kill all infants fathered by the previous leader. He then mates with the females and begins propagation of his "dominant" genes, while assuring food for himself and his offspring. When overthrown by a once subordinate monkey, a Japanese macaque will exhibit stress by pulling out and eating some of its own fur. A low-ranking male baboon will often seek out another troop because of the difficulty of rising in the dominance hierarchy of his own troop. Subordinate barbary ape males often "borrow" a female's infant before approaching a dominant male. The infant's presence inhibits male aggression and thus protects the inferior male from attack.

Although most Old World monkeys travel in groups of 2–1000, all primates tend to socialize in one of three ways. In the first, the "multi-male" grouping, the troop is composed of several males, females, and their offspring. Primates in this group include the **black lemur** (*Lemur macaco*), **common langur** (*Presbytis entellus*), **howler monkey** (*Alouatta palliata*), **macaques** (*Macaca* spp.), **Mangabeys** (*Cercocebus* spp.), **ring-tailed lemur** (*Lemur catta*), **savannah baboon** (*Papio cynocephalus*), **sifakas** (*Propithecus* spp.), and the **vervet** (*Cercopithecus aethiops*), to name but a few.

In the second, the "single male" grouping, each troop is composed of one male, several females (one of which may be the troop leader), and the young. Other males generally form bachelor groups. Examples include the **colobus** (*Colobus guereza*), **gelada baboon** (*Theropithecus gelada*), **gorillas** (*Gorilla gorilla*), **guenons** (*Cercopithecus* spp.), **hamadryas baboon** (*Papio hamadryas*), most **langurs** (*Presbytis* spp.), and the **patas monkey** (*Erythrocebus patas*).

In the third category, the "family group" contains a single male, one female, and their offspring. This is seen in the **gibbons** (*Hylobates* spp.) and the **titi monkeys** (*Callicebus moloch*). These primates and humans are the only mammals that mate year-round regardless of where the female stands in her estrous cycle.

Certain primates, such as the **chimpanzee** (*Pan troglodytes*), do not fit into any of these groupings because their basic unit is in a continual state of change. They have been known to exhibit all the preceeding relationships at one time or another.

Breeding season in many Old World monkeys is often heralded by the "blossoming" of the female's genitalia and buttocks. This is particularly true of baboons, mangabeys, and some macaques. The most outstanding example is the female mandrill, whose genitalia and rear end swell considerably and become brightly colored, as do the face and genitalia in the males. It seems that the greatest swellings are most attractive to the males. In the **gelada** (*Theropithecus gelada*) female, there is a "necklace" of warty protrusions that become brightly colored when she is in heat. Female proboscis monkeys, on the other hand, preferentially mate with those males that sport the largest nose, although this may reflect that these are simply the older, more dominant males since the nose increases in size with age.

In most species, the females are promiscuous and are free to mate with any of the males. In the **Southern black-and-white colobus** (*Colobus polykomos*), many males and females mate together in succession as if participating in a large orgy. Fights in this species are extremely rare. In other species, such as the hamadryas baboon, female promiscuity is not tolerated. The males closely guard their female harems while attempting to steal away another's females. Males are quick to discipline their own females should they wander away or misbehave. The discipline usually takes the form of a bite on the neck, though unfaithful females have reportedly been beaten to death.

Birth is sometimes assisted by another female in the troop. In many, such as the crab-eating macaque, the female quickly devours the afterbirth after delivery.

Monkey mothers (and occasionally fathers as in the tailless **Barbary ape** [*Macaca sylvanus*]) take good care of their young as a rule. When a troop moves on, the young and females are generally found in the center while the males patrol the periphery. Colobus females often share their babies with other females in the group. This is quite unusual for primates. The closely related langurs will let other females "baby-sit" their young in what has been termed "aunt behavior." During these times, when not with their mother, the young suck their thumbs as a substitute for suckling on the mother. In some instances, mothers have allowed jealous females to steal their young without putting up much resistance. Macaques, on the other hand, jealously guard their young from other females. The latter will go to great lengths, such as long periods of grooming, in order to get a glancing touch of the baby. In the **chacma baboon** (*Papio ursinus*) and others, the young ride on the female's back jockey-style. Her tail is held erect at its base through fusion of the proximal vertebrae so that it serves as a "sissy bar" for added support. The **olive colobus** (*Colobus verus*) female carries her offspring in her mouth. This is probably because her fur is too short for the baby to get a good hold on. Males usually limit their paternal care to the disciplining of "naughty" youngsters; proboscis monkey youngsters, known to irritate the old males by running around tweaking their noses, usually meet with a quick reprimand.

It has been reported that female guenons have given their babies to other females or have hidden them beneath their own body when mortally wounded. Guenons also are known to use separate alarm calls depending upon whether an approaching predator is coming from the air or the ground, as do vervets (who have a third call for snakes). Most monkeys depend upon flight when danger approaches. The vervet monkey not only flees to the treetops but actually pulls branches around itself to better conceal it. Male patas monkeys (and some guenons) will lead threatening predators away from the young and females by way of a diversionary display. They are aided by the fact that they are probably the fastest runners of all monkeys, reaching speeds of 35 mph. They are even built somewhat like greyhounds. Patas monkeys occasionally stand on their hind feet while using their tail as a prop in the manner of kangaroos. Talapoin monkeys spend the nights in trees and bushes overhanging creeks and pools. If threatened or disturbed, they quickly drop into the water from heights of up to fifty feet. Proboscis monkeys also escape by jumping into the water.

A few monkeys are big enough to handle themselves against most predators. These are the ground dwellers, such as the generally bold baboons. Two or more are usually enough to keep leopards at bay, and at times they have even killed leopards. They have been known in rare instances to steal and kill human infants, and they have even attacked and killed adult humans. Chacma baboons post sentries to keep watch, and baboon troops often mingle with antelope, each taking advantage of the other's alert senses. The ungulates can smell approaching predators from upwind, while the primates' good vision can spot lurking predators in the open. Both are known to flee at the other's warning. When on the march, immature baboon males usually take up the vulnerable point and rear positions. They are the most expendable and are generally the ones picked off by a quick pounce from a leopard. When the troop is attacked, the females with babies fall back as the dominant males take up position between them and the attacker. Should a young male lag behind and come under attack, it will often be left on its own. Some baboons throw rocks at enemies, and gelada baboons will roll large rocks over cliffs or down hillsides in attempts to drive out intruders such as humans. They are also known to engage in "warfare" with the hamadryas, although the latter are usually the aggressor. "Battles" include surprise attacks and the taking of prisoners which are invariably female. Wounds are sometimes "dressed" with leaves.

**Family Pongidae
(apes—13 species)**

This is the family of true apes, consisting of the gibbons, orangutan, chimpanzees, and gorilla. It does not include humans, which belong to the Family Hominidae and are omitted from this book.

Gibbons (9 species)

The nine species of gibbon are sometimes placed in their own family, Hylobatidae. **Gibbons** (*Hylobates* spp.) are the smallest of apes and are the only primates other than humans that habitually walk upright on their hind legs. While doing so, they hold their extremely long arms overhead for aid in maintaining balance and to keep from tripping over them since their upper extremities are one and a half times as long as their legs. They are also the only mammals to regularly travel by brachiation (swinging hand over hand) throughout their lifetime.

Family Pongidae—Gibbons

They are known to make swings of ten feet or more and leaps of fifty feet are not unknown. They can travel up to 20 mph through the trees.

The agility of these animals is truly remarkable. They can seize a bird in mid-leap and transfer it to a foot without breaking rhythm while brachiating. Accidents do occur when the animals misjudge a distance or grasp a branch that will not support their weight. The resulting fall can lead to fractures of bones. In one study of gibbon skeletons, ten percent bore evidence of such fractures. This is substantially better than the results of a similar study done on proboscis monkeys, thirty percent of whom had sustained at least one fracture.

As with other great apes, gibbons are thought to possess color vision. In some species, there is sexual dimorphism, the males being black while the females are brown. The calloused pads on gibbons' buttocks (a feature they share with baboons and others) allow them to sleep comfortably while in tree forks and on branches.

Most primates swim well and have no natural fear of water. In fact, few mammals cannot swim naturally. The exceptions to this rule include the apes, some lemurs, the slender loris, a few bats, possibly the giraffe, and a few others. The gibbons,

Siamang (*Hylobates syndactylus*)

74

chimpanzees, orangutan, and gorilla would in all likelihood drown if they fell into deep water. Gibbons obtain their water by carefully dipping their hands and sucking up the water entrapped in the fur on the backs of their hands. Only a few apes, including the **white-browed gibbon** (*Hylobates hoolock*) and the **siamang** (*Hylobates syndactylus*), can swim.

The siamang has a booming voice audible up to two miles away, thanks in part to a special throat pouch that increases resonance. The siamang alters its call by using a cupped-hand slap to the mouth at various intervals. Most gibbons produce these loud calls not only to announce ownership of territory, but also, in males, to call for females and, in females, to warn others of her sex away. It is interesting to note that juvenile males are discouraged from calling by their parents since this might attract an unmated female, while young females are actively encouraged since their call reinforces that of the mother in repelling free females. Aggressive displays include acrobatics and the breaking of tree branches.

Gibbons are the only apes not to construct sleeping nests for nighttime use. Instead they sit on bare branches or, if the night is cool, huddle together. Their rears have thick skin pads to cushion them when sleeping.

Gibbons are thought to mate for life. Copulation often occurs as the pair hang from branches. A single offspring is produced at birth and is raised with much care. Many young are white and do not obtain full adult coloration until two to four years of age. Young gibbons (like young langurs and others) often play "blind," walking about with their eyes closed, bumping into objects, and falling down. They seem to laugh and experience much enjoyment from these antics.

Orangutan (1 species)

The **orangutan** (*Pongo pygmaeus*) has one of the most bizarre appearances of all primates, though with great individual variation. The word orangutan is Malay for "old man of the woods." Orangutans inhabit the forests, where they spend most of their time in the trees. They are the only truly arboreal members of the great apes (orangutan, chimpanzees, and gorilla) and are the heaviest tree dwellers on earth. Large old males spend much of their time on the ground. Both sexes test all

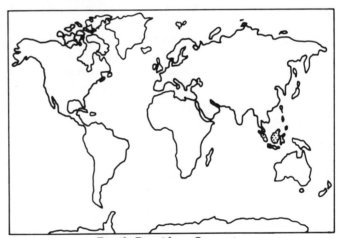

Family Pongidae—Orangutan

branches before committing their full weight to them. From studies, this appears to be an instinctive process, but accidents do occur, with up to a third of skeletons in one study exhibiting signs of fractures.

Since orangutans cannot swim, they avoid deep water. Drinking water is obtained by sucking rainwater from their fur, by squeezing water from tree mold, or by drinking from pitcher plants. They

75

will approach small streams to drink with cupped hands and will splash water over their faces. Some will even wade across small streams.

Orangutans are omnivorous but rely mostly on the tropical fruits that grow in their habitat. Mineral-rich soil is occasionally consumed. Like chimpanzees, they use tools in the form of sticks to get at termites. Some have also used sticks as back scratchers.

These red-furred apes usually build a new bed of vegetation each night; and may even construct an overhead roof. Some are known to cover their face with large leaves to block out light and facilitate sleep. During rainstorms, they will extend this to covering their entire body with leaves or will use a leafy branch for an umbrella.

Orangutans have a large pharyngeal sac enabling them to produce horrifying roars heard a half mile or more away, as well as continuous "groaning" sounds. When alarmed, they produce smacking sounds and loud, defiant "burps." Lip smacking is used for interspecies communication. As many as fifteen different vocalizations have been reported.

Males "serenade" the females prior to mating, the only time the two sexes come together. The song starts out low and gradually builds in volume to reach deafening levels. Premating behavior consists of wrestling, slapping, and mock biting, all accompanied by low grunts from both participants. Females may sometimes be forced into submission. Pairs generally mate face to face while suspended from tree limbs. The females give birth once every four to nine years since weaning may not take place until the age of seven. The very young are carried on the female's hip with added support from one of her arms. The female washes her young and even chews its fingernails to keep them short. When foraging with young the female may narrow gaps in the route by pulling vegetation together or allowing her body to be used as a bridge.

Orangutans will sometimes swing through the trees via hanging vines, generally, they use the vines only as an aid in climbing trees. This is not an innate behavior but instead is painstakingly taught from one generation to the next, as is much of their behavior except for nest building. Territorial intruders are occasionally greeted with a shower of vegetation. They may live for nearly sixty years, longer than any other nonhuman primate.

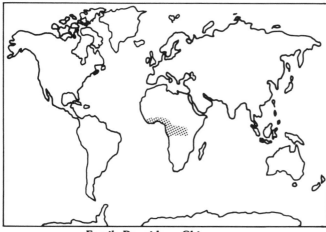

Family Pongidae—Chimpanzee

Chimpanzees (2 species)

Although **chimpanzees** (*Pan troglodytes*) are frequently considered small, lovable apes, they actually grow to heights of 5.5 feet, weigh up to 175 pounds, and can be quite aggressive. They are much stronger than humans. This is aptly illustrated by the fact that one female when upset was known to exert a one-arm pull in excess of a thousand pounds.

Chimpanzees are very closely re-

Chimpanzee (*Pan troglodytes*)

lated to humans, more so than any other mammal. It has been stated that they have approximately the same number of hairs as humans although theirs are obviously longer, coarser, and usually darker. Except for one subspecies (*P.t. schweinfurthii*), they become bald in old age, and as a group they suffer many human maladies, including dental caries, near- and far-sightedness, arthritis, and many of the same infectious diseases.

Chimps spend about half their lives in trees. Except for the infants, who sleep with their mother, each chimpanzee builds itself a small bed of vegetation aloft each night. Occasionally, the same beds will be used repeatedly.

Chimps are omnivorous and will feed on almost anything edible, but primarily on fruit, leaves, and roots. Their stomachs possess the same strong digestive acids as those of humans, requiring a rapid turnover of stomach cells. (In humans, it has been calculated that the stomach can replace cells at a rate of half a million per minute). Chimpanzees can pack away a lot of food. An average adult savannah chimp consumes about fifteen pounds of grapefruit daily. A large male can eat more than fifty bananas in a single meal. Both of these fruits are peeled instinctively. At times chimpanzees will band together to track down large prey including antelope, bushpigs, and even monkeys. Food is frequently shared among individuals. Although a dominant may not share with a subordinate, the reverse is almost always true. Food is solicited by extending the arm with palm up. Young chimps are known to beg for and even consume fecal material.

Next to humans, chimpanzees are probably the most extensive "tool-users" in the animal world. These behaviors are learned and are not instinctive. Chimps obtain termites by inserting carefully trimmed sticks into the tunnels penetrating large termite mounds. After a brief pause, they withdraw the sticks to consume all the termites which have inadvertently crawled onto or attacked the "invader." Sticks have also been employed to steal honey from the hives of bees and to capture ants, while rocks are occasionally gathered to crack open nuts. Leaves are also important tool material. At times they are rolled into "cones" and used as drinking cups, while others are carefully chewed to form primitive "sponges" to absorb water. Leaves are also used to wipe dirt, blood, and excrement from their bodies. Other ingenious methods are used by chimps to get water. Some will dip their extremities into streams and suck out water trapped in the fur. Others will sometimes get water by sucking it from the fur of conspecifics. Intelligence and tool-using insight are clearly demonstrated in the well-known example of the chimp who piled boxes together in order to reach suspended food.

Chimpanzees greet each other in many ways. If there has been a long separation, the greeting can be an elaborate ceremonial ritual. They may run towards each other with one hand raised high in the air, meeting in a long embrace. If both are male, this is usually accompanied by mutual erections. They may greet each other by reaching out both hands to each other, followed by vigorous handshaking. If one is much inferior, it may give an appeasement gesture in the form of gently touching the other's head, arm, or genitalia. The superior may extend his arm to an inferior who responds by pressing his lips to it often at the dorsum of the hand. To show affection, chimps will often kiss each other on the lips. Generally, hierarchies are not well established in chimpanzee groups, as they are in other social primates, and dominant animals don't possess much in the way of special rights.

Chimpanzees occasionally engage in ritualistic forms of behavior. At times a troop will participate in what has been called a "carnival" or "kanjo." At this time, day or night, the adults burst into a display of screaming and drumming which may last hours. Although the exact reason for these outbursts is unknown, many feel that they are related to aggression or temporary possession of territory since they often take place when two different bands meet. During rainstorms, males perform what appears to be a "rain dance." They run up and down hillsides, climb up and down trees, break and wave tree branches, and then repeat the hill running. The cycle may continue for as long as half an hour.

When in heat, the females display a large, red swelling of their hindquarters. Mating is promiscuous, at times appearing even brutal, with the males physically abusing the females. Females usually bear a single young every three to five years. At birth, the females readily consume the placenta. Family groups consist of a mother and her offspring, but not the father since their system of communal mating prevents a male from knowing which young are his. Should the mother die, the very young will usually be "adopted" by an older sibling or another female. Baby chimps, frequently observed to suck their thumbs, display a white tuft of fur on their tails. This serves to inhibit aggressive action toward it by others of the group. When the light fur is replaced with dark its status changes, and it is then disciplined when exerting itself beyond accepted behavior. Until that time, however, it may misbehave and even throw "temper tantrums." These are completely ignored by the adults. Puberty does not begin until about the age of nine years.

Chimpanzees band together to ward off enemies. They will often defend themselves successfully from snakes, leopards, and humans by throwing sticks and stones at the intruder. These are often thrown overhand with a definite measure of aim taken. Their distress calls can be heard up to two miles away.

Recently it has been found that large chimpanzee troops may make "war" on smaller troops. "Warriers" from the large troop periodically raid and kill members of the smaller group, sometimes wiping out the latter. Overcrowding and other stresses have been known to lead to cannibalism and may play a factor in these intraspecific battles.

Gorilla (1 species)

The **gorilla** (*Gorilla gorilla*) is truly the giant of the primates. Gorillas grow to six feet in height, have an arm span up to nine feet, and can weigh close to 500 pounds. In captivity, they reach weights

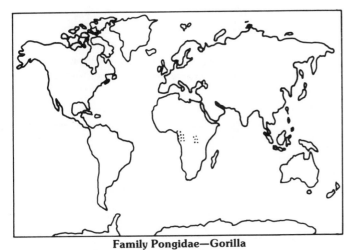
Family Pongidae—Gorilla

over seven hundred pounds due to overfeeding and unhealthy, excessive fat. They may live up to fifty years of age.

Despite their alleged ferocity and aggressive nature, gorillas rarely if ever have been known to take the life of a human, and what few documented reports of attacks exist can be attributed to self-defense on the part of the ape. The fact is that these animals have been shown to be quite gentle and peace-loving. They are extremely powerful, however, having been known to upend medium sized trees, pop footballs under their arms, bend rifle barrels and two-inch steel bars when angry, remove boulders with one arm that a strong man could not even budge, and kill leopards by grabbing them and snapping their necks with a quick twist. It has been suggested that a gorilla is capable of lifting up to 1800 pounds. Clearly, such a strong animal should not be carelessly harried or pressed.

Gorillas cannot swim and are naturally afraid of deep water. They are strict vegetarians except for an occasional snail or worm and will even feed without concern on stinging nettles. Some gorillas have climbed up to 120 feet into trees to feed on leaves. Adults eat about 60 pounds of food daily. They obtain most of their drinking water from their food. On those occasions when they do drink water, they suck it from the fur on the back of a hand carefully dunked in shallow streams or puddles.

Like other apes, they usually sleep in a new spot each night. They take about fifteen minutes to construct a primitive bed of moss, soil, and foliage. In such construction, they are known to tie true knots. Beds are sometimes constructed for midday siestas as well. In some areas, the beds are built predominantly (97 percent) on the ground while in other areas mostly (80 percent) in trees. Ground-nesting gorillas have been observed to roll out of bed and down slopes for as much as twenty feet without even waking until morning. They will not leave their nest during the night even for the purpose of defecating. This is often done inside the nest (unlike the chimpanzee, which keeps its nest clean). This probably reflects the fact that a gorilla can lay on its feces without soiling itself due to its course, fibrous wastes, whereas a chimpanzee would quickly soil its fur should it lie in its excrement.

Gorillas (and also other primates, dogs, and humans) threaten by staring. Other emotional states are displayed in easy-to-understand expressions. For instances, they bite their lips if uncertain, frown if annoyed or upset, show submission by nodding or folding their arms, shake their head to show they pose no threat, and throw temper tantrums if frustrated. They are known to communicate various messages through the vocalization of more than twenty different sounds. They belch when content and bark when curious or alarmed.

Although gorillas can appear quite human in the way they pant, cough, yawn, belch, hiccup, and scratch themselves, they are not as closely related to humans in behavior or biochemical tests as are the chimpanzees. The well-known chest beating accompanied by hooting serves several purposes. It helps to space out troops, even though aggression rarely appears when two different troops intermingle. It also acts as a cohesive stimulus within a group and serves as a displacement activity and/or threat when a gorilla is confronted by another gorilla or human. Rather than the bipedal form of locomotion usually depicted, both gorillas and chimps almost always walk quadrupedally using their finger knuckles (middle phalanges) for weight support.

A gorilla group's behavior is frequently determined by the character of its leader, this usually being a large "silverback," the name used for large, mature males whose back fur takes on a silver-gray appearance with age (about 12 years). Should he be killed, the remainder of the troop will often submit to any consequence without resistance, even to the extent of being clubbed to death by human hunters.

Males rarely if ever fight over females. Despite numerous novels, movies, and folklore that attribute great sexual prowess to male gorillas, scientific studies have shown them to be the "least sexy of all apes" in terms of intercourse rates. In some studies, it was found that they only mated once or twice per year. In addition, the male's penis measures only 2–3 inches when fully erect. Females produce young once every three or four years. Immediately after birth, the female will ignore its offspring unless its "mothering instinct," which is actually passed on from generation to generation, has been stimulated by prior observation and experience with young. She must support the baby with her arms for the first month since it is so helpless that it is unable to cling to her fur with enough strength to support itself. Frequently, "aunt behavior" is observed in that a childless female will babysit for a mother who goes off to feed or sunbathe. Interestingly, the large silverback leader will also babysit older young and will even take care of them should their mother die. Baby gorillas occasionally suck their thumbs or even their big toes. Infanticide by newly dominant male leaders brings the females back into heat, eliminates potential rivals, and hastens the spread of his genes.

When the gorilla's territory is invaded, the lead male takes up a position between the intruder and the other members of the troop. He then acts out an elaborate display that includes chest beating (this is done with cupped hands rather than clenched fists as is usually supposed), hooting, breaking branches, throwing objects into the air and at the invader, sometimes striking other nearby gorillas, and an occasional charge which usually proves to be a bluff. Since natives know that they usually carry through an attack only when the intruder flees from the false charges, anyone injured by gorillas is generally held up to ridicule and disdain.

Primate Intelligence

The largest primate brain is the human's, which measures 1500 cubic centimeters, comparing quite favorably to the 700 cubic centimeter brain of the gorilla. The primates with the largest brain size relative to weight are the common marmoset and squirrel monkeys, whose brains comprise over five percent of their weight. The human brain comprises only 2–3 percent of body weight.

The high intelligence of primates is well known, although many of their actual accomplishments are not. Some have been alluded to in the previous sections, but many of the better examples of primate ingenuity are artificial in that they were observed in or taught to captive specimens.

Humans have manipulated the primates and tested many thoroughly to see just how smart they

80

really are. Even thousands of years ago, various macaques and baboons were used to retrieve fruit, flowers, and coconuts. Chacma baboons have learned to herd sheep and work as train signalmen, rhesus monkeys to drive tractors, and chimpanzees to play games and communicate with humans via computer and sign language. They have learned to communicate with more than 150 "signs" and have even coined appropriate new words via gestures. Some experiments imply that chimps are capable of communicating with each other through very subtle signs at first mistaken for possible telepathy.

Other caged primates have also exhibited marked insight and intelligence. Capuchins and orangutans have used bread and corn kernels to lure ducks and chicks within reach of their cage. The unsuspecting animals were quickly grabbed and consumed. One captive guenon would habitually malinger (fake signs of illness) in order to obtain certain medication that apparently gave it a high. Chimps have taken up smoking and will even light a new cigarette from the burning butt of another. Orangutans have been taught to dress themselves in human clothing, drive nails with a hammer, and to ride tricycles. Gibbons can easily be taught to eat at the table with silverware and to make their bed. In the wild, gorillas are known to play with frogs and pet baby antelope. The list could go on and on, but these few examples are sufficient to illustrate the remarkable accomplishments of the primates.

At one time tool-making and -using were said to be the major criteria for separating humans from the other animals, but this, primarily thanks to the primates, no longer holds true. As we have seen, the chimpanzees exhibit these characteristics in a number of ways including the use of crumpled leaves as sponges, sticks for weapons and termite traps, stones for cracking nuts, and even small twigs to clean each other's teeth. Most probably, the primates exhibit the greatest intelligence compared to all other animal groups. However, as we shall see, some feel the cetaceans come closest to humans, and a few even say they surpass humans in some ways.

7

ORDER EDENTATA
(29 SPECIES)

The Edentata includes three families of primitive New World mammals. Although the name itself means "no teeth," only the anteaters are completely without teeth. Both the armadillos and the sloths possess primitive grinding teeth composed of dentine but no enamel. Edentates generally maintain low body temperatures in the 86–96°F range. The males' testes are internal.

Family Myrmecophagidae (anteaters—4 species)

Of the four species of anteater, the **giant anteater** (*Myrmecophaga tridactyla*) is the strangest in appearance. It grows to seven feet in length and can weigh over 100 pounds. Its long, saliva-laden tongue entraps the ants and termites upon which it feeds and attains a length of two feet (though only about a foot of this can be protruded from the mouth). It walks on its knuckles to keep the long, sharp claws from becoming blunted. With its double-jointed hips, it thus appears crippled. The six-inch claws help it tear through the tough walls of termite nests. The large, bushy tail protects the animal from the sun's hot rays

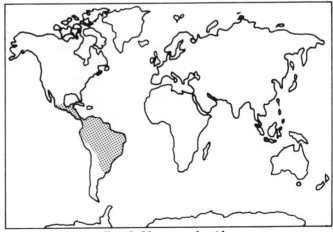

Family Myrmecophagidae

Tamandua (*Tamandua tetradactyla*)

83

Giant Anteater (*Myrmecophaga tridactyla*)

while taking its mid-day nap. It can also keep the animal dry during a tropical rainburst by serving as an umbrella. At such times, it is said to make the animal look like a large pile of hair. It is reported (but disputed) that the tail can also serve as a broom to sweep insects together to make eating more efficient. It also brushes away any pests that find their way onto the animal's body. The **tamanduas** or **lesser anteaters** (*Tamandua* spp.) and the **pygmy** or **silky anteaters** (*Cyclopes didactylus*), unlike the giant anteater, have prehensile tails that let them be as much at home in the trees as on the ground.

All anteaters feed primarily on termites and ants. In the giant anteater the tongue is pushed in and out up to 160 times per minute, trapping up to 500 ants or termites with one pass inside a nest and consuming as many as 30,000 daily. The tongue is liberally coated with sticky saliva secreted by the largest salivary glands, relative to size, of all mammals. The glands extend from beneath the tongue all the way to the sternum. Anteaters sometimes feed by laying the tongue out over an ant trail. After a short time, the ant-infested tongue is retracted, the ants swallowed and the tongue extended back for more.

Giant anteaters have poor sight but can smell termites more than 100 feet away. Since they are toothless, they must depend upon their tough stomachs to grind up the hard exoskeleton in much the way a bird's crop "chews" its food. The stomach is aided by the presence of sand and small stones swallowed by the anteater. Ant meals are often brief since bites, stings, and noxious chemicals can discourage the predator. Thus the nests can be repeatedly visited and cropped without annihilating the prey. Tamanduas occasionally eat bees and their honey.

Anteaters are solitary and only come together to mate. The female giant anteater gives birth from a standing position. The single young is carried on the mother's back for the first few months to a year of life. It has been found that the young align their body stripe with their mother's to enhance the appearance of a single animal. In the pygmy anteater, the young are carried until almost full grown, and there are reports that the parents of this species feed their offspring regurgitated masses of insects. This is common in birds but very rare in mammals.

When harassed, the lesser anteaters are capable of emitting an unpleasant odor. Giant anteaters produce a loud roar and can put up a savage fight while standing on their hind legs. They have been known to hold their own against such formidable foes as jaguars with the help of their powerful forearms and claws. There are reports of their having killed people through a fatal "embrace" and clawing. The pygmy anteater is often overlooked by predators because it resembles the opened seed pods of the silk cotton tree (*Ceiba* spp.) it frequents.

Family Bradypodidae
(sloths—5 species)

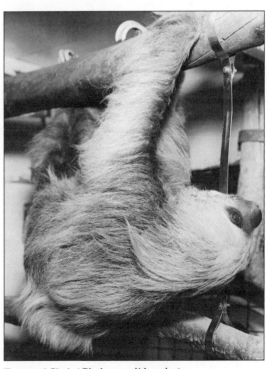

Family Bradypodidae

Sloths literally live in an upside-down world. These animals eat, sleep, mate, and give birth all while hanging upside down from tree branches. What is not well known is that they will frequently support their weight with their backs by resting on a branch or in a tree fork, rather than hang suspended by their claws for long periods of time. Some rare individuals have been thought to spend their entire lifetime within the confines of the tree in which they were born. They are physically designed for an arboreal life with their large, hooked claws; if placed upright on the ground they must literally and clumsily drag themselves about.

Sloths are best known for their lethargic nature. Truly the slowest mammals alive; they sleep for up to 18–20 hours per day. They even digest their food and sneeze slowly. A meal takes about a week to digest, but it sometimes requires more than a month. Nearby gunshots elicit a slow head turn and a blink of the eyes although this may reflect poor hearing. One sloth in a hurry was clocked at five yards per minute. In all fairness to the sloth, it should be mentioned that this was while the animal was on the ground, and that they can travel up to forty yards per minute (about 1.4 miles per hour) while in the trees. Surprisingly, sloths are good swimmers and can make their fastest times while in this medium. They can stay submerged for up to half an hour.

There are two types of sloth, the **ais** or **three-toed sloths** (*Bradypus* spp.) and the **unaus** or **two-toed sloths** (*Choloepus* spp.), where "toe" actually refers to the fingers since all sloths have three toes on their hind legs. Neither type can move the digits independently, and both kinds collect a green algae (*Trichophilus* spp. and *Cyanoderma* spp.) in their fur during the rainy

Two-toed Sloth (*Choloepus didactylus*)

85

season. This algae serves at least two functions. First, it gives the sloths a greenish hue to provide an effective camouflage; secondly, it serves as food for small pyralid moths, more specifically the **South American snout moth** (*Bradypodicola hahneli*), which inhabit the sloths' fur. Recent study has shown sloth fur to be "home" to nine moth species, four scarab beetles, and six tick species. A single sloth may carry as many as 978 beetles in its fur. The fur, incidently, is never cleaned or groomed and lies in a direction that is opposite to that of other mammals. This is an added adaptation to its upside down world in that it allows rainwater to be shed more efficiently. Further camouflage comes from the back hair, which is black with surrounding white and yellow resembling the flowers of *Cecropia* trees.

Sloths exhibit a few additional differences from most other mammals. The three-toed variety has nine (sometimes eight) cervical vertebrae, rather than the seven of almost all other mammals, including the giraffe. These sloths can rotate their heads through a 270 degree arc in both directions. The two-toed species have only six cervical vertebrae (some individuals have seven or even eight), a characteristic they share with the **African manatee** (*Trichechus senegalensis*). Again, all other mammals have seven although they may be variably fused in some of the cetaceans and jerboas.

Sloths are also among the few mammal species incapable of closely regulating their body temperature. Instead, it is known to fluctuate somewhat with the environment. This means that they must be cautious of direct sunlight or other temperature extremes. In fact, the two-toed sloth's body temperature varies more than that of any other mammal excluding hibernation changes; from 74–92°F.

Sloths feed on leaves, stems, and fruit. All are slowly chewed with teeth composed of dentine only (though they grow continuously throughout life). The three-toed sloths feed almost exclusively on the leaves and flowers of *Cecropia* trees. Stomach food can account for up to 1/3 the animal's body weight. Sloths obtain their water from the leaves they eat or by licking dew drops from the vegetation. They defecate and urinate at approximately weekly intervals. At this time, the three-toed sloths descend the tree-trunk tailfirst, dig a small hole with their hindlegs and/or tail, and then bury their feces. In so doing, large mounds of fecal material are built up because they often defecate in the same spot repeatedly. The two-toed species descend headfirst, have no tail, and do not bury their excrement.

Some researchers think that sloths are capable of communicating via ultrasonic utterances, although as a rule they are solitary and come together only for the purpose of mating. If forced together, as in a small zoo enclosure, they are known to fight, occasionally to the death. The actual act of mating in two-toed sloths takes place with both participants facing each other suspended by their arms only. Females carry their young who cling tenaciously to her body. Later she aids them by pulling branches together or by allowing them to use her body as a "bridge" for climbing between branches. They excrete all their wastes at weekly intervals into her fur. The mother promptly sweeps these away. She may nurse her young for as long as two years.

A sloth's primary defense is its camouflage. This is enhanced when the animal curls into a ball to resemble a wasp nest or dried leaves. If pressed, they will lash out with their long claws. It has been claimed that sloths are the hardiest mammals alive. The evidence is convincing when one considers the amount of trauma individual animals have sustained and still survived. Sloths have pulled through injuries, poisonings, and electrical shocks that would almost certainly have killed any other mammal.

86

Family Dasypodidae
(armadillos—20 species)

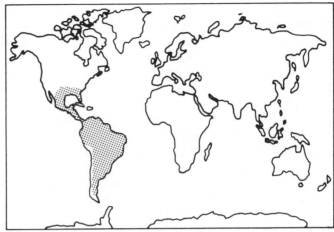

Family Dasypodidae

The armadillos are the only mammals enclosed within a shell. The shell is actually composed of true bone, not keratin. It can, surprisingly, be punctured relatively easily and bleeds readily. The largest of the clan is the **giant armadillo** (*Priodontes giganteus*). It grows to almost five feet in length and weighs up to 130 pounds. Its mouth contains more than 100 teeth, usually 104—more than any other land mammal—but many are lost as the animal ages. Its third claw is elongated and is reportedly the largest claw in the living animal kingdom, measuring up to eight inches in length. Armadillos have poor sight and on occasion have actually collided with humans while foraging at night.

Armadillos are omnivorous, feeding mostly on insects, plants, and small vertebrates. This includes lizards, snakes, fire ants, toads, scorpions, and "tarantulas." The **nine-banded armadillo** (*Dasypus novemcinctus*) can smell worms and insects buried up to eight inches deep. It can lick up seventy ants at once and makes a meal out of as many as 40,000. Giant armadillos, at times, will dig open fresh graves in order to feed on the buried corpses. The **peludos** or **hairy armadillos** (*Chaetophractus* spp.) are known to occasionally feed on snakes. To dispatch them, the animal crushes and cuts them in pieces with the lower edges of its armor plates. Some species burrow into animal carcasses, possibly to eat maggots and other invertebrates. The scats of some, including the nine-banded, are marble-sized clay spheres because they inadvertently consume much dirt with their prey.

The nine-banded armadillo, whose bands really vary from seven to ten, readily takes to the water

Nine-banded Armadillo (*Dasypus novemcinctus*)

and is capable of walking across the bottoms of ponds and streams, staying submerged for up to six minutes at a time. More frequently, it swims across the surface but first swallows large amounts of air so that it will be buoyant enough to float. Females at times carry nesting material by holding it in their forelegs and hopping on their hindlegs.

Armadillos are generally solitary animals but sometimes live in groups, usually of one sex only. Males mark their territory with urine and court the females, who respond by lying on their backs for mating. After fertilization, the egg usually lies dormant for three and a half months (delayed implantation). The females produce but one egg at a time, but after implanatation the egg divides and some, like the **seven-banded armadillo** (*Dasypus septemcinctus*), can produce up to twelve genetically identical offspring, all of the same sex. This process is known as polyembrony. Unfortunately, female armadillos bear only four teats at most. In the nine-banded species, the female almost always gives birth to identical quadruplets. In many species, only one or two offspring are produced. The shells of the young do not fully harden until they are almost fully grown.

Some armadillos sunbathe while resting on their backs. This practice can be fatal, for it can allow hungry predators to sneak up. **Pygmy armadillos** (*Zaedyus pichiy*) are known to hibernate where they inhabit colder regions.

When alarmed or frightened, armadillos rely on a number of defensive mechanisms. Some, such as the nine-banded, will suddenly leap into the air (a fatal jump if an approaching car is what scared it) and then take off on the run. The hairy and nine-banded species will dig their way to safety or head for a nearby briar patch or cactus clump. The nine-banded armadillo can bury itself in less than two minutes if the ground is soft, while the giant armadillo can do so in less than ten minutes in rock-hard ground and is even known to burrow through asphalt. Armadillos hold their breaths when digging. If only partially buried, many can "expand" their plates to wedge themselves in.

Only the **three-banded armadillos** (*Tolypeutes* spp.) can completely roll up into the well-known protective ball. The nine-banded armadillo can do so only incompletely, while others, such as the hairy armadillos, are incapable of it. It is not a foolproof method of defense, however, in that predators such as foxes and maned wolves can overcome it by rolling the armadillos into the water. Three-banded armadillos sometimes leave a "gap" in the "ball" and snap this shut when touched. This seems to be effective against natural predators.

The **fairy armadillo** or **lesser pichiciego** (*Chlamyphorus truncatus*) protects itself by quickly running into its burrow or by digging one and then plugging the entrance with its special anal plate, which acts like a cork. It digs with its hind legs while supporting its rear weight with its tail. When threatened, the **naked-tailed armadillos** (*Cabassous* spp.) occasionally take to the water. Some others are known to feign death.

Armadillos are frequently persecuted by humans because many believe that they feed primarily on bird eggs. The truth is that 80–90 percent of their diet consists of insects, particularly termites, making them in reality beneficial to humans. In fact one million armadillos can dispose of 100,000 metric tons of insects yearly. They can be repelled with mothballs.

8

ORDER PHOLIDOTA
(7 SPECIES)

This one-family order is ecologically the Old World equivalent of the edentates.

Family Manidae (pangolins—7 species)

The **pangolins** or **scaly anteaters** (*Manis* spp.) are the only mammals covered with scales. The scales are actually composed of "compressed hair" or keratin, the substance of human hair and fingernails. Their scaly skin can account for up to 30 percent of their total weight and individual scales are periodically shed and replaced. The scales found on the tail of some species are used as an aid in climbing vertical tree trunks. The tail of the **long-tailed pangolin** (*M. tetradactyla*) contains 46–47 caudal or tail vertebrae, on average the most of any mammal. In some species the tail is prehensile. The **giant pangolin** (*M. gigantea*), the largest species at up to six feet in length, and others will occasionally walk or run erect on their hind legs with tail extended up and back to counterbalance the body weight.

Pangolins feed primarily on ants and termites, up to 200,000 each day. In fact, the **Cape** or **common ground**

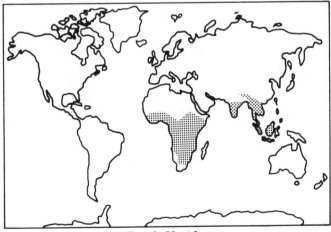

Family Manidae

Giant Pangolin (*Manis gigantea*)

pangolin (*M. temmincki*) even feeds on the dreaded army ant. Pangolins often look for termites in cow patties. They occasionally hold the dung up with their feet while lying on their backs, eating the termites as they fall onto the stomach. Prey is quickly snatched up with long, sticky tongues, which can be protruded up to sixteen inches. The tongue is actually much longer than this, as it is anchored to the sternum's xiphoid, which reaches almost to the pelvis. When not in use, the tongue is folded into a special throat pocket.

The nostrils and ear openings can be closed to keep out agitated insects. The eyes are protected by thick, transparent membranes. When under attack by a large number of ants, some pangolins retaliate by forcing their way through small openings so that their scales become pressed together to squash the attacking pests. Since none of the pangolins have any teeth, their stomachs do the "chewing" with the aid of small stones the animals have swallowed. In addition, the stomach contains "horny teeth" composed of stratified epithelium to help this masticatory function.

Pangolins have been observed to take "ant baths." Some experts believe the formic acid given off by the ants stimulates the skin and is essential to the health of the animals. It may in fact keep them free of vermin, and some reports state that they lift their scales to allow entry by the ants, which they can later rid themselves of by submerging in water.

The young are generally born singly with soft, pliable scales that harden within a few days. The baby rides on its mother by grasping the scales at the base of her tail. Should danger threaten, the baby quickly finds itself inside the protective "ball" of its mother since she folds both her head and tail beneath her.

The pangolins use many defensive behaviors. When curled about a tree limb sleeping, they effectively resemble pine cones, causing them to be overlooked by passing predators. When molested, they are known to hiss, take to the water, emit foul-smelling fluids from glands near the lower base of the tail, and roll into a protective ball. The ball can actually be pulled apart by a strong man, but this does require considerable force. Some pangolins have been known to curl into a ball and then roll away down a hillside. While curled into a ball, they can also perform several active defensive maneuvers. The scales can be erected and then quickly clamped down to "bite" a probing snout. This ability to erect the scales is reportedly also used to help break a fall and to flick off attacking insects.

While in the ball form, these animals will sometimes use the tail as a "barbed" whip. The scales have been known to cause lacerations to unwary predators. In the cape species, the tail is continually slapped against the body in attempts to trap one of the limbs of its attacker. If successful, the tail is then engaged as a saw. It reportedly actually severed the leg of an assailant in some instances. There are rare reports of pangolins using their tails to strangle enemies to death. If forced open from the ball position, they are known to urinate and defecate on their attacker, and some can put up a good fight. The giant pangolin is extremely strong and has been known to twist apart iron bars and tear through concrete to make good its escape.

9

ORDER LAGOMORPHA
(65 SPECIES)

Lagomorphs include the pikas, hares, and rabbits. Once thought to be closely related to the rodents, they are distinguished from rodents by having an additional pair of upper incisors located just behind the larger, anterior pair. Serologic tests have recently shown them to be more closely related to the ungulates. In those species with hair on their soles, these hairs are rectangular in cross section.

Family Ochotonidae (pikas—19 species)

Pikas (*Ochotona* spp.) are also known as **calling hares, conies, haymakers, mouse hares, rat hares, rock rabbits, slide rats,** and **whistling hares.** They inhabit cold areas such as high mountain rock slides and steppes. They can be seen to be active on days when the temperature is as low as O°F. The **Mount Everest pika** (*O. wollestoni*) can be found at altitudes reaching 20,000 feet, thus rivaling the yak as the highest living mammal in the world.

Pikas can close their nostrils in bad weather. They possess three coats of fur for warmth, and their feet are covered with fur for insulation and

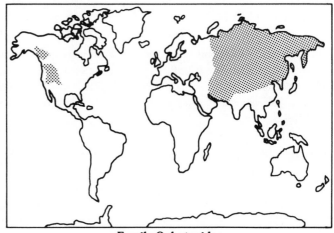

Family Ochotonidae

added traction on the bare rock and icy surfaces upon which they roam. Their whistles and chirps have a ventriloquial quality which helps to keep their location concealed from predators. This probably results from the animal's jumping upward and forward with each call.

Pikas do not hibernate. Instead, all species other than the **royal pika** (*O. roylei*) and **large-eared pika** (*O. macrotis*) gather and lay out to dry various grasses in summertime. These are then stored for the lean winter periods, which accounts for the common name of haymakers. Stacks are often marked with urine and facial gland secretions to show ownership. Sometimes unusual items such as paper or marmot droppings are gathered and placed on the stacks. Occasionally, the latter are even eaten. Should a rainstorm break while the hay is out to dry the pikas quickly get to work and bring it "indoors" to prevent it from getting soaked. They sometimes work well into the night with such activity. The hay is brought back out when the sunshine returns. They are known to leave the hay out during light showers, and some species leave their piles out all winter long regardless of weather.

Pika (*Ochotona princeps*)

Haystacks are often walled in or held down with rocks to keep them from being blown away by the wind. Stores of hay will often be two feet high, three feet across, and weigh up to forty-five pounds. Hay piles are jealously guarded from conspecifics, and each animal may have up to six stockpiles of hay. When the opportunity presents itself, they will steal each other's caches.

Like rabbits and hares, the pikas re-eat certain of their feces (see refection in the next section on rabbits and hares). Territories are marked out with cheek gland secretions and hierarchies exist among colonies. The male's testes are abdominal except at breeding time, when they descend to the base of the penis. At times, two pikas that suddenly come into view of each other have been observed to fall off their rocks because of the resulting "surprise." The resident usually manages to chase the intruder out quickly after recovering.

Pikas usually inhabit large dens. They construct these themselves and often live there socially. Separate chambers are usually constructed for defecation. The **Pallas pika** (*O. pallasi*) builds rock piles three feet long at the bases of cliffs. The small spaces are chinked closed with camel and horse dung as well as vegetation. At the approach of danger, pikas will warn each other by producing shrill whistles.

Family Leporidae (rabbits and hares—46 species)

The words rabbit and hare are often used interchangeably although there is a difference between the two types of animal. **Hares** (*Lepus* spp.) are generally larger and have longer hindlegs and ears. A hare's young are precocial (they are born fully furred with open eyes and teeth and can hop

96

immediately), while those of the **rabbit** (8 genera) are altricial (they are born naked, with closed eyes and no teeth and are unable to walk). These differences probably reflect the unequal gestation periods in the two groups. Rabbits are generally more gregarious, while hares are usually solitary. Hares generally have black-tipped ears, while no rabbits are so marked. Actually, the word rabbit was originally used for the young of the animals known as conies. Thanks in part to the Bible, cony has been transferred to the hyrax so that the original conies became rabbits. To

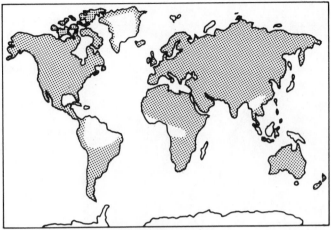

Family Leporidae

make matters even more confusing, well-known common names are often used inappropriately, so that the **jackrabbits** (*Lepus* spp.) and the **snowshoe rabbit**, more correctly called the **varying hare** (*Lepus americanus*), are hares, while the **scrub hare** (*Poelagus marjorita*), **red hares** (*Pronolagus* spp.), **hispid hare** or **bristly rabbit** (*Caprolagus hispidus*), and **Belgian hare** (*Oryctolagus cuniculus*) are all rabbits. In leporids, the males are known as bucks, the females as does, and the young as leverets or kittens.

Large hares can run up to 45 mph and can jump up to 25 feet. The snowshoe rabbit can accelerate from zero to 35 mph within one second when on solid ground. Some hares have been seen to make vertical leaps up to seven feet high. Many make vertical jumps to survey their surroundings checking for the possible approach of predators. Rabbits take up to twenty short naps daily.

Snowshoe hares are probably one of the best known animals that change coat color with a change in season. The individual hairs do not actually change color but instead fall out little by little as new colored hairs grow in. Most animals that exhibit seasonal color changes do so in this manner, although a few, such as the **Arctic fox** (*Alopex lagopus*), are known to change each individual hair's pigmentation. In hares, the actual change is triggered when hormones are released in response to changes in day length. Those at lower latitudes remain brown all year long. Snowshoes have adapted to life in the snow by growing long tufts of hair on their large feet. These serve the same function as snowshoes (i.e., they give additional surface area for weight support). They also aid in giving the hare better traction on slippery ice, increase conservation of body heat, decrease the amount of scent left in the tracks, and are responsible for its name.

Most rabbits and hares produce few noises. When captured by predators, they will sometimes let out a loud scream. This is often the only noise they are heard to make. Many do produce "growls" during fights, copulation, and suckling; and occasionally the females will scream if the males become too forceful during mating. The **volcano rabbit** (*Romerolagus diazi*), the only rabbit with a tail not externally visible, and the red hares are exceptions in that they are quite vocal even when just resting. The former produces sounds similar to those of the pikas, and the latter often let out loud screams.

Varying Hare (*Lepus americanus*)

Rabbits and hares are for the most part strict vegetarians, and all of them have a cleft upper lip. This is where the term "hare lip" comes from for those people with congenital cleft lips. Lagomorphs eat all forms of vegetation, including poison ivy (cottontails), toxic belladonna (many species), cactus (carefully consumed around the spines by desert-living hares), and even kelp (the **Arctic** hare [*Lepus arcticus*]). The Arctic hare has markedly bucked teeth. At times, it and the snowshoe rabbit are known to eat carrion and other meat items. Snowshoes are even known to feed on the dead bodies of conspecifics. Some hares have caught and eaten voles, and rabbits occasionally eat snails and insects including ants. Lagomorphs' incisors, like those of rodents, never stop growing and must thus be kept worn down through grating with the lower teeth and through eating. Many also have cheek pouches where food can be temporarily stored.

Rabbits, hares, pikas, shrews, sportive lemurs, and a number of rodents including the house mouse, kangaroo rat, and beaver are known to excrete two types of feces. The first is greenish and moist due to a covering of mucus. These are known as caecotrophs. They are promptly re-eaten directly from the animal's anus and are thus exposed to further digestion and absorption. This process of refection is necessary because the caecotrophs contain many essential nutrients and vitamins such as B_{12} resulting from bacterial action within the caecum. Unfortunately, these nutrients can only be absorbed in the small intestine proximal to the caecum. The leporid caecum has ten times the capacity of the stomach. Its bacteria secrete important enzymes, such as the cellulases capable of digesting cellulose, the main constituent of plant material. These bacteria are originally obtained by the young through eating some of their mother's feces, a behavior which if prevented will lead to death.

After the second trip through the digestive tract, the feces appear dry and much darker in color. These are discarded and never eaten. It is also interesting to note that when leporids vigorously clean their ears they ingest special body oils in the process. One of these, ergosterol, is a precursor to Vitamin D and prevents the development of the disease known as rickets.

Most lagomorphs obtain adequate amounts of water in the vegetation they consume so that they rarely need to drink. Some hares and **cottontails** (*Sylvilagus* spp.) will eat snow. One cottontail, the **marsh rabbit** (*S. palustris*), requires relatively large amounts of drinking water and will actually die if forced to go for more than a few days without it.

The hares that inhabit deserts must depend upon their large ears to radiate away excess body heat. Since the ears comprise up to a fourth of their total body surface area, this can be accomplished by controlling the amount of blood flow through them. They cannot rely on evaporation since water is too

98

scarce to support this mode of cooling. Jackrabbits occasionally use their huge ears as umbrellas during rainstorms.

As a rule, hares and fast-running rabbits build nests known as forms above ground, while slow-running rabbits take over or dig out dens known as warrens or buries. Warrens can become quite large and are themselves sometimes taken over by foxes or badgers when abandoned. Although most lagomorphs do not dig their own dens, a few species that are good diggers and are known to do so include the **European rabbit** (*Oryctolagus cuniculus*) (from which more than 50 domestic breeds have come), the **pygmy rabbit** or **Idaho cottontail** (*Sylvilagus idahoenis*) (the only species of cottontail to dig its own den), the hispid hare, and the endangered volcano rabbit. The European rabbit is the champion digger. Its warrens may be ten feet deep and extend lengths of one hundred and fifty feet. One large colony of four hundred and seven rabbits maintained a warren containing over two thousand entrances.

Rabbits and hares are quite diversified, having taken up residence in areas ranging from the Arctic to the desert and even swamps. Both the marsh rabbit and the **swamp rabbit** (*Sylvilagus aquaticus*) are excellent swimmers. So are most lagomorphs, but these two actually seem to enjoy lounging about in the water and often feed on aquatic plants. Needless to say, they are found in wet habitats such as swamps and will quickly take to the water to avoid predators, but their fur is not waterproof. Many species take dust baths to help retard insect attacks, and some will stretch out to take sun baths as well.

The holding of territory by individuals and groups serves to space them out and helps to prevent overpopulation with its resultant consequences. Many rabbits mark their territory by "chinning." To do this, they rub a scent produced by glands in the skin of the chin area onto rocks, grass, the ground, and even their mates and offspring. Cheek gland secretions are also carefully groomed into the animal's fur and are thought to aid the animals in recognition as well as to make them more attractive to the opposite sex. Feces and urine are also used for marking territory, as are anal and nose gland secretions. Many periodically trim branches and move obstacles from their well used "runs" inside their territory. In hares, mass migrations by the thousands have taken place many times and seem to be a result of food shortages and even inclement weather.

Rabbits are the proverbial breeders, a reputation they often live up to. It has been calculated that one pair could theoretically produce over a million descendents in as little as four years. Another reported calculation has the **Eastern cottontail** (*Sylvilagus floridanus*) producing two and a half billion descendents in five years. In some instances, females are known to produce 35 young within a year. This is no mean feat when one considers the natural controls that limit population growth in this family. Even though there are distinct breeding seasons, the females have no estrous cycle *per se*. When females come into "heat," they remain receptive until they are either impregnated or pseudo-impregnated since the process of copulation is what stimulates ovulation. (This is also true of many other mammals including cats, ferrets, mustelids, shrews, and some squirrels.) Both rabbits and hares show double pregnancy or "superfetation," in whch females become impregnated one to five days before giving birth. In these instances, both uterine horns are used simultaneously. This is not a common occurrence, although females often mate again within one to two hours after giving birth. The females usually chase the males away after mating since they are known to kill the young.

Male lagomorphs are unusual in that the testicle-containing scrotum is located anterior to the penis, not posterior as in most other male mammals except marsupials and tree shrews. The testicles drop down into the scrotum only during the breeding season, at all other times, they ascend into the abdomen. The males or bucks exhibit erratic actions during the breeding season. They are known to kick, bite, leap on, and even box each other from standing positions. In some instances, this has been fatal to one of the participants, usually due to the powerful hind leg kicks. On one occasion, a hare was even observed attacking a dog while in such a state. This erratic behavior has given rise to the saying "as mad as a March hare." It is thought to be hormone-induced aggressiveness designed to develop a hierarchy among the males in a given area. Some carry these antics right into courtship and occasionally injure females severely. In fact some recent studies have found "boxing" to occur only between male-female pairs.

In cottontails, named for their cottonball-like tail, one of the pair will jump high into the air while its mate runs underneath. Some male rabbits and hares (as well as agoutis, maras, and porcupines) will urinate on the females as a prelude to mating. Male hares are known to be very rough with their does and will box them into position for mating. Some rabbits will at times precede copulation with mutual licking of each other's head and ears. At times, a buck may attack and attempt to dislodge a copulating male so that he can take its place. After mating, female cottontails often turn on the males and drive them away. They may even tear mouthfuls of his fur out in the process.

Does, generally larger than bucks, construct a nest in anticipation of the coming young. In hares, the young are known as leverets and in rabbits, kits. The nest is often lined with some of the female's fur. It is taken mostly from around her teats, thus exposing them for the young to find more easily. The hair is mixed with grass to form an insulated bed for the babies, born after a gestation of about thirty days in rabbits and forty-two days in hares. The females can be sexually receptive one hour after giving birth. Female hares will frequently divide their young among two or more forms above ground and will even cover them with loose vegetation mixed with fur to form a warm, camouflaged blanket. This provides better odds for the survival of at least a few of her offspring. Like fawns, the young are practically odorless. The females often make their way to and from the nest via long roundabouts, often going through puddles and marshes to help conceal scent trails. They are also known to make large leaps into and out of the forms to prevent leaving a telltale scent right up to the nest.

Rabbit does conceal their nest chamber, known as a stop or stab, within the warren. This is usually accomplished by blocking the chamber's entrance. This helps them protect the young both from predators and from bucks that may kill the offspring on occasion. One Eastern cottontail female raised her young in a nest nine feet high in a willow tree. Female rabbits nurse their young for about three minutes once every 24 hours. Some female lagomorphs are known to chase off cats, dogs, snakes, and even humans in protection of their offspring.

Overpopulation in rabbits and hares leads to a number of compensating processes. It decreases fertility, inhibits breeding, promotes aggression, and lowers resistance to infection and disease. Most of these changes are mediated through hormonal imbalance as a result of the stresses imposed. At these times, many females go through what is known as intrauterine absorption, a process common also in pikas. About midterm in their pregnancies, the embryos stop growing and are instead reabsorbed by the mother. Complete absorption takes about two days. This feature is not that uncommon, particularly in overcrowded areas. Its incidence is 30–60 percent of all pregnancies in some studies, mainly in subordinate females and overpopulated colonies. The ten year population fluctuation in snowshoes may result from toxic metabolites (phenolic resins) formed by plants in response to overbrowsing by the hares.

Lagomorph mortality is often high, mostly because they are preyed on by so many predators, including snakes, raptors, and numerous other carnivores. In one study of snowshoes, it was found that 70 percent of all adults die yearly. Approximately two percent of the population lives to five years of age, even though they are known to live in excess of eight years. In cottontails, known to live to ten years of age, 85 percent die in their first year of life. In times of high mortality, the thinning of the population results in the reversal of the measures outlined above.

Leporids use various means to protect themselves from predators. Some hares lay down multiple false pathways near their forms. These distract and confuse predators, allowing the hare to spot them first and avoid a surprise attack. They are also known to look out for danger by standing on their hind legs or by making periodic "observation leaps" every fourth or fifth jump while on the run. Marsh rabbits walk on their hind legs at times, and Arctic hares will even flee by hopping away on their hind legs kangaroo-style in order to keep an eye on the danger. Some experts feel that hares stand on their hind legs to better enable themselves to pick up the scent of enemies.

Many rabbits and hares will give warnings when alarmed. Like deer, goats, sheep, shrews, wallabies, and others, many species will sound an alarm by thumping the ground with their hind feet. This has even been known to frighten away some intruders. **Cape hares** (*Lepus capensis*) warn others by grating their teeth. Some leporids are known to signal by flashing the white area beneath their short, stubby tails. This is also used to signal babies to follow their mother or to lure predators away from the young. The **white-sided jackrabbit** (*Lepus callotis*) has a strange way of alerting others to danger by performing an amazing, instantaneous "color change." Its ears and lower sides are completely white while the rest of its body is brown. It can, however, shift its skin greatly over its body with specialized skin muscles, so that one side will appear white while the other is brown. This shift takes place repeatedly so that the hare actually flashes out a conspicuous signal as it runs away at full speed. It may be a display to show ownership of its territory or even a way of luring danger from the form or young. It is thought by some that it confuses a pursuing predator into miscalculating shifts of direction or may serve as a learning signal signifying "you can't catch me." The latter has been documented in some domestic dogs that learn to ignore this hare after several unsuccessful chases.

Rabbits and hares use various defense mechanisms in coping with danger. Their usual tactic is to freeze or lie flat against the ground with their ears back for as long as fifteen minutes. If approached too closely, they will flee in a zig-zag pattern with occasional backtracking. Some seek shelter in a briar patch, while marsh rabbits, swamp rabbits, and others are known to take to the water. They can submerge all but their nostrils and remain under until the danger passes. Some have escaped by running across ice that would support their weight but not that of the pursuing predator. Strangely, some have been known to hide by climbing trees. These include the **brush rabbit** (*Sylvilagus bachmani*) and the **desert cottontail** (*Sylvilagus audubonii*). When chased, the eastern cottontail is known to run in a huge circle and then take a long jump sideways to break from its circular scent trail. Large hares will occasionally stand their ground and fight. Some have been known to injure foxes with their powerful hindleg kicks. One doe, in protecting her young, was seen to kick a stoat sixteen feet through the air. Red hares will take shelter in a den and will even barricade the entrance with piles of vegetation. Arctic hares and **white-tailed jackrabbits** (*Lepus townsendii*) are known to hide on leeward sides of rocks and even burrow under the ice and snow during storms for protection. In

101

addition to female **snow** or **blue hares** (*Lepus timidus*) and a few others that burrow tunnels up to six feet long for their young, these are the only times that any of the true hares do any burrowing. Some rabbits will urinate when alarmed and will even spray their attacker with it. When caught, many may produce a loud, terrible scream; a few have been known to die of shock when captured, particularly if they have been living under crowded conditions.

After shooting a rabbit or hare for eating, one should carefully clean the carcass, particularly if there are any cracks or sores on the hunter's hands. The meat should then be thoroughly cooked, since it can transmit the disease tularemia through contact or consumption. Leporids are also known to be reservoirs of brucellosis, bubonic plague, Q fever, and Rocky Mountain spotted fever. They are also useful to humans in that some felts are composed mostly of rabbit and hare fur.

One strange and unexplained social behavior noted in white-tailed jackrabbits is their formation of large circles that gradually close down as all the hares move towards the center. When a certain point is reached, they all bound away in different directions. After a few minutes have passed, they will reappear to repeat the performance. Some specialists have attributed these antics to "play."

10

ORDER RODENTIA
(1,686 SPECIES)

The rodents comprise the largest of all mammalian orders. Their tremendous number of species can probably be attributed to the fact that their awesome reproductive potential makes more likely the production of new characteristics. With many, frequent, large litters and short generations, they offer natural selection an ideal opportunity to generate new species.

One of the main distinguishing features of rodents is their incisors. These teeth grow continuously throughout life, but only the leading edge is coated with enamel. Use thus wears them naturally to a razor-sharp edge, since the posterior portion of dentine erodes much more readily. Should a mishap occur, through either genetic deformity or environmental trauma, so that the teeth are poorly aligned, the rodent will be unable to keep them worn down. Continued tooth growth will then eventually prevent eating; the lower incisors may even grow up through the palate and perforate into the brain. Both produce fatal results.

The rate of tooth growth has been carefully measured in some species. In the common rat, the incisors grow six inches per year. In the same amount of time, a guinea pig's will grow ten inches, and the pocket gopher's a whopping twenty inches. To keep this growth under control, most rodents gnaw on hard wood,

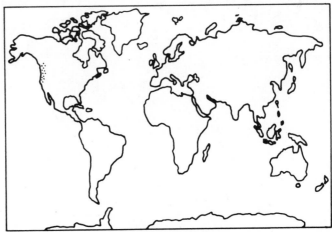

Family Aplodontidae

Ground Squirrel (*Spermophilus* spp.)

antlers, bone, and rocks, and they will often sit and grind their teeth together.

Suborder Sciuromorpha (squirrel-like rodents)

Family Aplodontidae (mountain beaver—1 species)

The **sewellel, mountain beaver,** or **boomer** (*Aplodontia rufa*) occupies a family of its own. Its cheekteeth as well as its incisors grow continuously, enabling it to chew on coarse, abrasive vegetation such as grass without wearing away the molars (which would lead to starvation). They are good swimmers and fast runners, and have been seen to brachiate through treetops. Each of the four legs is prehensile. They constantly drink water because their kidneys are incapable of concentrating urine. Mountain beavers take frequent baths and are even known to divert small streams into their tunnels, through which they then swim. Their sleep is so sound that they have been picked up without awakening.

In some areas, mountain beavers are known as "haymakers" because of their habit of cutting and drying grass before taking it below into their burrows. The hay is either eaten right away or stored for winter since they do not hibernate. During winter, they travel through tunnels under the snow. Their burrows have chambers for food storage, wastes, and mountain beaver "baseballs," large stone or clay balls used to block off tunnels or chambers as well as to gnaw on to sharpen teeth. Burrow entrances are often sheltered by blackberry, thimbleberry, or other brush. They mark their territories with urine and are known to defecate in fixed areas. They use refection, and at times they will even take balls of freshly excreted feces into their mouths and spit them onto the top of a dung heap. When threatened, mountain beavers roll over onto their back, grate their teeth, and strike out with the clawed feet. They also have the dubious distinction of being host to the world's largest flea, up to a third of an inch long.

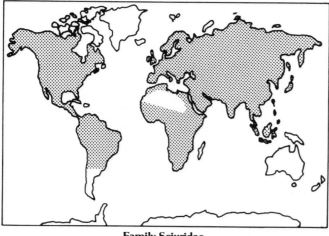

Family Sciuridae

Family Sciuridae (squirrels—251 species)

The squirrels range in size from the three inch (lengths exclude the tail), half ounce **pygmy squirrel** (*Myosciurus pumilio*) to the twenty inch, six and a half pound **giant squirrels** or **tree dogs** (*Ratufa* spp.), capable of making leaps of twenty feet from tree to tree. The tree dogs are the largest arboreal species. The two foot, almost twenty pound **Alpine marmot** (*Marmota marmota*) is the largest "land" squirrel and heaviest member of this family. "Land" squirrels such as the **woodchuck** or **groundhog** (*Marmota monax*) are occasionally found in trees.

Squirrels are clothed in a variety of colors. sometimes varying with season. The **bush squirrels** (*Paraxerus* spp.) rank among the most colorful with their green backs and yellowish bellies. The **fire-bellied squirrel** (*Callosciurus erythraeus*) is greenish above but as its name implies, has a bright red belly. (Others of this genus are tricolored, white, or otherwise beautifully marked.) In contrast, the belly of the **oil palm squirrel** (*Protoxerus strangeri*) is nearly naked. It is sometimes called the "ivory eater" from its habit of gnawing on elephants' tusks.

The **pine squirrel** (*Sciurus vulgaris*) is one of the few that changes colors with the seasons. The fur color varies from dark brown to light brown to red to cream as the seasons progress from spring to winter respectively. In the Siberian varieties of pine squirrel, the tails vary in color not with the season but with the type of forest the animals inhabit and therefore with diet. In pine forests the tails are red, in fir and larch they are brown, and in cedar forests the tails are usually black. This adaptation is known as polymorphism and is common to other mammals including the **antelope ground squirrel** (*Ammospermophilus leucurus*), some jirds, pocket mice, and others. The **Olympic marmot** (*Marmota olympus*) often changes color from spring to fall, becoming a pale yellow. This is thought to result from the ammonia given off by the urine in its burrow. Generally, the more humid a mammal's environment, the darker is its fur coloring. Certain populations of **gray squirrels** (*Sciurus carolinensis*) consist solely of albino individuals. One such group has brought fame to Olney, Illinois, where a population exceeding one thousand lives within the city limits. Melanism is also common in this species.

One of the most attractive aspects of squirrels, mainly the tree varieties, is the large bushy tail. It is actually a versatile tool used for balance, braking, turning, and communication of messages dealing with dominance and sexual matters. In addition, the tail can serve as a crude parachute in jumps, an umbrella and sunscreen to protect the animal from rain and heat, a blanket on cold nights, and a misleading target to foil attacks from birds of prey. In some species, such as the gray squirrel, the animals huddle together when cold and will use their tails to cover each other. On rare occasions, this has led to such an entanglement of tails that failure to resolve the matter has led to starvation. Seven individuals were once found bound together in just such a predicament.

During the 19th century, gray squirrels were known to make massive lemming-like migrations for unknown reasons. Hunting and loss of habitat have now made such eruptions less spectacular, but they still persist with instances reported as recently as 1968 and 1979. In 1819, a migrating mass 130 miles wide was observed in Ohio. At that time, many thousands of squirrels drowned in attempts to swim across lakes and large rivers. Actually, in smaller rivers and streams, many species, such as the **Arctic ground squirrel** (*Spermophilus undulatus*), woodchuck, and **red squirrel, chickaree,** or **barking squirrel** (*Tamiasciurus hudsonicus*), are good swimmers and will actively take to the water. In these cases, the tail serves an added function as a rudder.

Most mammals, including squirrels, tolerate only narrow limits of body temperature. The **Harris's antelope ground squirrel** (*Ammospermophilus harrisii*), however, is an interesting exception. They have shown no signs of distress despite body temperatures in excess of 110 degrees Fahrenheit. They have adapted physiologically to withstand the intense heat of their desert homes in a number of ways. They lose very little body water needlessly, as exemplified by the fact that their urine

Flying Squirrel (*Glaucomys sabrinus*)

is so thick it verges on being solid. They lose some water in their breath, but they do not sweat. To stay cool, they rely on radiative and conductive losses of heat by lying prone in shady areas or by burrowing into the cooler ground. Body temperature has been observed to fluctuate a full 7°F within just two to three minutes. Should their temperature become excessive, the squirrels drool copiously and use their forepaws to spread the saliva over the head and shoulders, where evaporation aids cooling. (Jerboas, kangaroos, kangaroo rats, rats, wallabies, and others use the same process.)

Squirrels produce various sounds, ranging from the soft chattering of the red squirrel to the booming of the oil palm squirrel and the ultrasonic projections of flying squirrels (possibly a primitive form of echolocation). Many of the better known squirrel calls are alarm calls. The **Douglas squirrel** (*Tamiasciurus douglasii*) has been called the "mockingbird of squirrels" through its varied productions of barks, screams, chirps, and chatterings.

Squirrels are found almost worldwide and in many different environments, even including the air. There are thirteen genera of **flying squirrels** (including *Aeromys* spp., *Glaucomys* spp. *Petaurista* spp., and *Pteromys* spp.). None exhibit powered flight like birds or bats, but there is a report of a **South Asian flying squirrel** (*Hylopetes lepidus*) flapping its membranes and rising three feet in a twenty-foot flight. Some can produce extraordinary glides. All have a membrane or patagium stretching between the front and hind legs. Gliding direction is controlled much as parachutists control their descents with guidelines. The tails of most have thick side hairs to increase surface area and thus improve their rudder effect. With the proper use of wind currents, some flying squirrels can end flight at a higher altitude than the take-off point or by banking can return to it. Many can perform loops, right-angled turns, dives, and even spiralling ascents. It is not unknown for some to outmaneuver attacking owls. Various species glide in excess of two hundred yards at a time, while the **great flying squirrel** (*Petaurista* spp.) has flown for distances of fifteen hundred feet with the aid of wind currents. Some flying squirrels such as the **southern flying squirrel** (*Glaucomys volans*) roost during the daytime, occasionally by hanging from their hind legs bat-style. Most species are nocturnal.

The vast bulk of a squirrel's diet is vegetation, particularly seeds and nuts but often including thistles, poisonous toadstools, and mushrooms with seemingly no ill effects. Red squirrels feed on the poisonous fungus known as fly agaric, while the **European red squirrel** (*Sciurus vulgaris*) will pick, but then lay out to dry before eating, poisonous toadstools (*Amanita* spp.). Almost everyone has seen a squirrel sitting on its haunches eating a small morsel held in its hands. What is not commonly known, however, is that squirrels (and most other rodents, with the notable exceptions of beavers, coypus, kangaroo rats, and a few others) are unable to pick objects up with the forepaws. They must do so with their teeth, and then later drop the objects into the hands after sitting up. Some flying squirrels may at times stuff themselves until flight is impossible.

Squirrels and most other rodents include fresh meat in their diets. Most species will consume small animals including insects, snails, frogs, snakes, birds (as well as their eggs and nestlings), mice, shrews, and carrion. Prairie dogs occasionally leap into the air to catch insects. Red and gray squirrels are known to eat rabbits and rats, while some ground squirrels will even eat each other during periods of population explosions. The **long-nosed squirrel** (*Rhinosciurus laticaudatus*) has long, forceps-like incisors used to grasp the insects which surprisingly make up the bulk of its diet. Oil palm and palm squirrels include pollen in their diets and in so doing become pollinators of certain plants.

Squirrels are often mentioned as the classic example of food storers, even though countless other animals including bears, big cats, foxes, shrews, shrikes, weasels, and many other rodents, including beavers, gophers, hamsters, lemmings, mice, muskrats, and more, are also known to do so. Squirrels do spend considerable amounts of time burying nuts and seeds. This is an innate drive that can even be observed in captive specimens that go through the motions symbolically. Some, such as the red squirrel, place stores in one or two large caches measuring up to 32 quarts each. Others, like the gray squirrels, hide many small supplies in numerous places, even poking nuts into tree holes drilled by woodpeckers. Still others, like the **chipmunks** (*Tamias* spp. and *Eutamias* spp.), use both methods, frequently keeping a handy store underneath their sleeping quarters. One **eastern chipmunk** (*Tamias striatus*) stored a bushel of nuts in three days. As a rule, squirrels cannot remember where all their caches are located and leave many unrecovered. This aids in the reforestation of new areas. It is interesting to note that burying is a prerequisite for the sprouting of certain nuts, including butternuts, hickory nuts, and walnuts. Although most squirrels dig out areas where they gently drop their booty, the flying squirrels will occasionally "hammer" their nuts into cracks and crevices, using their front teeth. Red squirrels sometimes impale mushrooms on small branches for later use.

Squirrel methods of transport are interesting and sometimes hotly debated. Some, such as the chipmunks, have special cheek pouches capable of holding up to 35 sunflower seeds. The pouches are crammed so full at times that the chipmunk gains a lovable but grotesque appearance. They avoid scratching the insides of these pouches by carefully biting off the sharp edges and points of acorns and nuts. Another method often relegated to fantasy is the tale of marmots pulling other grass-loaded marmots by the tail. This just may be true after all. Documented evidence from Switzerland describes how one marmot will gather a load of grass beneath its belly, grab hold with all four feet, roll over on its back, and then submit to being pulled to the burrow by its tail. Similar transport methods have been reported for the Virginia opposum, house mice with sugar cubes, and rats with eggs. Still, many authorities laugh these stories off as pure bunk. (See Family Muridae for another possible explanation of this behavior.)

Caches are recovered by both memory and smell. Squirrels will dig up any store they find, whether they buried it themselves or not. The **fox squirrel** (*Sciurus niger*) can smell nuts buried two inches below ground or up to twelve inches beneath snow. Gray squirrels can also smell nuts through a foot of snow, and they can bite open walnuts with teeth capable of exerting a force up to eleven tons per square inch. Red squirrels can smell pine cones buried up to a foot underground, and one Douglas squirrel was known to dig through twelve feet of snow to get to its cache. Beneath the feeding stations of red squirrels, piles of shucked pine cone scales have grown to thirty feet in length and three feet high. These are likely the work of several generations. Such piles of empty shells and husks are called middens.

Most squirrels require some water intake. If this is unavailable, red, fox, and gray squirrels will substitute maple sap, chinking holes in the bark of maple trees. Should the sap freeze, the squirrels are known to spend time in the morning licking on their sweet "popsicles." Red squirrels also eat snow when their water supply freezes over. Flying squirrels drink huge amounts of water. Should a human drink an equivalent amount he would have to drink about two gallons per day. Red squirrels are also avid drinkers and are known to take frequent baths, at times using fresh moss as "washcloths." They are also known to wipe their noses on bark after meals.

As a rule, "land" squirrels hibernate and tree squirrels do not, although there are exceptions. The antelope ground squirrel is not known to hibernate, while the red squirrel is known to do so in northern latitudes. Chipmunks go into a state of torpor during bad weather but don't truly hibernate, while some flying squirrels will huddle together in groups of up to 50 to keep warm in cold weather. Marmots plug all den entrances with earth and stones and hibernate in small groups of up to 15 during winter. They often carry all manner of objects into their dens for insulation, including vegetation, rags, and even dry dung. During hibernation, their body temperature falls from a regular 95–104°F to 40–55°F, their pulse drops from 90–140 to a mere 2–3 beats per minute, and their breathing rate is only once every 3–5 minutes. At this time the animals can be handled without awakening them. During such deep sleep, woodchucks have been known to drown when their burrows become inundated by heavy rains. Many do awaken on their own about once every three to four weeks to urinate and defecate. They will also awaken should the ambient temperature fall to dangerous levels. Activity at these times will keep the body temperature at an acceptable level. The **hoary marmot** (*Marmota caligata*) and the **barrow ground squirrel** (*Spermophilus parryii*) are known to hibernate up to nine months of the year, the longest hibernation period in mammals. Most of the hibernating animals lose from a fourth to half of their body weight while in this winter sleep. Hibernation ends when the mammal's endogenous circannual alarm clock goes off, since it cannot depend on external factors such as temperature, day length, etc., when it is insulated from these in its burrow system.

The old tale of how a groundhog that sees its own shadow on February 2 will return to its burrow to sleep through six more weeks of bad weather is of course pure fantasy. Yet many animals do seem to have an uncanny sense of approaching weather conditions. It is also true that clear days in early February are generally associated with additional cold weather; cloudy conditions often accompany the approach of warmer days.

In addition to hibernating, many **ground squirrels** or **susliks** (*Spermophilus* spp.) also estivate during hot spells. Estivation is thought to be triggered by lack of water rather than increase in temperature.

Squirrel homes range from the single, stick drey of the pine squirrel to the large "towns" of **prairie dogs** (*Cynomys* spp.). In between are the red squirrels and others that build underground tunnels in addition to treetop runways and dreys. In winter, the tunnels are built under the snow, where temperatures can be fifty degrees warmer. A typical squirrel drey has no permanent entrance or exit in that these are torn open and patched closed every time the squirrel comes and goes. They are usually

110

constructed by the females, although in some species the males are known to help. Dreys usually consist of three layers. The outer core is composed of coarse sticks, the middle is a waterproof layer of leaves, and the inner part is made up of soft grass, feathers, and fur. Nests are usually built in a tree fork, out on smaller limbs, underground, in deserted woodpecker holes, in tree stumps, under rocks, and even over an old hawk or crow nest. The South Asian flying squirrel is known to build its nest inside coconuts.

Marmots are good diggers and can remove seven hundred pounds or more of dirt in making their burrows. In New York state, woodchucks turn over an estimated 1.6 million tons of soil annually. A woodchuck can bury itself out of sight in less than a minute in loose ground. It and others can close off their ear canals while digging to keep out dirt and debris. Their burrow entrance is sometimes dug between close boulders or tree trunks. This prevents large predators from widening the opening to gain entrance. Like so many other burrowers, marmots construct special latrine areas. These may be closed off when full or regularly cleaned of their deposits by taking the wastes to the main entrance and burying them outside. Ground squirrels and chipmunks carry out dirt in their cheek pouches and scatter it away from their burrows. The entrance is usually camouflaged with leaves, moss, and rocks and often plugged shut to keep out predators and the cold. One **African ground squirrel** (*Xerus inaurius*) is known to share burrows harmoniously with suricates and mongooses. Their young are even reported to play together.

While most of the five prairie dog species are solitary or only partially social animals, the **black-tailed prairie dog** (*Cynomys ludovicianus*) lives in large groups known as "towns." The largest of these covered almost 25,000 square miles and housed in excess of 400,000,000 residents. Towns are further divided into geographical areas known as "wards" which can be further split into numerous "coteries." A coterie consists of a "family" group: a male, several females, and their young, even though the individuals are not necessarily related.

These prairie dogs build extensive burrow systems. They serve a multitude of beneficial functions in addition to housing and protecting the animals. They prevent floods, increase aeration of soil, stop the encroachment of forests onto prairies, lead to continued large-scale turnover of vegetation, and supply dust bathing areas for larger mammals such as bison. The systems are well engineered to serve the prairie dogs' needs. The entrance mound is built up above ground level to prevent the entry of water in the case of flooding. The typical mound is one to two feet high and six feet in diameter. Should flood waters rise high enough to pass over this barrier, the dogs still remain dry thanks to the ingenious air locks built into the system. The locks are constructed by digging straight down for a number of feet and then burrowing back up at an angle. At the very least, this slows the water sufficiently should the animals find it necessary to dig their way to safety.

The large entrance mounds also serve as lookout posts. The surrounding vegetation is removed for bedding and food as they eat twice their weight in food each month. This also removes cover from near their burrows to prevent predators from mounting surprise attacks. Special "listening chambers" are built just inside the mound where the "dogs" can sit and listen for predators overhead. The mounds serve a third purpose in that they set up convection currents, drawing air out of the tunnel system while pulling air in through various ventilation shafts. This ensures the animals of proper levels of oxygen and carbon dioxide within their homes. Inside, the tunnels provide cooler temperatures in summer and warmer ones in winter as compared to the outside. There is also a higher level of humidity all year

111

round, aiding in the important process of water conservation. The intricate tunnel systems are often cut short by earthen plugs appropriately sealing off chambers containing wastes or the remains of conspecifics. Abandoned burrows or even peripheral sections are often taken over by owls, snakes, insects, and many other creatures. Adults will usually give up their burrow system to their young which are very aggressive and importunate.

The members of a coterie keep together through "kissing" or "tongue touching" rituals. When two "dogs" meet they drop down and crawl up to each other on their bellies. After meeting and determining whether the other is friend or foe through kissing, they will either tolerate each other's presence or fight. If rivals, the intruder is almost always driven away. If familiar, they may continue kissing for ten seconds or more. This is usually accompanied by mutual grooming, caressing, and often feeding together. All of these behaviors serve to strengthen the "friendship" bonds between them. An important exception to all of these standards of behavior occurs in times of emergency, when prairie dogs from one coterie are allowed to take refuge in the burrows of another coterie without provoking a fight. Pups are almost universally accepted by any coterie and will be suckled by any available female.

Prairie dogs' eyes, like those of rabbits, possess an orange pigment to protect them from the bright glare of the sun. As alluded to, they are known to bury their dead, but at times a group has been observed to mob and kill one of their own kind for unknown reasons. Cannibalism has also been reported.

Many of the squirrels mark out their territories with special secretions and/or urine. In addition, prairie dogs verbally mark out their region. In producing their territorial calls, they fling their upper torso into the air. The young frequently end up on their backside when learning to perform this maneuver, as may excited adults.

During the breeding season, the males of many species, including gray and fox squirrels, are more aggressive since their male hormones are highest then, the testicles being shrunken and located inside the body cavity at all other times of the year. Male **striped palm squirrels** (*Funambulus* spp.), **tree squirrels** (*Sciurus* spp.), and **Belding's ground squirrel** (*Spermophilus beldingi*) fight each other for mating rights to the females. Occasionally, the females sit by and peep as if to urge the combatants on. Male ground squirrels are sometimes killed in these fights, and most are hairless from mid-chin to mid-chest by the end of the mating season as a result of numerous bites, scratches, and kicks. Males are even known to jump rivals during copulation to fight. Male **yellow-bellied marmots** (*Marmota flaviventris*) are known to collect harems of up to 31 females.

It is interesting to note that during courtship squirrels (like dormice, hamsters, humans, and others) are known to "baby talk." As the adults cuddle, they produce sounds that mimic the calls of the young. A few species, including the fox squirrel and some marmots, are thought to mate for life. The females of some species, such as Belding's ground squirrel, are receptive for only three to six hours shortly after hibernation ends. Male chipmunks may chase females for up to nine hours before mating. In many squirrels (and in cats, shrews, and weasels), mating is preceded by what appears to be a vicious fight, even though neither usually becomes overly aggressive. This premating spat is thought to induce ovulation to ensure a successful pregnancy. Others believe that a waxy plug secreted by the male into the female's vagina during copulation both stimulates the female to ovulate and increases the likelihood that he is the progenitor of her offspring, since it prevents entry of other males' sperm.

112

After mating, most males leave of their own accord or are driven away by the females since they have been known to kill and even eat the young, as in the case of Belding's ground squirrel. At times, the mother of a large litter, as in the **thirteen-lined ground squirrel** (*Spermophilus tridecemlineatus*), may eat some of her own young. The females will use their own tail hairs to bind nesting material together. When danger threatens, the female will usually transport her babies by grasping the loose fur on the belly and carrying them one by one in her mouth to a new, safer nesting area. Female tree squirrels cover their young with nest material when they leave them alone. At weaning time, the young of some, such as the **African ground squirrels** (*Xerus* spp.) (like many Artiodactyls), are known to eat dirt before advancing to solid food. By doing so, they obtain some of the intestinal flora (bacteria) required for proper digestion. The female woodchuck is known to prepare new dens for her young. She leads each to its new home at weaning time.

Most species give some sort of warning when danger approaches. In fact, many are said to post "sentries." Actually, no particular animal stands "watch," but at any given time one or more will be keeping an eye out and will chatter out warnings at the first sight of danger. Prairie dogs signal danger by barking, erecting their tail, and by lying flat against the ground. They and ground squirrels apparently produce separate calls to distinguish predators approaching from the air and the ground. The former call provokes an immediate dive by all to cover. The latter prompts freezing and scanning of the area prior to taking evasive action. Some ground squirrels even produce a special warning for snakes. The antelope ground squirrel and others give warning by flashing the white bottom side of their tail, while woodchucks and others do so by chattering their teeth and marmots may whistle. Squirrel warnings are also headed by other mammals in the vicinity, including numerous species of antelope, deer, and sheep. Many also give an "all clear" signal; in the prairie dog, this consists of standing erect and holding the tail downward. This particular signal is used as a sign of territory ownership as well.

Evasive maneuvers to foil predators are usually successful. Marmots and other ground squirrels construct "dive tubes" and "flight tunnels" into which they scamper when danger threatens. If enemies are encountered below ground, as in the case of rattlesnakes, prairie dogs and others will seal off that portion of the burrow. Tree squirrels readily retreat to the safety of the trees where they quickly maneuver themselves to the opposite side of the tree from the predator. They will also quickly shinny up the tree and slither out onto a branch hugging closely to the upper surface to keep from being spotted. If pursued they head for the outer limbs and will even jump as far as twelve feet to a neighboring tree. When chased by nimble predators such as fishers, they are known to head for the uppermost branches and then jump straight to the ground as the fisher closes in. Some squirrels have jumped as far as 30 feet in just such an escape. One **Mexican tree squirrel** (*Sciurus* spp.) leaped 550 feet to a ledge when chased and landed without apparent harm. When traversing large distances on the ground, tree squirrels frequently stop at intervening trees, climbing part way up to get a good look around before progressing to the next "lookout post" on the journey. They have sharp claws, needed for a good hold. Since these are not retractile, they are kept extended upwards when running along the ground to prevent dulling the sharp tips. On rare occasions, female gray squirrels have attacked and bitten humans who threatened their young. An evasive tactic used by many flying squirrels is scampering around to the other side of a tree immediately after touchdown. This occasionally proves lifesaving when a predator such as an owl is "hot on its heels." All marmots are capable of emitting smelly secretions from their anal glands when agitated. In some species, such as Belding's ground squirrel, harsh weather is the chief cause of mortality.

There are many stories of red squirrels and gray squirrels being enemies and of their exterminating each other in various localities, but these tales are unfounded in fact. Natives believe the bite of some African ground squirrels to be poisonous. Since these squirrels have salivary glands filled with streptobacillae their bites readily cause infection and occasionally septicemia with death a result. However, there is no venom.

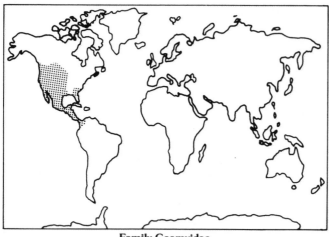

Family Geomyidae

Family Geomyidae (pocket gophers—30 species)

As is obvious from their name, **pocket gophers** (many genera) have cheek pouches. In them, they carry food but never dirt, unlike chipmunks and other "pocketed" squirrels. Most people do not realize that the cheek pouches of pocket gophers (and of pocket mice, pocket rats, and kangaroo rats) are externally located and open to the outside. In addition, they are fur-lined, not skin-lined like the internal pouches of chipmunks and hamsters. The pockets must be periodically cleaned. At such times, they are manually everted in the same manner as they are emptied. After cleaning, they are quickly retracted into place by the action of special muscles.

Pocket gophers exhibit many anatomical adaptations to their burrowing mode of life. Like beavers and many other rodents, pocket gophers' lips close behind the large incisors to prevent dirt from entering the oral cavity. As a result, their front teeth are always visible. These incisors grow up to twenty inches per year, a record for tooth growth rate, but are kept worn down through the processes of burrowing and tooth grinding. Stiff paw bristles prevent dirt from clogging the areas between their toes. The color of their fur generally varies to match the color of the soil they inhabit, and it is somewhat greasy to repel moist dirt. Like moles, their hairs can lie in any direction. All species are equipped with functional eyes and ears. The eyes are kept moist and free of dirt through the secretion of thick tears, and the ears can be closed to keep out soil. Their bodies are resistant to the high carbon dioxide levels in their burrows and are adapted to the lower oxygen concentrations. They do not need to drink since they get all the water they need from the vegetation they eat.

A single pocket gopher can dig out a 300 foot tunnel in one night's work. Digging is done primarily with the feet, although the teeth are brought into play against roots and stones. During winter, excavated dirt is deposited in tunnels dug in the snow. When the snow melts in the spring the dirt remains as long (up to 40 feet) caterpillar-like mounds referred to as "gopher sausages." Gophers can yearly bring up to ten tons of earth per acre to the surface through their digging efforts. Burrow entrances are marked by fan-shaped mounds, not the round mounds of moles. Mole tunnels descend vertically while gopher tunnels slant at an angle. Entrance ways are usually blocked closed to discourage predators and to maintain proper temperature and humidity levels. When moving backwards, the tail is used for guidance and is lined with special tactile organs for just this purpose. Pocket gophers can run almost as quickly backwards as they can forwards. The upper tunnels are used for feeding (i.e., to pull plants down by the roots), while the deeper tunnels are for nests, storage, and wastes. A few, such as the **mountain pocket gopher** (*Thomomys monticola*), occasionally nest above ground. Some species are good swimmers. Rarely a gopher may consume a mouse or dead conspecific.

During the breeding season, the males wander out of their tunnels or dig long straight extensions to search for females, who remain inside their own tunnels. After mating, the male returns home to resume his solitary existence. If two gophers of the same sex meet, they may fight to the death. They are, however, somewhat protected by their loose skin, which thickens around the head and neck. The female's small, fused pelvic bone is widened during pregnancy. This is completed through the secretion of special hormones capable of dissolving away part of the pelvis to facilitate the passage of the young during the birth process.

When alarmed, pocket gophers will hiss. One of their main enemies is humans, who poison them because of the damage their burrows do to crops and gardens. In the wild, however, these burrows serve to loosen, aerate, and add organic debris to the earth as well as to help prevent flooding. Some gophers are the only host to certain parasites such as the **gopher biting louse** (*Geomydoecus scleritis*).

Family Heteromyidae (kangaroo rats and pocket mice—63 species)

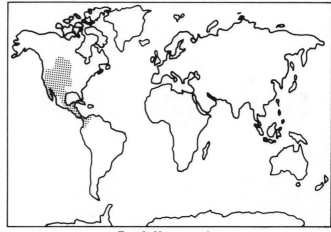

Family Heteromyidae

Kangaroo rats (*Dipodomys* spp.) and **kangaroo mice** (*Microdipodops* spp.) are appropriately named for their anatomic and locomotive peculiarities. Hopping on the hind legs is physiologically the most efficient means of traveling in sandy areas, where these mammals are often found. They can make leaps up to 15 feet and with the aid of their tail can leap in any direction and even change direction in mid-air. **Pocket mice** (*Perognathus* spp.) and **spiny pocket mice** (*Liomys* spp. and *Heteromys* spp.) differ in their mouse-like appearance; they travel on all fours. Many species in this family have tails tufted at the tip for greater efficiency in steering and maintaining balance. The tails of kangaroo mice are unusual in that they are thickest in the middle rather than the end. This is where fat is stored for future use. They do not have tufted tailtips. The spiny pocket mice are covered with flattened, bristly hairs thought to give them protection not only from cactus spines but also from insect bites. Kangaroo rats must dust bathe regularly to prevent the development of skin sores and to keep their fur from becoming matted by body oils secreted from a gland on the back. External cheek pouches are also present in this family.

With a few exceptions, such as the **chisel-toothed kangaroo rat** (*Dipodomys microps*) and the **Mexican spiny pocket mouse** (*Liomys irroratus*), most of the rodents in this family (and many other mammals, including the addax, oryx, and gazelle antelopes; desert mice; fennecs; gerbils; ground squirrels; jerboas; jirds; koalas; and pocket gophers) obtain all the moisture they need from foodstuffs and the oxidative breakdown of fats. Actually, in most mammals, including the camel and some of those just listed, more water is wasted through respiration than can be produced through oxidation of fat so that they must rely on other mechanisms, mainly a high water content in their food. Water is lost through respiration because exhaled breath is humidified and usually warmer than the ambient air so that its water content is higher. Kangaroo rats, however, possess special nasal passageways that not only cool their breath to as much as 50°F less than the rest of their body, but also absorb enough water to allow their food metabolism to produce positive water balance. Increasing their activity increases water production through stepped-up oxidation. If their food has a high water content, they are often much more lethargic, possibly because of the decreased need for oxidative water production.

Kangaroo rats conserve body water in many ways. The animals do not sweat and avoid heat, often by remaining in cool, humid (30–40 percent relative humidity, compared to 10 percent outside) underground tunnels. They produce very dry droppings and an almost solid urine through water absorption by the bladder and very efficient kidneys. Similarly efficient kidneys are also found in gazelle, jerboas, kangaroos, pronghorn antelope, and the **spinifex hopping mouse** (*Notomys alexis*), which produces the most concentrated urine of any rodent species. If human kidneys worked as well, we could drink sea water without problem, but as it is, we can only concentrate our urine about one

Merriam's Kangaroo Rat (*Dipodomys merriami*)

fifth as well. If kangaroo rats become overheated they will coat their forebody with saliva to cool down with the help of evaporation.

Most of these rodents hoard food. Often the seeds are first dried under a one-inch layer of sand to prevent germination after they are buried. Grasses and seeds are often placed in separate piles of their own kind. Pocket mice can collect and store three thousand seeds or more in an hour. When filling their cheek pouches, the hands move so fast that they become a mere blur. One **Merriam's kangaroo rat** (*Dipodomys merriami*) stored a cache of fourteen bushels. The **giant kangaroo rat** (*Dipo-*

116

domys ingens) buries small seeds in holes near its burrow entrance. One specimen dug almost nine hundred such storage holes for its goods. Some members of this family are known to steal from the caches of others rather than work to stockpile their own. In the chisel-toothed kangaroo rat, one that requires supplemental free water, the animals carefully avoid the outer, salty layers of leaves by scraping them away with their incisors before consumption. In hard times, the metabolism of pocket mice falls to conserve energy, but this is not a true hibernation/estivation. The **little pocket mouse** (*Perognathus longimembris*) can remain in such torpor for up to one week.

Many species in this family place their burrow openings near vegetation for stability and camouflage. The **banner-tailed kangaroo rat** (*Dipodomys spectabilis*) builds mounds up to 3 feet high and 15 feet across with up to 12 openings. Kangaroo rats and pocket mice generally keep burrow entrances closed. This is done not only when they are inside but after they leave as well to maintain decreased temperatures and increased humidity inside. Sometimes they are observed bursting through what appears to be solid ground as they exit these sealed entrances. This is also observed in other rodents, including the **Oldfield mouse** (*Peromyscus polionotus*).

Kangaroo rats are solitary for the most part, and they are known to fight to the death if forced together. If their territory is infringed on by a conspecific, pocket mice squeal and kangaroo rats thump the ground to announce ownership. Kangaroo rats fight and defend themselves with their powerful, clawed hind feet. These feet are haired not only to improve traction on sand but also to kick sand more efficiently into the face of attacking predators, including rattlesnakes. The powerful hind legs of the banner-tailed kangaroo rat allow it to make jumps in excess of ten feet. Some also drum their feet to attract females at mating time.

Female **Ord's kangaroo rats** (*Dipodomys ordii*) are known to pursue the males when in heat. After copulation, female spiny pocket mice form vaginal plugs. When threatened, female kangaroo rats carry their young away by holding them in their arms. Litter size and frequency both fall off in times of drought because of a decrease in estrogenic substances found in the green plants they normally feed on. There is also a lowering in the urge to mate at such times.

The primary enemies of this family are snakes and owls. In one study, fewer than five percent of kangaroo rats lived greater than a year, even though they are capable of lifespans lasting five years. It is interesting to note that the hearing of kangaroo rats has been found to be most sensitive in the range (1,000–3,000 Hz) of those sounds produced by the rustling of owls' feathers (1,200 Hz) and the friction of scales in rattlesnakes coiling to strike (2,000 Hz). This enables them to make more escapes with dodges at the last fraction of a second so that fewer are lost to predators. In fact, their middle ear is larger than their brain and is proportionately the largest of all mammals. Those in the kangaroo mice are the biggest in this respect, enabling them to hear more than four times as well as humans. Enlarged bullae are also present in jerboas, gerbils, desert antelope, foxes, gundis, and bats.

Family Castoridae (beavers—2 species)

Beavers (*Castor fiber* and *Castor canadensis*) are well adapted to life in cold water. They are kept warm by the presence of two fur coats. The inner one is silky and warm while the outer coat or "kemp" is coarse and made waterproof by secretions from the perianal glands. They can stay submerged for up to fifteen minutes at a time and can swim up to half a mile underwater at speeds exceeding five mph.

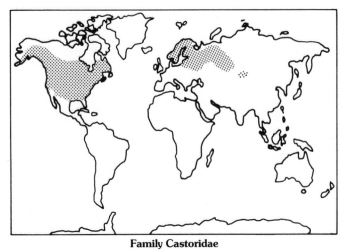

Family Castoridae

The ears and nostrils are closed off by special valves prior to making a dive. (Other animals with valved nostrils include desmans, dugongs, hippos, manatees, mouse-tailed bats, otters, platypus, seals, some shrews, and whales; other animals that can close off their ears include hippos, muskrats, platypus, and otters.) Beaver eyes are protected by special transparent lids.

Beavers are the only rodents possessing a cloacal fold. This closes over to protect the ano-urino-genital area and makes the sexing of a beaver difficult. Water and debris such as mud and splinters are kept out of the oral cavity by flexible lips that can be closed behind the ever-growing, orange incisors, even when the mouth is full with a large load of vegetation. The teeth are embedded in large bony areas that also serve to prevent splinter penetration. Their large, webbed feet increase swimming efficiency, and the hind feet's first and second claws are grooved for grooming the fur and can be used to pick debris from between the teeth. The characteristic large, flat tail is used for balance, fat storage, propulsion, shedding of excess heat, steering, support, and warning of danger, a feat accomplished by slapping it against the surface of the water to produce a noise audible for half a mile or more.

Beavers are best known for their seemingly industrious nature and their dam building. The average dam measures about one hundred feet in length and five feet in height. Some go as high as eighteen feet and are up to twenty feet wide. The record dam was 4,000 feet long and was located on a lake near Berlin, New Hampshire. The largest dams are the work of one to three beaver families and numerous generations. It takes about a week for a beaver family, which rarely exceeds ten members, to build a dam thirty feet long. They occasionally work in pairs. One does the cutting while the other keeps watch for danger. At intervals they trade places.

A beaver's large incisors can bite out chips of wood six inches long. A single beaver can fell a five-inch diameter willow tree within three minutes. Although they rarely cut trees greater than a foot in diameter, they have been known to cut some measuring five feet or more in diameter and in excess of one hundred feet high. These would often require several nights work before coming down. One beaver pair cut down 266 trees in fifteen months. Beavers gnaw until crackling signals the imminent fall. At this time, the beaver runs for cover, usually by diving into nearby water. Contrary to popular belief, beavers do not always drop the tree towards the river; riverbank trees naturally fall this way because that side receives more sunlight and has more foliage. Since they cannot really control the direction of fall, beavers are sometimes found crushed to death by the very tree they cut down.

Even though beavers can climb, trees that are cut down six feet above the ground were actually

118

taken when there was snow present to that level. At times, a tree is intermittently gnawed at when the snow is at different levels producing multi-layered cuttings on the tree (up to eleven layers in some instances). Beavers usually don't take trees that are more than two hundred yards from the water. In some cases, special waterways up to twenty inches deep complete with locks to guide the direction of the logs are built. Some of these have measured up to seven hundred fifty feet or more in length. At other times, only a pathway to the river is cleared, and the logs are dragged to the water. Mud slideways make for easy passage through the locks. One dam in North Dakota was actually built with coal and another in New Jersey with corn stalks.

The first logs to be installed in a dam are either cross-braced against existing structures such as trees and rocks or thrust with considerable force into the river bottom at acute angles so that the force of the river's water will anchor them even further. Secondary brush is often weighted down with rocks weighing as much as thirty to forty pounds, again a remarkable feat for an animal that weighs only about 50 pounds. Large adults have weighed up to 90 pounds or more, with the record being 115, but these are exceptional.

Mud used for dams and nest building is not carried on the tail but in the forepaws held against the chest. When so loaded the beaver walks on its hind legs using its tail for support. Usually the mud is obtained from an area just upstream of the dam. This results in the formation of a pit which further slows water flow taking some of the pressure away from the dam structure. Should the pressure build up to dangerous levels, the beavers will often go downstream and build accessory dams. Thus the downstream water level rises to counteract some of the upstream pressure and ensure the safety of the dam. Sometimes, particularly when flood water is present, spillways are cut to relieve excess pressure. Damage is continually repaired as the beavers are alerted to the need by the sound of rushing water. Most dam work is done at night.

Dams are useful in that they conserve water; prevent flooding of lowland areas; and lead to the production of lakes and marshes. The latter are important because they are habitats required for the existence of other animal species.

The nest or "lodge" measures up to 25 feet in diameter. It is equipped with a vent for air-conditioning, as many as seven underwater entries, and a floor located above the water level. When the water level rises, the floor is built up by chinking out part of the ceiling of the lodge. Later, mud and sticks are used to reinforce the ceiling. The lodge is sturdy and can support the weight of several humans, and dams have been ridden over on horseback. Lodges occasionally house other mammals, such as muskrats, simultaneously with the beavers. In winter, the mud coating becomes concrete hard and prevents enemies from digging through. In some areas, beavers cut dens into river banks rather than build a lodge.

Beavers are vegetarians thriving on bark, saplings, water scum, and other vegetation. They occasionally hold and eat the bark from small branches in much the way humans eat corn on the cob. Their scat consists of oval pellets of sawdust. It has been calculated that adult beavers use about nine thousand pounds of wood and bark annually in eating and building. Since they live up to twenty years, this adds up to a lot of wood per beaver.

Winter food is not gathered until the leaves turn yellow. At this time, the trees have stored the most nutrients beneath their bark, which both helps beavers get through winter and keeps the branches from rotting in storage. Small, green branches are stored at the bottom of the pond for winter use, a necessity since the winter frost freezes the tree sap so solidly that not even powerful beaver incisors can overcome larger branches. They may store 2–4,000 pounds of vegetation for winter use. Beavers are among the few species of mammal that actually gain weight during winter. This often applies mainly to the young, since many of the adults are known to fast through much of the winter, sustaining themselves on stores of fat. During winter, they frequently breathe from air trapped under the ice when they go swimming. Sometimes holes are cut in the dam during winter to widen the gap between the ice above and the water surface below.

Beavers can emit a foul-smelling substance known as castoreum from the castor or scent glands, located anterior to the perianal glands which also discharge into the cloacal region. This serves to mark territory and to advertise for a mate. Castoreum has been used in the past as a cure-all. The fact is that it probably did some good since it contains salicylic acid obtained from willow bark. Salicylic acid is closely related to aspirin. Beavers mate back-to-belly, but occasionally belly-to-belly, both underwater and on the ground at the water's edge. They are thought to mate for life (as are badgers, coyotes, foxes, gibbons, hooded seals, marmots, mongooses, etc.).

The young or kits are known to take to the water inside the lodge within hours of birth. It is impossible for them to leave the lodge at such an early age because they are too buoyant to dive and are thus unable to get through the exit holes. They learn quickly, however, and are said to be competent swimmers at one week of age. The female will nurse from a standing position but will often fold her tail beneath her body, allowing the young to sit on a warm support while nursing. The female carries the young in her mouth or on her back when in the water, depending upon how old they are. When on land, she will sometimes carry them on her tail or in her front paws while waddling about on her hind legs. The young stay with their parents for up to two years, longer than any other rodent. At the end of this period, the female forces them out of the area and must usually use force to do so, on rare occasion even killing them. At other times, they and the father are forced out of the lodge temporarily when the female is about to give birth again. They live up to 23 years.

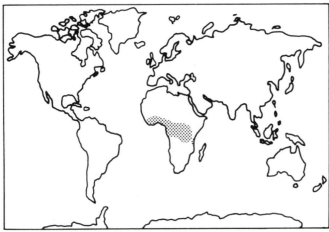

Family Anomaluridae

Family Anomaluridae (scaly-tailed squirrels— 7 species)

With the exception of the **flightless scalytail** (*Zenkerella insignis*), all **scalytails** (*Anomalurus* spp. and *Idiurus* spp.) are capable of gliding flight, like in the flying squirrels. Some cover distances measuring 750 feet. The flying membranes are attached to long elbow spurs (not the wrist spurs of flying squirrels) that can be used to tighten the flying membrane. Their name comes from

120

the sharp, pointed scales located on up to one third of the undersurface of the tail. These are used to gain traction when climbing vertically up tree trunks. Scalytails are quite nimble in trees but do not move well on flat ground. Since they are generally nocturnal, they spend the day sleeping, clinging to branches and tree trunks or bunched together inside hollow trees. When disturbed, a "cloud" of squirrels takes to the air in many directions. Their color blends in well, giving them the appearance of bark. Scalytails usually live in family groups, but some, such as the **pygmy scalytail** (*Idiurus zenkeri*), live communally in large numbers of a hundred or more, often in the company of **free-tailed bats** (*Tadarida leonis*). Pygmy scalytails occasionally sleep by clinging to the side of their drey with all four feet.

Family Pedetidae (springhaas—1 species)

The **springhaas, springhare,** or **Cape jumping hare** (*Pedetes capensis*) looks like a rabbit-sized kangaroo with long ears and a bushy tail. It is said to make an interesting pet and has been known to follow after its owner like a dog. Leaps of 36 feet have been attributed to it, but lengths of 5–10 feet are probably more typical, although still remarkable for its size. The animal is usually bipedal when traveling or resting but quadrupedal when feeding. The

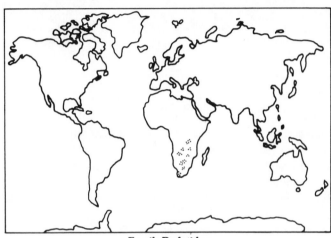

Family Pedetidae

springhaas is a vegetarian but is known to frequently consume mineral rich soil. It is an excellent digger capable of tunneling up to ten feet in a few hours. The ear canals can be voluntarily occluded with a special "tragus" to keep out dirt and debris when digging. Springhaases exit their burrows with large leaps to foil any predator that might be lurking near the entrance. When inside, the entrance is often blocked closed. The animal usually sleeps while standing.

Suborder Myomorpha (mouse-like rodents)

Family Cricetidae (American mice, gerbils, hamsters, lemmings, muskrats, pack rats, voles, etc.—578 species)

We can only scratch the behavioral surface of the cricetid grab-bag. It is the largest of all mammalian families in total number of species (this claim sometimes belongs to the Family Muridae, of which the Cricetidae are sometimes considered a subfamily).

One of the best known members of the family is the **common hamster** (*Cricetus cricetus*). When swimming, it adds needed buoyancy to its body by inflating its internal cheek pouches with air. However, pet **golden hamsters** (*Mesocricetus auratus*) can die if exposed to damp conditions. (All

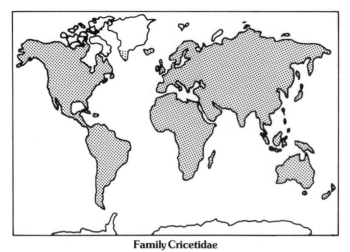

Family Cricetidae

pet golden hamsters are the descendants of a single family captured in 1930; the species is probably extinct in the wild.) Wheel-running in pets is thought to fulfill the needs of exploratory migration.

The **snow** or **collared lemmings** (*Dicrostonyx* spp.) and the **Asiatic hamster** (*Cricetulus* spp.) are the only rodents to turn white in winter and brown in summer. During winter, snow lemmings also grow enlarged accessory claws on their third and fourth phalanges to help them travel on and dig in ice and snow. These are cast off after winter in what is the only known case of rodents shedding claws. Other **lemmings** (*Lemmus* spp.) and some **gerbils** (*Gerbillus* spp.) grow long hairs on their feet to increase traction in snow or sand and to help insulate them from cold or heat. Diurnal gerbils such as the **fat sand rat** (*Psammomys obesus*) and the **midday jird** (*Meriones meridianus*) are protected from the carcinogenic solar radiation of the desert by dense fur, thick skin, and extra skin pigment. They are exceptions to the general rule that densely furred animals possess thin skin and vice versa.

Pack rats, cave rats, trade rats, or **wood rats** (*Neotoma* spp.) are best known for their thefts of brightly colored objects, which they carry off to hide in their nesting areas. They do not, however, purposely exchange one gift for another. Instead, should they, while carrying one item to their nests, come across a second with more appeal, they will drop the first and take the new one home, giving the impression of bartering. They often just steal what they please without leaving anything in exchange. They are known to take just about any object they can carry—including bones, bullets, coins, dynamite, false teeth, glasses, knives, mousetraps, nails, shells, silverware, rags, tin cans, and so on. These objects are often incorporated into the nest structure, as are cow and horse dung. Giant rats and others make similar thefts.

One of the great animal leapers is the **Indian gerbil** or **antelope rat** (*Tatera indica*). It is credited with jumps of sixteen feet, not bad for a mouse-sized animal. Like jerboas and kangaroo rats, they travel about by bounding on their hind feet. This is thought to be the most energy efficient means of travel in dry areas with short food supply. It is also said that this animal escapes dogs by leaping onto its pursuer's back.

The **muskrat, mudcat, muskbearer,** or **musquash** (*Ondatra zibethicus*) is the largest member of this family, and the best known aquatic member. Its vertically flattened tail offers increased effectiveness in propulsion and steering. Its middle ear is protected when diving by closing a special skin flap. The feet are only partially webbed, but they are fringed with bristles. It can stay submerged up to seventeen minutes and can swim forwards and backwards. Its name comes from the odoriferous secretions emitted from glands near its tail. During the winter it maintains breathing holes when the surface of its marsh or river freezes over. The breathing holes are kept open with plants placed by the muskrat and changed daily; these plants also provide a protective camouflage. Adults often feed on rafts of reeds, which are also occasionally used by females for giving birth while floating in the middle of the pond safe from many predators. These rafts can measure several feet across. The young are reported to swim before their eyes even open.

Housing arrangements vary considerably in this large family of rodents. The muskrat's nest is constructed by piling up a mound of rushes and reeds as much as five feet high and eight feet in diameter and then digging out a central cavity. When winters are lean, the inhabitants feed on the inner walls of their homes. The entrance and exit tunnels are often underwater. Pack rats build large stick nests six feet or more in diameter and height. They resemble beaver lodges in shape. Such nests may be continually added to by successive generations and total more than a ton in weight. They are frequently built around cactus

Muskrat (*Ondatra zibethicus*)

plants, and the openings are often well lined with cactus thorns to discourage intruders. At weaning time, the females often leave the large nest to the young while they go off to build another.

Many cricetids are aboreal. The **golden mouse** (*Ochrotomys nuttalli*) builds spherical nests as much as thirty feet high in trees. Their prehensile tails aid in balance and support. The nests of **dusky-footed woodrats** (*Neotoma fuscipes*) can be either fifty feet up a tree or on the ground, where they can measure up to eight feet in diameter. Such ground structures may house many of these partially social creatures along with other animals including scorpions, spiders, frogs, lizards, snakes, tortoises, wrens, and mice. **Red tree vole** (*Phenacomys longicaudus*) females rarely leave the trees and nest up to 150 feet high in Douglas firs. Their nests are cemented to the branches with urine and feces. The males dwell in ground burrows. **White-footed mice** (*Peromyscus* spp.), also known as **wood mice, deer mice,** and **vesper mice,** frequently dwell in an abandoned bird or squirrel nest. During winter, they have even been found inside occupied beehives, although most of the worker bees have died and the hive is inactive. These mice will often move to a new nesting area when their current one becomes too soiled with their wastes.

Other unusual housing arrangments have been reported. The **red-nosed mouse** (*Wiedomys pyrrhorhinos*) often nests communally inside termite mounds or in the hanging nests of birds such as the **thorn bird** (*Anabates rufifrons*). The **marsh rice rat** (*Oryzomys palustris*) occasionally builds its nest inside a muskrat's lodge or anchored in cattails above the waterline. It also constructs feeding platforms by bending over vegetation. The **Florida mouse** (*Peromyscus difficilis*) is known to take up residence with turtles (*Pseudemys floridanus*) inside the latters' burrows. One subspecies of **white-footed mouse** (*Peromyscus leucopus tornillo*) lives only in the darkness of remote areas in Carlsbad Cavern where it feeds almost exclusively on **cave crickets** (*Ceutophilus* spp.).

Many of the rodents in this family dig burrow systems. The **woodland vole** (*Microtus pinetorum*) can dig at a rate of fifteen inches per minute. The **great gerbil** (*Rhombomys opimus*) has built systems with almost 2000 feet of tunnels. **Mole voles** (*Ellobius* spp.) have tiny eyes and atrophied ears. They spend their lives digging underground and, like moles, have fur that can be flexed in any direction to facilitate backward movement through their tunnels. **White-throated wood rats** (*Neotoma albigula*) and others line their burrow entrances with cactus spines to discourage intruders. During winter, when there is heavy snow cover, the **meadow mouse** or **meadow vole** (*Microtus pennsylvanicus*) tunnels out ventilation shafts to prevent harmful build-up of carbon dioxide. The **water vole** or **water rat** (*Arvicola* spp.) has at least one underwater exit in its burrow system. The females in this species will often take over abandoned bird nests, and the males are known to build small, twig platforms upon which they regularly sunbathe. If they are prevented from taking frequent water baths, they are known to develop an eye disease.

Most of these rodents are omnivorous but feed primarily on vegetation. The meadow vole is said to consume its own weight in seeds, leaves, and grasses every day. The **long-clawed mole vole** (*Prometheomys schaposchnikowi*) appears to prefer poisonous **buttercups** (*Ranunculus elegans*) as its staple. The golden mouse eats poison ivy and sumac, while the **pygmy mouse** (*Baiomys taylori*), the smallest of all rodents, eats so much prickly pear fruit that its urine and mouth are often stained red. The red tree vole feeds almost solely on Douglas fir needles. The **salt-marsh harvest mouse** (*Reithrodontomys raviventris*) is one of the very few mammals able to drink saltwater without ill effect.

A few species are primarily carnivorous. **Grasshopper mice** (*Onychomys* spp.) track down their prey, mostly grasshoppers and other insects, but also scorpions, lizards, and other mice (including grasshopper mice), by scent much as do hunting dogs. In fact, once they pick up a fresh scent they are said to get excited and utter sharp squeaks detectable up to fifty feet away. The **fish-eating rats** (*Ichthyomys* spp.) feed primarily on fish, spearing them with their upper incisors. White-footed mice avoid the irritating bristles on gypsy moth caterpillars by peeling the insect and eating only the insides. The Indian gerbil feeds regularly on eggs, birds, and even its own young on occasion (as has the muskrat). **Field voles** (*Microtus agrestis*) are known to eat the bodies of dead conspecifics, and muskrats have been reported to eat their own tails during hard times.

Most members of this family are active during winter and are known in colder habitats to dig numerous tunnels under the snow where the temperature can be as much as 50°F warmer. The **fat-tailed gerbil's** (*Pachyuromys duprasi*) tail becomes swollen and club-shaped in times of plenty to serve as a built-in food supply for leaner times, usually winter. At times the tail becomes so fat that it cannot be lifted in the air by its owner. The **snow vole** (*Microtus nivalis*) and **high mountain vole** (*Alticola* spp.), and a few others, are haymakers like the pikas. Pack rats have stored up to 90 pounds of pinon nuts, and hamsters have been recorded to store up to 200 pounds of food, mostly corn and

124

potatoes, in their winter pantry. The food is transported in cheek pouches able to hold 42 soybeans; the head can then reach a third the size of the remaining body. They are one of the few species to hibernate but are known to awaken frequently to feed on their foodstuffs. When hibernating, their pulse drops from 400 to four while their respiration rate falls to two breathes per minute. **Harvest mice** (*Reithrodontomys megalotis*) reduce their metabolism up to 25 percent in cold weather. They conserve heat by huddling together when it is freezing outside.

Most of these rodents are solitary and will fight conspecifics should one wander into their territory. When male hamsters meet, they often threaten each other by inflating their pouches. If the intruder does not back off, a fight will erupt. The victor spreads its scent over the opponent's body. When next the loser comes across the scent of the winner, it quickly retreats. Many of these animals mark their territory with their scent. Water voles do so with special side gland secretions that are wiped onto the hind feet and from there to their territory. Gerbils do so with secretions from glands on their belly. Some mark their areas with urine and feces.

Pack rats within a family group have the habit of licking each other's lips, faces, and heads. This is thought to promote good feelings between adults, as well as to play an instrumental part in transmitting antibodies from the mother to her offspring. The males both court the females and transmit alarm by stamping the ground with their feet. The female may give him a "cold shoulder" in the form of an injured ear or tail if he is not careful. She may allow him to return to the nest after the young are born, a behavior unusual to rodents. Male **cotton rats** (*Sigmodon* spp.) often fare even worse, since the females have been known to attack, kill, and eat suitors. In **rice rats** (*Oryzomys* spp.), the females actively seek out the males but drive them away after mating. They are also unusual in that the females again go out "mate-hunting" within hours of giving birth. These animals, like most rodents, are extremely prolific. The females are ready to start their own families by the age of seven weeks. They are also aggressive animals and do not shy away from cannibalism, a practice that helps to keep their numbers down.

Almost all mammals, including bats, mate in the dorsal-ventral position referred to as *more canum* (the dog's way). A few, such as the beavers, dugongs, hamsters, humans, orangutans, pottos, pygmy chimpanzees or bonobos, sea otters, two-toed sloths, and whales, are known to mate ventral-ventrally or *more humanum* (the human way). The current record holder in the mating department is the male **Shaw's jird** (*Meriones shawi*). On one occasion a male copulated 224 times within two hours. In captivity, male golden hamsters should be removed from the females after mating or they will probably be killed by the female. The males are much larger but are fully inhibited from harming the females in any way. When sexually excited, male common hamsters chatter their teeth, a signal also used in displays of threat. The gestation period of golden hamsters is a quick fifteen days, the shortest of all non-marsupial mammals. Other species in this family also have short gestation periods, usually three to four weeks. In many female rodents, such as the house mouse, a vaginal plug is formed shortly after mating.

Even though some rodents, such as the **bank vole** (*Clethrionomys glareolus*), use delayed implantation, most don't. The majority are extremely prolific. To give an idea of how fast these rodents can "be fruitful and multiply," the meadow mouse female, most prolific breeder of all mammals, can produce up to 17 litters of 4–13 young each year. Fortunately, they usually die of old age at about one year, although aged ones have reached eighteen months. Only 2% surpass the 100 day mark.

The **common vole** (*Microtus arvalis*) is one of the most fecund of all mammals. Female sucklings have been found to be sexually mature and can mate at twelve days of age, allowing them to give birth by five weeks of age. One female produced 127 young in 33 litters during her lifetime. The female meadow vole can reproduce when twenty-five days old. Many species in this family are sexually mature at eight to twelve weeks of age. A female hamster can given birth to as many as twelve young. She has only four teats, however, and her excess offspring are occasionally eaten by her and/or the surviving young. Female Indian gerbils are also known to eat their young at times. In deer mice, grasshopper mice, and others, the males will sometimes act as "midwives." They not only aid in delivery but clean the young, cover them with nesting material, and eat the placenta. In some species including the **prairie vole** (*Microtus ochrogaster*), the males aid the females in caring for the young. Female meadow mice and others mate again within 24 hours of giving birth.

Norwegian lemmings (*Lemmus lemmus*) exhibit population explosions at intervals of about four years or so. Most people have heard of their "mass suicides." What happens is that conditions favor a rapid population growth that leads to a "panic" causing many of the animals to emigrate or irrupt. During such periods, areas of up to two square miles can be filled with animals. They head out in all directions, inevitably leading many to the sea. These lemmings are actually good swimmers, normally being able to swim fjords 2.5 miles wide, but the stress imposed upon them at this time leads to exhaustion and death. In fact, many die without reaching any water at all. Those that do reach water are known to run up and over boats that lie in their path.

Research has shown that the blood vessels in the emigrating lemming's brain contain abnormal concentrations of certain adrenal hormones. These hormones are thought to be responsible for many deaths. Many more deaths are due to predation, for migrating lemmings are easy prey to all the local predators as well as to some animals, such as caribou, which normally feed mainly on vegetation. Some predators are so attuned to the cycles that snowy owls lay more eggs and foxes have more cubs in anticipation of the lemmings' abundance. During irruptions, living lemmings will even feed on their dead conspecifics.

Other rodents, including the common hamster, **steppe lemming** (*Lagurus lagurus*), squirrels, and common or brown rats, are also known to irrupt. Cotton rats have irrupted by the millions. During these times, they have been known to feed or gnaw on books, boots, clothing, furniture, lead, leather, paper, pewter, shot, wood, and even the hooves of domestic cows, as well as on sleeping humans. They eventually reach a state of cannibalism and eat each other or die prematurely from stress. It is interesting that stress affects most mammals in a physical way. In humans, it shows as hypertension, headaches, ulcers, heart disease, gastrointestinal problems, and kidney disease. In many rodents, primates, and others, it can lead to abortion in females and low testosterone levels in males. It can cause early death in these animals in addition to numerous forms of abnormal behavior.

Most cricetid young are born blind, naked, and deaf; but they mature quickly in two to four weeks. When hamster young are very small, the female protects them by placing them in either her cheek pouches or her oral cavity. During this time, they go into what is known as "carrying paralysis" and thus put up no resistance. The young of some, such as pack rats, grasp onto their mother's nipples for transport. Water vole young are at first extremely wary of the water. They often kick and scream when first thrown in by the female. Sometimes she must resort to bribery with food to entice them back in later.

126

The rodents in this family defend themselves in diverse ways. Pack rats emit warnings by thumping the ground with their hind feet and vibrating their tails. White-footed mice stamp their front feet when alarmed. Gerbils stamp their hindfeet and, if cornered, fight from a stance similar to the defensive posture of kangaroos, kicking with their front legs and biting with their teeth. Bank voles chatter their teeth when disturbed. **Brush mice** (*Peromyscus boylii*) escape by climbing trees. The Indian gerbil will jump onto the backs of attacking dogs, while **European water voles** (*Arvicola terrestris*) are known to group together to ward off predators. They are even known to take turns "tag-team" style, so that a fresh member is always on the attack. **Richardson's water vole** (*Arvicola richardsoni*), **water rats** (*Scapteromys tumidus*), muskrats, and others will quickly take to the water to escape predators. If followed, the Richardson's water vole produces an effective "smokescreen" by churning up mud along the bottom. European hamsters will sometimes defend themselves by forcefully expelling the contents of their cheek pouches directly into the face of the enemy, and some mice (including several species of deer mice) make a quick getaway because of their ability to shed the outer skin on their tails without suffering ill effect. **Maned rats** (*Lophiomys imhausi*) can part their long back hairs to expose two white stripes of shorter hairs giving them the appearance of the skunk-like zorilla and affording them the same protection since the erected hair exposes a scent gland which exudes a stifling odor. Some authorities believe that erecting its hair also gives the maned rat the appearance of a porcupine, which wards off some predators. Pack rats are 130 times as resistant to snake venom as mice and can thus survive the bite of most rattlesnakes.

Most species in this family live a year or less, although many have been kept in captivity for as long as eight years. The average lifespan in the wild for the **clawed jird** (*Meriones unguiculatus*) is three to four months, and one study of wild meadow voles found lifespans to average only one month.

Family Spalacidae (Mediterranean mole-rats—3 species)

The **Mediterranean mole-rats** (*Spalax* spp.) have no external eyes or ears. Instead, they have functionless, rudimentary eyes located beneath the skin and a small, cartilaginous opening leading to the auditory canal. They find their way about with tactile bristles on their snout. There is no sign of a tail, and the erect fur bends freely as the mole-rat moves in either direction. They differ from the true moles in that they dig with their teeth rather than the forefeet. Like many other rodents, they can close off the oral cavity behind their incisors with flexible lips.

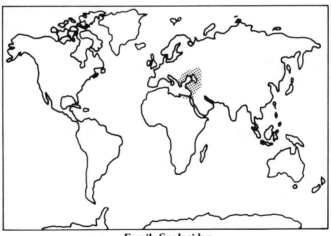

Family Spalacidae

To avoid flooding, the mole-rats build large mounds, sometimes over three feet high and nine feet

Palestine Mole-rat (*Spalax giganteus*)

in diameter. "Restroom" chambers within their burrows are sealed off after being filled. The **lesser mole-rat** (*Spalax leucodon*) also builds nuptial chambers. These, too, are sealed shut after being used. Other female mole-rats build breeding mounds three to eight feet in diameter and 16–36 inches high. Each one is usually surrounded by 15–20 smaller mounds built and inhabited by the males during the breeding season. Mole-rats store food, carefully biting off the germinal tips of corms and tubers to prevent germination.

Family Rhizomyidae (bamboo rats and African mole-rats—6 species)

The rodents of this family look very much like pocket gophers but lack cheek pouches and are very similar to spalacids. The **bamboo** or **root rats** (*Rhizomys* spp.) dig their tunnels with both their incisors and their forefeet. Bamboo is a staple of their diet. The **African mole-rats** (*Tachyoryctes* spp.) (see also family Bathyergidae) are extensive burrowers and produce so many tunnels in some areas that passing horses

Family Rhizomyidae

cannot take a step without falling in. They use primarily their incisors in constructing these mazes. Both groups of animals can close their lips behind their incisors. The bamboo rats are known to produce a repertoire of grunts, groans, and mumbles so loud they can be heard up to 100 yards away. To keep warm they defecate beneath their nest. Bacterial decay then generates considerable heat.

Family Muridae (rats and mice—469 species)

Murids range in size from the **dwarf** or **harvest mouse** (*Micromys minutus*), which weighs 0.1–0.33 ounce, to the **cloud rat** (*Phloeomys cumingi*), which can be 21 inches long minus the tail and weigh up to 4.33 pounds. Any member of this family less than six inches long is usually called a mouse. Larger animals are rats. An additional distinguishing feature is that the **true mice** (*Mus* spp.) tend to stink; the **true rats** (*Rattus* spp.) do not.

128

Many rodents are beneficial to humans, for they destroy countless insects and help aerate soil with their burrowing. However, no animals have been so devastating as the rats and mice in this family. Rats alone, primarily the **sewer, wharf, Norwegian,** or **brown rat** (*Rattus norvegicus*) and the **ship, roof, house,** or **black rat** (*Rattus rattus*), cause more damage in the U.S. than all other mammals combined. Their atrocities range from direct damage to cables, cement, furniture, pipes, and wires from gnawing (they are unable to gnaw through glass or steel) to extensive damage, often estimated at hundreds of millions to billions of dollars each year in the U.S. alone, caused by consumption and fecal contamination of foods. Rats and mice have been estimated to consume one fifth of the world's crops yearly amounting to over 30 billion dollars worth of damage. Some reports say that these animals consume as much as fifty million tons of rice yearly and that corn shipments from certain sections of the U.S. are less than five percent free of rat and mouse contamination. This explains why some of your cornflakes are "burnt" only in isolated parts of the flakes. It is also something

African Mole-rat (*Tachyoryctes splendens*)

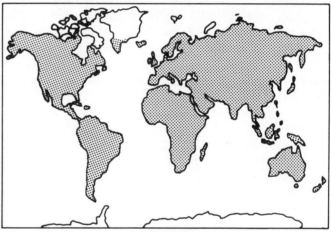

Family Muridae

to think about next time you are tempted to kill one of their natural predators, such as snakes and birds of prey.

If you are still not convinced, consider that in one study state chemists found rat hairs in more than 25 percent of the canned goods investigated. Rats have started fires by gnawing on matches and caused floods by tunneling through dams. And rats serve as reservoirs and transmitters of many serious diseases, including amoebiasis, leptospirosis, plague, rabies, rat-bite fever, salmonellosis, tapeworm, trichinosis, tularemia, and typhus. In this way, they have been responsible for more human deaths than all the wars of history. Actual confrontations with humans usually lead to injury and occasionally death to the latter. About 14,000 Americans are bitten by rats every year. Rats are

House Mouse (*Mus musculus*)

known to feed on the dead bodies of humans and their own kind. At times, they have attacked or begun to eat the still bodies of sleeping humans, and they have even viciously attacked small infants.

These rodents are not all bad news. In times gone by, the **house mouse** (*Mus musculus*) was considered a panacea (cure-all) for arthritis, baldness, cataracts, constipation, epilepsy, goiter, snakebite, and other ills. It is the most widely distributed of all mammals. Millions of white rats (albino brown rats) and mice have given their lives in service to humanity as experimental lab animals. In the course of some research, a few peculiar breeds of house mouse have been developed. These include the "waltzing mouse," which habitually runs in circles, and the "singing mouse," which can be heard chirping like a bird as far away as fifty feet. "Waltzing mice" actually have defective balance organs.

Many species in this family exhibit rather extraordinary features. The **multimammate rat** (*Praomys natalensis*) female has up to twelve pairs of teats, even though she has not been known to give birth to more than 22 young at a time. This is exceeded only by the domestic pig in which some females have been known to bear fourteen pairs of teats and the common tenrec which may have as many as 29 teats. The **pouched rats** (*Cricetomys* spp., *Beamys* spp., and *Saccostomus* spp.) are named for their cheek pouches, used for the temporary storage of food. The **spiny mouse** (*Acomys cahirinus*) wears sharp-pointed spines interspersed with its fur to help discourage predators. It is interesting to note that they often nest communally, with individuals piled on top of each other.

The tails of the murids are generally sparsely haired and scaly. The single exception is the **bushy-tailed cloud rat's** (*Crateromys schadenbergi*) long, hairy tail, resembling that of a squirrel. House mice, **field mice** (*Apodemus* spp.), and many other mammals, including the genera *Acomys, Dryomys, Eliomys, Glis, Graphiurus, Muscardinus, Perognathus,* and *Zyzomys*, are capable of shedding their tail skin or even portions of their tail in order to escape from an enemy. The harvest mice (and others, including *Melomys, Pithecheir, Pogonomelomys,* and *Pogonomys,* and some Cricetids such as *Brachytarsomys*) have a prehensile tail which helps support their weight in bushes and tall grass. The **thick-tailed rat's** (*Zyzomys* spp.) tail gradually thickens with age.

Many rodents, including the house mouse, are known to communicate in ultrasonic squeaks. In fact, the house mouse can be killed by audiogenic seizures brought on by intense ultrasonics. (Ultrasonic sounds can be produced by rattling keys together or by running one's finger over glass, although such sounds are not loud enough to cause audiogenic seizures in this mouse.)

Murids are intelligent and quite agile. A rat is capable of walking along pipes and wires placed vertically as well as horizontally. Some can climb up string with ease, and a brown rat can jump five feet. A foot-long rat can work its way through openings less than one inch square and survive a fall of fifty feet without injury. Most are good swimmers and can cross open stretches of water measuring more than half a mile wide. Probably the most aquatic of all rodents in this family is **Monckton's rat**

130

(*Crossomys moncktoni*). It has waterproof fur, large, paddle-like webbed feet, tiny eyes, apparently no external ears, and a fringed tail to increase its effectiveness for thrust and steering.

Rats and mice feed on virtually anything they can swallow, whether it's digestible or not. They have been observed to eat beeswax, books, cloth, garbage, glue, lead pipe, leather, paint, paper, paste, soap, sewage, and more. Many seem to prefer animal to vegetable food, becoming predators on fish, baby birds, and small mammals such as house mice. At times, rats have grouped together to attack larger prey, including chickens, ducks, geese, lambs, and pigs. Since their incisor teeth grow 4–5 inches per year, they must be kept worn down through gnawing. This is done mostly by chewing on concrete, metal, and wood, or by grinding their teeth together. This also helps keep them sharp.

Much of these animals' ingenuity and intelligence shows in their feeding behavior. Black rats of the Tobriand Islands go down to the water's edge, squat on some flat coral, and dangle their tails into the water. If the rat is patient, a crab will often latch onto this "bait." The rat immediately leaps shoreward, turns, and consumes the would-be predator. Rats allegedly work together to carry off large food objects such as eggs. One rat wraps its body around the egg, while the second then drags it by the tail to the nest. This behavior has been reported on countless occasions down through the centuries, but its motivation is still hotly debated. Some experts feel that the first rat is merely attempting to protect its find from the second, which is doing its best to take it away.

Single rats are known to carry large objects in their forepaws and hop away on their back legs with the aid of their tail. Another interesting technique is the "conveyor belt" carrying method. In this, a rat lays on its back, picks up an object such as an egg with its hind legs, passes it to the front legs, and then pushes it over its face. It then gets up and runs to the front to repeat the process over and over. Upon discovering an abundant food source, larger, older rats will often allow smaller, younger ones to feed first. If the former, more experienced rats come upon a food source they suspect is poisonous, they will warn other rats by urinating and defecating on it. In this way, rats are capable of transmitting the information of poisons to future generations. Thus, the best rat poisons act slowly so that they do not become associated with the bait.

Some species have taken to certain specialty food items. The **marmoset rat** (*Hapalomys longicaudatus*) feeds almost exclusively on a single species of bamboo (*Gigantochloa scortrechinii*). It easily scales the slick stems with the help of sticky secretions on its feet. It even lives inside the bamboo stems. The **giant pouched rats** (*Cricetomys* spp.) exhibit coprophagy. The **yellow-necked mouse** (*Apodemus flavicollis*) has a particular liking for honey which has led to its taking up residence within beehives. It seems that these mice are immune to bee-stings and occasionally feed on the bees as well as their honey. **Beaver rats** (*Hydromys* spp.) feed primarily on fish and aquatic insects but are also known to take spiders, frogs, turtles, and bats.

Bandicoot rats (*Bandicota bengalensis*), **pest rats**, (*Nesokia indica*), and others often build up huge stores of corn or other foodstuffs for lean times. These are occasionally discovered by humans, who promptly confiscate the loot for their own use. The **fat rats** (*Steatomys* spp.) put on large amounts of fat during the fall, causing them to become grossly obese but allowing them to sit lethargically through winter. None of the species in this family exhibit true hibernation. Another aspect of feeding discovered in experimentation with rodents and thought to apply also to humans is that infrequent feedings and low temperatures predispose to increased life spans.

Australian **native mice** (*Pseudomys hermannsburgensis*) dig up small pebbles and place them in the area surrounding their desert burrows. The pebbles collect enough dew to supply the mice with their water requirements. Some **hopping mice** (*Notomys* spp.) can survive without drinking water.

Rats and mice dwell in almost any type of shelter, ranging from underground burrows dug out by the animal to people's homes and garages. As mentioned, the yellow-necked mouse can sometimes be found in beehives, and the **pencil-tailed tree mouse** (*Chiropodomys gliroides*), like the marmoset rat, lives in the stems of bamboo, sometimes in social groups. **Groove-toothed rats** (*Pelomys* spp.) burrow into and construct their nests within the stems of large plants. Some native mice occasionally rest and even shelter below flat "cow pies," and the house mouse has been found living inside meat being held in cold storage. **Tree mice** (*Dendromus* spp.) will take over weaverbirds' and sunbirds' nests. Those that dig their own burrows include the bandicoot rats, pouched rats, and **wood** or **long-tailed field mice** (*Apodemus sylvaticus*). **Gerbil mice** (*Malacothrix typica*) pile excavated dirt away from their tunnels to prevent leaving tell-tale signs for predators. Pouched rats sometimes dig their burrows in termite mounds (as do some **grass mice** [*Arvicanthis* spp.] and spiny mice), where they are known to collect odd items such as chalk, cloth, coins, earrings, and pens in the manner of pack rats. Pouched rats also possess parasitic, wingless earwigs (*Hemimerus talpoides*) in their fur which feed on skin flecks or fungi on their skin.

The **stick-nest rat** (*Leporillus conditor*) builds large stick nests up to three feet high and four feet in diameter. In some areas, the nests are anchored with rocks piled both on top and within to prevent them from being swept away by high winds. The harvest mouse builds one of the most perfect bird-like nests of all mammals. The wood mouse has the curious habit of erecting miniature "altars" (two-inch cubes of totally unknown function) near the entrance to its burrow. Wood mice are good climbers and frequently move into owls' nests and squirrels' dreys. The stick-nest rat's nest can be connected to the burrows of penguins and shearwaters. In one instance, a single burrow opening was used for shelter by a tiger snake, penguin, bandicoot, shearwater, and stick-nest rat. The most unusual case is probably the behavior of the **long-tailed climbing mouse** (*Vandeleuria oleracea*). It is known to raise its young in the middle of an occupied spider's web.

Rats and mice mark out their territories with urine, feces, and, in some species, glandular secretions. This is often why the cages of such pets are immediately soiled after being cleaned. Rats live in "family" groups numbering up to two hundred or more. They do not tolerate the presence of stranger rats, killing them if they refuse or are unable to leave. The rats from a common group establish a dominance hierarchy through ritualized shoving matches. Male tree mice are known to fight to the death. Dominant mice often bite off the facial hair and whiskers of subordinate mice, thus branding them as lower class "citizens." Dominant male mice excrete pheromones that prevent the maturation of the testes in other male mice within the same group. Should the dominant male be killed or driven off, another male will rise to dominance. At the same time, all the currently pregnant females in the group will self-abort, come into heat, and then mate with the new leader. The self-abortion is actually a resorption of the embryos. It is apparently triggered by the presence of hormones in the male's urine and only occurs in females during the first few days of pregnancy. A dominant female's urine contains pheremones that prolong estrus in other females and even increases the latter's incidence of pseudopregnancy.

The reproductive capabilities and hardiness of these animals make their extermination totally impossible. It is currently estimated that there are about three rats for every human on earth. Rats can produce up to twelve young at a time with up to ten litters per year. A single pair could theoretically have 15,000 descendants in a year; if all the progeny survived and reproduced, that initial pair would become about 350,000,000 in a mere three years. Fortunately, very few rats reach two years of age. Studies have shown that bandicoot rats have a 33 percent mortality per month, and that up to 99 percent of brown rats die before reaching adulthood. Should an area become overpopulated, the rats will either migrate away or become cannibalistic (the latter response to overpopulation is also seen in bullfrogs, herons, herring gulls, kestrels, lions, magpies, octopuses, pelicans, rabbits, many rodents, shrikes, spider crabs, scorpions, terns, voles, wagtails, and many others). In some, such as the house mouse, the stresses of overcrowding cause the females to become infertile and the incidence of intrauterine absorption of embryos, increases. In some population explosions, there have been found to be up to 82,000 house mice per acre.

Female rats have, in a single episode of heat lasting six hours, mated up to 500 times with various males in the same group. Male rats sing ultrasonic, post-ejaculatory songs after copulation. In most species, gestation lasts about three weeks (pouched rats have the longest murid gestation, six weeks). Delayed implantation occurs in some hopping mice. In all except the **vlei rat** (*Otomys irroratus*), the **karroo rat** (*Otomys denti*), and the spiny mouse, the young are born hairless and blind. The exceptions are born fully furred with open eyes and from a standing position rather than with the female lying down as in all the others. The smallest litters (single births) are seen in the karroo rat. In spiny mice, the maternal instinct becomes so acute just before birth that the females will sometimes steal and suckle the young of other females or will even grasp another adult and hold it baby-fashion while it licks its body. When the birth finally comes, the female is often aided by another female (midwifery such as this is also seen in dogs, dolphins, elephants, tamarins, and others). The midwife in this case often steals and eats the afterbirth and may kidnap the newborn. This theft is temporary, however, since after one to three days the young are raised communally, being nursed by any lactating female; they are weaned at about one week of age.

Hopping mice and others are known to mark newborns and group members with their scent gland secretions. Murid females frequently mate again the day after giving birth. Rats and house mice are also known to raise their young communally. In giant pouched rats, there are communal nurseries known to take in orphaned young. The females of many species, including deer mice, hopping mice, jerboa mice, native mice, pencil-tailed tree mice, stick-nest rats, swamp rats, and wood mice, transport their young while the latter are nursing at the teats, thus limiting the move to a single trip. Some females produce fecal pheremones that prevent the young from wandering. At weaning time, its production is halted.

Rodents in this family fall prey to numerous predators. Their main protections lie in explosive reproduction and keeping concealed. They hide in burrows, under leaf litter, in logs, and in numerous crevices, including those inside buildings. In fact, an adult house mouse can squeeze through a hole as small as three eighths of an inch in diameter. The spiny mice are aided by their spines; some species can shed the outer portion of their tail; and a few gain protection by nesting in inaccessible places such as beehives. The fat mouse escapes predators through quick tunneling, blocking with the freshly dug material as it goes. Vlei rats escape predators by taking to the water. Pouched rats can blow out their cheek pouches when alarmed, producing a muffled sound that sometimes startles pursuers. Mice have often been observed to take a "prayer-type" posture when cornered by enemies such as a cat. People have misinterpreted this behavior as the mouse pleading for mercy when it is in fact the aggressive stance taken by a mouse preparing for battle, even though its chances are slim at best. Interestingly, it has been found that rats and others grind their teeth when relaxed but stop immediately if alarmed. This is known to warn nearby conspecifics of possible danger.

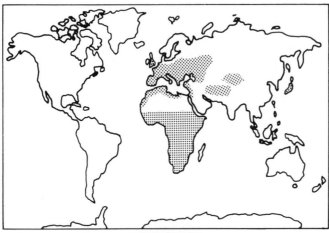

Family Gliridae

Family Gliridae (dormice—16 species)

Dormice (seven genera) are best known and even named for their profound sleep, both during hibernation and daily. They are also known by the common names of **sleepmouse, sleepy mouse,** and **dozing mouse.** (Dormouse comes from the Old French "dormir," meaning "to sleep.") The **edible dormouse** (*Glis glis*) spends half or more of each year in hibernation, during which it can be rolled around like a ball without being awakened. Hibernation seems to be initiated both by overeating and by weather factors such as external temperatures (below 60°F). Prior to entering sleep they are known to double their weight, although almost all the surplus is lost during hibernation. It has been found that 50–80 percent of the hibernating animals are lost to predators and severe weather conditions each year. Their body temperature falls from 97°F to 33°F. The **hazel dormouse** (*Muscardinus avellanarius*) holds its hibernating nest together with sticky salivary gland secretions.

The edible dormouse has

Edible or Fat Dormouse (*Glis glis*)

cushioned foot pads that produce sticky secretions to increase its climbing prowess, while some **African dormice** (*Graphiurus* spp.) have flattened skulls and bodies enabling them to escape predators by crawling into narrow crevices. (A similarly flattened skull is also found in the **rock**

134

mouse [*Petromus typicus*] which also hides in crevices.) In **garden dormice** (*Eliomys quercinus*), the tail skin can be shed if pulled on by attacking predators. Later, the animal will completely bite off that skinned portion of tail. Some species are reportedly the only mammals to regenerate lost tails.

Some dormice species nest or hibernate in unusual places. The edible dormouse will occasionally hibernate inside beehives and outhouses, while the **pygmy dormouse** (*Graphiurus nanus*) dens in the large nests of communal spiders (*Stegodyphus* spp.). The hazel dormouse frequently builds its nest near those of birds of prey, particularly the black kite. This is actually one of the safest areas in which to nest, since the hunting drive of most predators is suppressed in the immediate vicinity of its nesting area.

For most of the year, male garden dormice get along well. However, during the breeding season they are known to fight to the death. The victor is even known to consume the body of the vanquished. Other dormice, including some of the African dormice, are also known to fight to the death.

In edible dormice, the female stimulates her young to void by licking their anus. She quickly consumes both their solid and liquid wastes. She also presents them with some of her saliva which they quickly lap up. It probably contains beneficial substances such as antibodies. Garden dormice females and their young form caravans as described for some of the shrew species.

Family Platacanthomyidae (spiny and Chinese dormice—2 species)

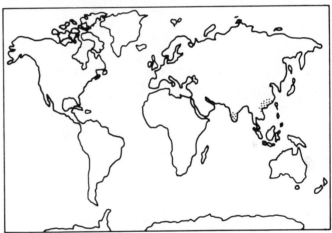

Family Platacanthomyidae

These animals are similar to the other dormice except that they have a brush-like tail with a scaly base. The **spiny dormouse** (*Platacanthomys lasiurus*) has quill-like bristles scattered throughout its fur. It also seems to have a yen for pepper plants and fermented palm juice. The **blind** or **Chinese pygmy dormouse** (*Typhlomys cinereus*) is almost blind. It spends its lifetime primarily in rotting logs hunting for insects. It is claimed that cats refuse to eat the Chinese pygmy dormouse.

Family Seleviniidae (desert dormouse—1 species)

The **desert dormouse** (*Selevinia betpakdalensis*) enjoys several adaptations to the desert way of life, including enlarged auditory bullae to help it detect approaching predators. Its hair does not come off alone when it molts, but in patches along with the accompanying skin. It becomes dormant in cold weather.

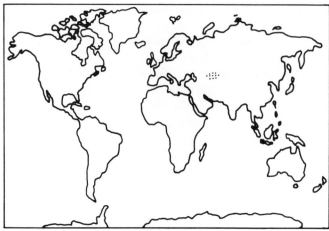

Family Seleviniidae

Family Zapodidae (birch and jumping mice— 11 species)

The **North American meadow jumping mouse** (*Zapus hudsonius*) weighs less than an ounce but can make jumps up to 10 feet long. The importance of the long tail in maintaining balance is shown by the fact that if the tail is lost the animal involuntarily somersaults with each attempt at a long jump. They can easily climb and swim without problem. They sometimes feed on an underground fungus (*Endogone* spp.) whose spores would not germinate if not exposed to the rodent's digestive juices. They are known to attack and eat small birds as they put on large amounts of fat for hibernation. Some eat 1½ times their body weight daily. All species in this family hibernate, some for up to eight months out of the year, during which time they lose up to half their weight. When alarmed, some may thrash the ground with their long tails.

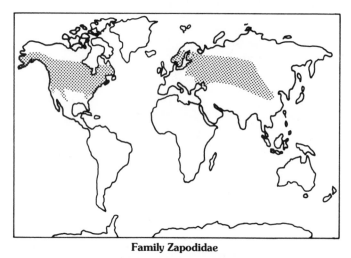

Family Zapodidae

Family Dipodidae (jerboas—29 species)

Jerboas have the appearance of small kangaroos. The hind legs, about four times as long as the front, enable them to make leaps up to ten feet long and to reach speeds of thirty miles per hour. They sometimes make vertical jumps of several yards to grab large bushes and trees with their teeth and then forelegs as they procure food in the form of blossoms, shoots, and twigs. Their soles are haired to increase both traction on sand and digging ability. They get around by hopping, a method both faster and more efficient than running.

Several species sport huge ears that are often well haired to keep out sand. The **big-eared jerboa** (*Euchoreutes naso*) sports 1.5 inch ears on a 3 inch body with a half inch head. Other long-eared jerboas, such as the **four-toed jerboas** (*Allactaga tetradactyla*), actually fold the ears up when running at high speed. The **dwarf jerboas** (*Salpingotas* spp.) have huge tympanic chambers, causing

136

their heads to be as large as the rest of their body but giving them an extremely keen sense of hearing.

Jerboas' tails are quite long and are often tufted at the end to maximize their efficiency as a rudder. The animals often use the tail kangaroo-style to prop themselves up. **Fat-tailed jerboas** (*Pygeretmus* spp.) have bulky tails in which they store fat.

Jerboas use their noses, teeth, and forefeet to construct burrow systems that, for some species, can reach depths of nine feet to avoid the desert's heat. To keep their nostrils from becoming clogged with sand, some **desert jerboas** (*Jaculus* spp.) use a fold of skin on the nose to cover the nares. The powerful hind legs are used to kick away loose sand. The **rough-legged jerboa** (*Dipus sagitta*) can dig a four-foot tunnel in 10–15 minutes. **Comb-toed jerboas** (*Paradipus ctenodactylus*) build their burrows with the entrance facing into the wind so that they automatically close over when the wind is blowing. It must be dug open whenever entering or exiting, although the animal itself never closes the opening. Other species are known to plug the entrances themselves after returning from a night of foraging. This serves to keep out predators and hot, dry air. In emergencies, they are known to burst through these plugs in spectacular exits. This has even been known to startle predators lurking nearby.

As previously mentioned, these small rodents can go indefinitely without any source of free water. Most species are known to hibernate.

Woodland Jumping Mouse (*Napaeozapus insignis*)

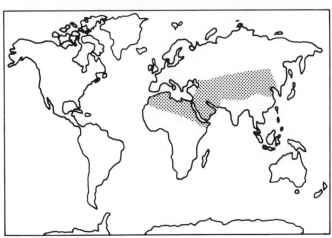

Family Dipodidae

In courtship, male desert jerboas clout the females' snouts with their own. The glans penis in some species (and in cavies and pacas) is adorned with special spines. Many female jerboas line their nests with some of their own fur. In the **common desert jerboa** (*Jaculus jaculus*), the females use camel hair.

Suborder Hystricomorpha (porcupine-like rodents)

Family Hystricidae (Old World porcupines— 11 species) and Family Erethizontidae (New World porcupines—10 species)

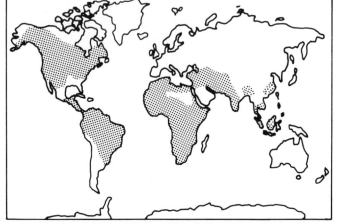

Desert Jerboa (*Jaculus jaculus*)

Old World porcupines often have, in addition to a set of short, barbed quills, many long (up to fourteen inches), barbless spines. They dig burrows and rarely climb trees. The New World porcupines have shorter, usually barbed quills; do not burrow; and have nonskid foot pads, for they spend a great deal of time in trees. Although the two families are not closely related, they are grouped together here for the sake of convenience and their many similarities.

The common denominator of the two families is their quills. Actually, one Old World species, the **long-tailed porcupine** (*Trichys fasciculata*), and one New World species, the **thin-spined porcupine** (*Chaetomys subspinosus*), lack true quills; they are instead covered with stiff, bristly hair. All true quills are actually derivatives of hair. Hollow, they increase buoyancy greatly when the animals take to water. Quills come in many varieties. In Old World species, they have a variable number of longitudinal furrows, giving them curious kidney, hourglass, and even star shapes in cross section. Quills, like hair, continually fall out and regrow. Quills that crack when growing will subsequently grow back together through a process known as cicatrization.

An average **North American tree porcupine** (*Erethizon dorsatum*) carries about 30,000 quills, each one covered with small barbs. The barbs lie flush with the shaft until the heat and moisture from penetrated tissues causes them to expand and pop out. This prevents withdrawal and actually causes the quill to work its way in more deeply, at rates up to one millimeter per hour or about one inch per day. This has led to the demise of such formidable predators as eagles, foxes, leopards, and mountain lions when the quills have worked their way into vital structures such as the heart or brain or because the pain of the

Family Hystricidae and Family Erethizontidae

138

quills in the mouth leads to death from starvation. If unlucky enough to get stuck, cutting the exposed end of a quill to relieve the air pressure inside aids removal. Occasionally this is of no use since some quills are not hollow. In the case of barbless quills, most fall out on their own.

The tails of most porcupines are heavily armed with quills. Exceptions include the **prehensile-tailed porcupines** (*Coendou* spp.). They spend most of their lives in trees, using the tail as a fifth hand. They differ from other prehensile-tailed mammals in that they grip with the upper surface of their tail. Other New World porcupines use their tails as a

Crested Porcupine (*Hystrix* spp.)

brace in tree climbing, much as woodpeckers use their stiff tail feathers. It is reported that some of the **brush-tailed porcupines** (*Atherurus* spp.) and the long-tailed porcupine have tails that can easily break away in emergencies, like the tail of a lizard, though they cannot be regrown. Several Old World porcupines, including the **crested porcupines** (*Hystrix* spp.), the **Malayan porcupines** (*Acanthion* spp.), and the **thick-spined porcupines** (*Thecurus* spp.), have large, hollowed "quills" mixed in with regular ones on the tail. When alarmed they shake their tails to produce a rattling sound and warn off potential attackers.

Porcupines are usually black or some shade of brown. However, albino North American porcupines are not rare. As in many other rodents, the incisors are bright orange. Most species have poor sight but excellent senses of hearing and smell. Some species also have sensitive vibrissae that allow them to feel their way through the dark, particularly on black nights or when in caves. They produce many sounds including ghostly cries. Some, such as the **Indian crested porcupine** (*Hystrix indica*), are extensive burrowers.

Although they are primarily vegetarians, porcupines are known to eat carrion. At times, they strip so much bark from trees that they kill the trees. While feeding, tree porcupines drop many branches and twigs that are later eaten by deer, rabbits, and others. During harsh winters, such dropped vegetation may even form the mainstay of these other animals' diet. Tree porcupines are also known to feed on mistletoe, and can eat opium and prussic acid without ill effect. Porcupines often smell like fermenting wood. Many porcupines crave salt and will go to great lengths to obtain it. They have been known to eat boots, car tires, gloves, gun butts, oars, saddles, tool handles, and even glass for the little bit of salt present in these objects, sometimes in the form of dried human sweat. They gnaw antlers and bones for their calcium and phosphate content and to help keep the incisors worn down. Some

Prehensile-tailed Porcupine (*Coendou* spp.)

porcupines crave soap.

Porcupines are generally solitary creatures. Some, such as the prehensile-tailed porcupines, are aggressive towards each other at chance encounters, being known to try to stick each other with their quills. Male North American tree porcupines fight by standing on their hind legs and boxing with their forelegs like kangaroos.

People often wonder how porcupines go about the mating process. The truth is that they mate much like most other mammals, if necessarily more carefully. The female usually takes the initiative and picks her mate. In some, such as the North American species, the male acknowledges by spraying her with his urine, an act thought to stimulate her further. (This is also observed in agoutis, cavies, coypus, hares, maras, and rabbits.) A pair may stand on their hind legs and "dance" prior to copulating. To facilitate safe mounting, the female raises her tail and flattens her quills tightly against her body by contracting special skin muscles. Males have never been known to force mating (neither do most mammal males), which could have dire consequences. The males stand to mate and push their tails against the ground to aid their thrusts during copulation.

New World females generally produce only one ovum per year. It is viable for about five hours, and she only allows mating on that single day of the year. At other times the vaginal area is sealed by a membrane. The offspring are born with flexible quills that harden and become useful for defense in thirty minutes to ten days, depending on the species. The female is further protected by the tough birth sac that envelops the young during passage through the birth canal. **African brush-tailed porcupine** (*Atherurus africanus*) females shake their tails to signal the young to come and when they are angry or frightened. The male crested porcupine protects the young by placing them between his forefeet and then raising his quills and hissing.

Most porcupines rely on their quills for protection from predators. Threatened brush-tailed porcupines are known to panic, resulting in reckless, headlong flights. These occasionally lead to collisons with rocks and trees resulting in loss of consciousness. (Similar hysterics are seen in maras, rabbits, and others.) When crested porcupines and North American porcupines are faced with danger, they are known to turn their backs, chatter their teeth, growl, squeal, stamp their rear feet, and shake their tails. This tail shaking produces a rattling in some species. If these antics do not ward off an intruder, the porcupine may charge by running backwards and jabbing the luckless antagonist with its quills. Porcupines cannot shoot their quills. However, vigorous tail shaking can cause some quills to fly out randomly, occasionally with enough force to become embedded in wood. Despite their quills,

140

porcupines are preyed on by coyotes, eagles, fishers, horned owls, jaguars, and mountain lions. Fishers feed on porcupines so regularly that they have been used to control porcupine populations. They and jaguars are known to eat the animals quills and all. Some porcupine species, such as the prehensile-tailed porcupines, may roll into a ball when threatened and are said to produce obnoxious odors to aid in deterring predators. The taste of their meat reportedly has a turpentine-like flavor. One **Sumatran crested porcupine** (*Hystrix brachyura*) attained the greatest age of any rodent, 27 years and 3 months.

Family Caviidae (guinea pigs or cavies— 16 species)

Most people are well acquainted with **guinea pigs** (*Cavia porcellus*). They make fine pets, partly because they do not bite. Like humans and Indian fruit bats, they lack the ability to synthesize Vitamin C. They have been used in so many laboratory tests that their name now symbolizes the subjects of experiments. The species is extinct in the wild.

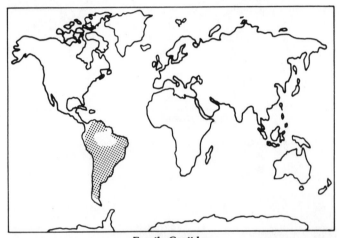

Family Caviidae

The animals in this family can often make do with little free water. **Wild cavies** (*Cavia aperea*) can survive with only the water in cactus pulp, while the **mara** or **Patagonian hare** (*Dolichotis patagonum*) needs only the water in the vegetation it eats and that obtained from fat metabolism. The same is true of the **moco** or **rock cavy** (*Kerodon rupestris*), a ground dweller known to climb trees to feed on tender young leaves. All the animals in this family have continuously growing molars, which allows them to chew large amounts of vegetation without wearing away their teeth. Interestingly, guinea pigs can eat strychnine without problem but die if fed penicillin.

Male **mountain cavies** (*Microcavia* spp.) may fight to the death in dominance battles. Male **cuis** or **yellow-toothed cavies** (*Galea* spp.) are thought to induce estrus in females through a chin gland secretion

Patagonian Cavy or Mara (*Dolichotis patagonum*)

they rub onto the females' rumps. Ovulation is induced by copulation. One female cui became pregnant at the age of nine days, a mammalian record for youngest pregnant female. Mara females

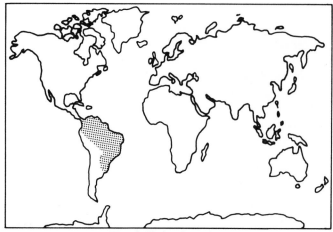

Family Hydrochaeridae

nurse their young while sitting back on their haunches. The young suckle from the same position standing at right angles to her. The female's 1.5 inch nipples resemble leeches and are located just behind the shoulder and in the groin. Female mountain cavies may suckle unrelated young, and guinea pigs can be weaned at 3 days of age. When threatened, maras signal each other by exposing white-haired regions on their rumps, rock cavies whistle, and cuis drum their hind feet.

Family Hydrochaeridae (capybara—1 species)

The **capybara** or **water cavy** (*Hydrochaerus hydrochaeris*) is the world's largest rodent, growing to four feet in length and weighing as much as 175 pounds. Its molars grow continuously, and its feet are partially webbed, revealing its aquatic nature. It is almost neutrally buoyant due to its fat stores and can often be seen standing belly deep in water feeding on aquatic vegetation. When content, these large rodents produce low clicking noises. Large grazing herds of up to 100 animals are sometimes encountered. The males take female

Capybara (*Hydrochaerus hydrochaerus*)

harems and are known to mate in water and on land. The young are often cared for in communal nurseries. Capybaras can escape predators through mad dashes towards and loud splashes into water where they submerge to swim underwater or hide among floating vegetation with only their nostrils exposed. Some experts feel that the loud splashes serve as a warning to other capybaras nearby, even though they are known to give out alarm barks most of the time.

Family Dinomyidae (pacarana—1 species)

The **pacarana** or **false paca** (*Dinomys branickii*) uses its forepaws for feeding; true pacas do not.

142

It is quite docile and if threatened will back into a hole or against a cliff to protect its backside. It may sleep in trees on occasion.

Family Dasyproctidae (agoutis, acouchis, and pacas—15 species)

The **agoutis** (*Dasyprocta* spp.) are extremely nervous animals known to jump from cliffs en masse at the sound of a twig snap. Some reports state that these small rodents are able to jump twenty feet horizontally or six feet vertically from a sitting start. They are aggressive towards conspecifics and have been known to stamp each other to death. They occasionally follow monkey troops to obtain discarded fruit. Food is held in the forepaws and is sometimes peeled before eating; excess food is buried. They dig "foxholes" about three feet deep and roof them over with vegetation for their den. In courtship, males urinate on females, which respond by performing a frenzied dance. Females give birth from a squatting position and bar the male from the nesting area while the offspring are young. Some female agoutis (*D. leporina*) and female **acouchis** (*Myoprocta* spp.) have membrane-closed vaginas during summer. Acouchis also form post-coital, vaginal plugs (also observed in viscachas). When alarmed, agoutis raise their hair on end and will often head for the protection of water while producing high screams or barks.

The **pacas** or **spotted cavies** (*Agouti* spp.) are sometimes placed in their own family, Cuniculidae. The **spotted pacas'** (*Agouti paca*) cheekbones are modified to form resonating chambers, a unique feature among mammals, even though the animals are only known to produce low-volume grunts. Pacas dig burrows, but their skin is so delicate that strips of it are often torn away when they rush

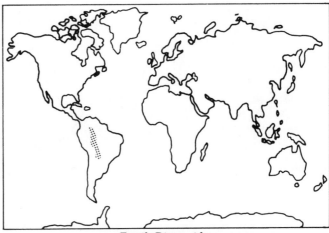

Family Dinomyidae

Pacarana (*Dinomys branickii*)

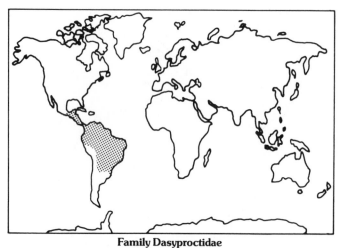

Family Dasyproctidae

Acouchi (*Myoprocta acouchy*)

through tangles of brush. The resulting wounds heal quickly. In courtship, paca pairs squirt urine on each other. They may also urinate on attackers. Should this fail to deter them, the pacas take for the protection of water, where they can swim strongly.

Family Chinchillidae (chinchillas and viscachas—6 species)

The tendency for animals of different or the same species to adapt to cold climates by growing smaller ears and shorter tails compared to others in warm climates is well illustrated in the chinchillas, whose soft, delicate fur is known the world over. In fact, the fur from the best known **chinchilla** (*Chinchilla laniger*) is composed of such fine hair that individual stands are barely visible to the naked eye. The number of strands, is so large (60–80 strands per hair follicle) that they are thought to pose a barrier to insect parasites. In **viscachas** (*Lagostomus* spp.) and chinchillas, the molars as well as the incisors grow throughout life. All members of this family dustbathe regularly. Chinchillas can make do with only the water obtained in their food along with small amounts of dew. They generally live communally in groups numbering up to 100. Pairs probably do not mate for life as once thought. The females grow to a larger size and are dominant to the males. Courtship is marked by a hair-pulling contest between the couple, each taking turns pulling tufts of hair from the other. After the young are born, the parents will often sit side by side to form a living, warm nest for the offspring, which squeeze in between them. When alarmed they may growl, chatter their teeth, or urinate.

Plains viscachas (*Lagostomus maximus*), whose males sport prominent "moustaches," live in large communal warrens known as viscacherias with as many as fifty members. Viscacherias are also known to lodge foxes, lizards, owls, skunks, snakes, swallows, toads, and other animals. The extensive network of tunnels caves in under the weight of a horse, which is thus in danger of breaking a leg. This viscacha collects various objects, ranging from cow dung to shiny keys, in the manner of pack rats. These are displayed as decorations on the mounds formed at the entrances to the burrows. The female plains viscacha produces 2–800 eggs at ovulation, the most of any mammal, even though the average litter size is two. To warn each other of danger, these animals produce a swishing noise

144

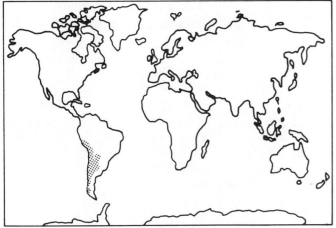

Family Chinchillidae

Chinchilla (*Chinchilla laniger*)

followed by a sound resembling a drop of water falling into a pond.

Family Capromyidae (coypus and hutias—14 species)

Coypus or **nutrias** (*Myocastor coypus*) resemble beavers with round tails. In cold habitats, the tail tips are known to freeze and break off. They are well adapted to their aquatic life. The hind feet are webbed, and they have two coats of fur, the undercoat being waterproof. The fur is groomed and coated with secretions from fat glands located at the corners of the mouth. They can stay submerged for up to five minutes. They frequently dip their food in water before eating, and often eat from a floating raft of vegetation up to two feet wide. Like rabbits, they exhibit refection. Entrances to their nesting areas are usually located below the water's surface. The female's nipples are found high on her side, above the waterline when swimming. This allows the young to feed while swimming with her, when they ride on her back, or when she is lying on her belly. The young can swim within 24 hours of birth. The mammary glands of the female **Cuban hutia** (*Capromys pilorides*) are located on her thighs, and both sexes have prehensile tails. Coypus are sometimes exhibited in sideshows, where they are billed as the "world's biggest rat." They do grow to two feet in length and weigh up to 25 pounds. They have killed cats by luring them into water and then drowning them. The **Jamaican hutia** (*Capromys brownii*) has the greatest number of chromosomes of all mammals—eighty-eight.

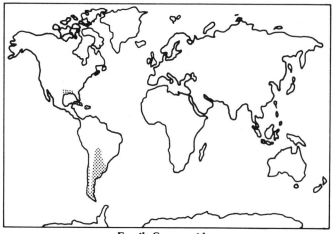

Family Capromyidae

Family Octodontidae
(octodonts—8 species)

These small animals, such as the **degu** (*Octodon degus*), resemble rats with coarsely haired tails. The tail is often carried erect when they run. If grasped by the tail, the skin can be shed. Later, the animal bites off that part of the tail. Their burrow entrances are often marked by piles of rocks, sticks, and fecal material. Degus occasionally consume fresh cattle and horse dung. Food is sometimes stored for winter use.

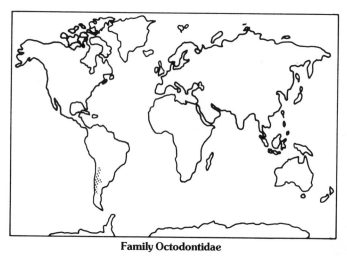

Family Octodontidae

Family Ctenomyidae
(tuco-tucos—38 species)

Tuco-tucos (*Ctenomys* spp.) look and live very much like pocket gophers but lack the external cheek pouches. They differ in that they live communally and have good eyesight. The fur is groomed with special bristles that grow from the bases of their hind claws. Their name comes from the bell-like sound they produce when alarmed. Burrows are almost never deeper than one foot and will collapse easily if trodden on by human or horse. *Ctenomys talarum*

packs its tunnels with the hind feet after urinating on the walls. This species is also reported to open and close tunnel entrances to keep the temperature of its burrows at 68–72°F. In dusty areas, it is said to use chopped grass to filter air coming through the entrances. Tuco-tucos often store vegetation and appear not to need any drinking water.

Family Abrocomidae (chinchilla rats—2 species)

Chinchilla rats (*Abrocoma* spp.) are South American rat-like animals with soft fur used in garment-making like the fur of their namesakes. Their claws are weak and hollow underneath. Burrows are occupied by colonies of these animals. Like chinchillas, octodonts, and tuco-tucos, there are stiff foot hairs forming combs to groom the fur and help remove loosened dirt when digging.

Family Echimyidae (spiny rats—52 species)

Most **spiny rats** (many genera) are covered with fur interlaced with very stiff hairs or spines. Some spiny rats (including *Hoplomys* spp. and *Proechimys* spp.) have true breakaway tails like lizards, though they cannot regenerate. This feature results from an anatomically weakened link at the base of the fifth tail vertebra. This point easily gives way with force. Almost half the animals caught in one study had lost their tails at just this point, showing the life-saving advantage of this defense mechanism. The **gliding spiny rat** (*Diplomys labilis*) lives in hollow trees and can make short glides utilizing a small gliding membrane.

Family Thryonomyidae (cane rats—2 species)

Cane rats (*Thryonomys* spp.) are large, bristly coated rodents known as pests on sugar plantations but also held in high regard as a food item. Some adults approach twenty pounds. They may shelter in termite mounds or aardvark or porcupine dens or dig out their own burrows. Males fight in nose-to-nose pushing contests. When alarmed, they often take to water to escape.

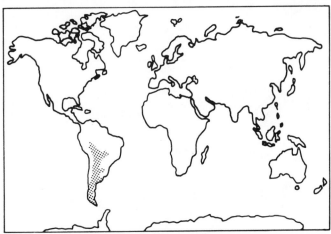

Family Ctenomyidae

Family Petromyidae (rock rats—1 species)

The **rock rat** (*Petromus typicus*) actually looks more like a thinly haired squirrel than a rat. It has quite flexible ribs and a flat skull allowing it to take shelter in very narrow crevices. Its bony flexibility allows it to be pushed flat without harm. The tail breaks away easily, and the two or three pairs of teats are located high at shoulder blade level.

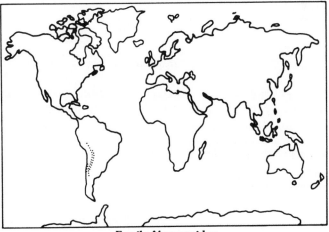

Family Abrocomidae

147

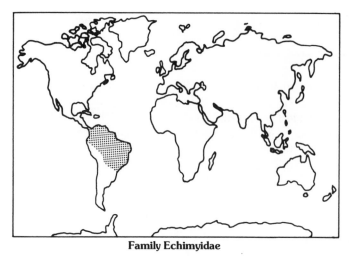

Family Echimyidae

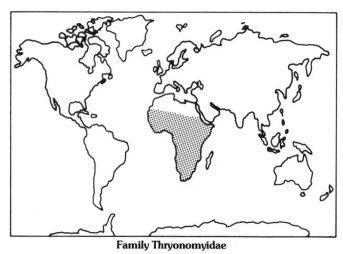

Family Thryonomyidae

Family Bathyergidae (African mole-rats—8 species)

These mole-rats resemble the spalacids and rhizomyids. All are burrowers. With the exception of the **dune mole-rats** (*Bathyergus* spp.) which primarily dig with their claws, the mole-rats tunnel with their incisors, the lower ones of which can be moved independently. One **common mole-rat** (*Cryptomys hottentotus*) maintained a burrow system in excess of 1100 feet, possibly a mammalian record. The extensive burrow systems of the **sand mole-rat** (*Bathyergus suillus*) can pose a threat to horseback riders and even train tracks, which have collapsed because of its diggings. The movement of many **blesmols** (*Cryptomys* spp.) is at least partially determined by air currents detected by the corneas of its eyes which have become sightless. The **Cape mole rat** (*Georychus capensis*) reportedly bites the "eyes" from bulbs and tubers to prevent their sprouting while in its storage chambers. A single individual can dig out 1000 pounds of soil each month.

One unusual member of this family is the **naked mole-rat** (*Heterocephalus glaber*). This species has almost no hair at all and is thus unique in the rodent world. It is also without sweat glands and has no subcutaneous fat. Since thermoregulation is poorly controlled in this species, it inhabits areas with warm soil. Its body temperature runs around 90°F. The burrow openings are cone-shaped and resemble certain volcanoes. When working near the entrance, the animals kick out excess dirt as streams of sand rising a foot high every few seconds. They do this in relays as they are colonial. Socially, they exhibit a caste system wherein a single dominant female, the queen, produces all the offspring. She is even known at times to kill other females to maintain her dominance. She is tended by sluggish, fat nonworkers of both sexes. The young are communally cared for and are known to beg adults for fecal material. This is eaten by the young, possibly for needed nutrients.

148

Family Ctenodactylidae
(gundis—5 species)

Gundis (four genera) are small rodents that are very susceptible to exposure when wet, since water clumps their fur and exposes bare patches of unprotected skin. Their hind feet have bristles above the claws for combing the fur, for aiding their climbing ability as they scale near-vertical rock surfaces, and as excellent adjuncts for digging in sand. Many species produce sounds that resemble bird chirpings, and most take time to sunbathe. Some species are desert dwellers and do not need to drink. Females often have a cervical pair of mammae in addition to the chest pair. When alarmed, some will thump the ground to sound a warning. A few **gundis** (*Ctenodactylus* spp.) will remain motionless or even feign death for up to twelve hours if threatened. One gundi species, *Felovia vae*, can autotomize its tail in extreme emergencies. Most react to danger by fleeing to crevices.

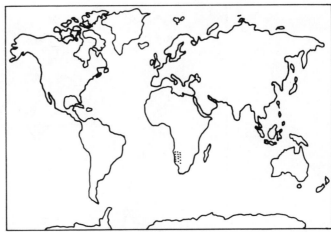

Family Petromyidae

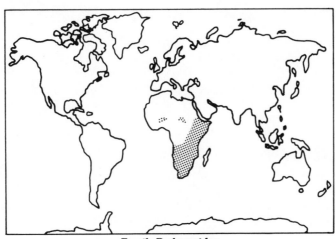

Family Bathyergidae

149

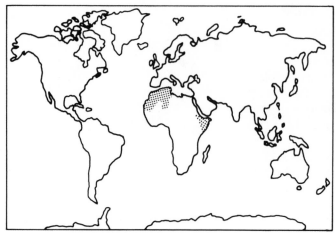

Family Ctenodactylidae

150

11

ORDER CETACEA
(79 SPECIES)

The cetaceans are generally split into two suborders, the **Odontoceti** or toothed whales and the Mysticeti or baleen whales. Because there are so many similarities both among families and between suborders, we will discuss the cetaceans as a whole, as we did with the bats.

The word "cetacean" refers to any species that is a member of the order Cetacea. The words "whale" and "dolphin" are a little more broadly used; as a rule, "whales" are animals in excess of twenty feet in length. The difference between "dolphin" and "porpoise" is less clear, although some specialists insist that there are definite differences between the two animals. Porpoises are smaller, being less than about six feet, while dolphins run about nine to fourteen feet. Porpoises' bodies are less streamlined; they have blunter noses and more broad, rounded flippers; and their dorsal fin is smaller and more triangular than the curved one of dolphins. Porpoise teeth are usually flat, while those of dolphins are conical.

All cetaceans inhabit the seas or the large riverways leading to them. Even the **Chinese lake dolphin** or **beiji** (*Lipotes vexillifer*), once thought to live exclusively in Dongtinghu (Tung Ting) Lake in China, now dwells primarily in the Chang Jiang (Yangtze) River and may even

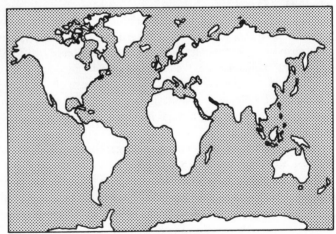

Order Cetacea

Sperm Whale (*Physeter catodon*)

153

Harbor Porpoise (*Phocoena phocoena*)

be extinct in the lake. At times even some of the big baleen whales, such as the **humpback whale** (*Megaptera novaeangliae*), will make their way some distance into harbors and up large rivers, but prolonged exposure to fresh water is harmful to the skin of many cetaceans. It is believed that some whales travel up freshwater rivers to rid themselves of skin parasites such as barnacles, which drop off when exposed to drastic salinity changes. Some cetaceans, however, including the genera *Inia*, *Lipotes*, and *Platanista*, do inhabit only fresh water, and some marine species such as the **Irrawaddy dolphin** (*Orcaella brevirostris*) can tolerate fresh water for extended periods without problem.

The smallest cetacean is probably **Commerson's dolphin** (*Cephalorhynchus commersoni*). In one study, the adults ranged from 51 to 77 pounds, and none exceeded five feet in length. Rivals for the title include the **finless porpoise** (*Neophocaena phocaenoides*), captive specimens of which ranged from 56 to 84 pounds in weight and four feet seven inches to five feet one inch in length; the **Amazon dophin** (*Sotalia fluviatilis*), averaging 80 pounds and four feet in length in one study; and the **harbor porpoise** (*Phocoena phocoena*), which rarely exceeds four and a half feet in length and usually weighs less than 100 pounds.

The largest cetacean is the huge **blue** or **sulphur-bottom whale** (*Balaenoptera musculus*), the largest animal ever to have lived on earth, reaching lengths of 110 feet and weights of almost 200 tons or more in large pregnant females. The tongue alone can weigh 8900 pounds, exceeding the weight of a full-grown elephant. The eye, the size of a large grapefruit, is the largest mammalian eye. Its liver weighs a ton, and the heart, comparable to the size of a Volkswagon "beetle," can weigh 1200 pounds. The cardiovascular system holds in excess of 9000 liters of blood (humans hold about six liters), and a child could easily crawl through its aorta, while full-grown salmon could swim through most of its major blood vessels.

In baleen whales, the females are larger than the males. The reverse is generally true in the toothed whales, although there are exceptions as in the **beaked whales** (Family Ziphidae), Chinese lake dolphin, harbor porpoise, **La Plata dolphin** (*Pontoporia blainvillei*), and the **susu** or **Ganges River dolphin** (*Platanista gangetica*). With the exception of the baleen whales, most cetaceans never stop growing; that is, epiphyseal or growth plates in their long bones never close. A rare occurrence, this is

154

seen in only a few mammals, such as elephants and the gray kangaroo.

Many cetaceans are long-lived if given the chance. **Fin whales** (*Balaenoptera physalus*) can reportedly reach 116 years of age, second only to humans. Blue whales have been credited with 110 years; **orcas** or **killer whales** (*Orcinus orca*) 100 years; and many, including **Baird's beaked whale** (*Berardius bairdii*), **Bryde's** (pronounced "breedas") **whale** (*Balaenoptera edeni*), the **gray** or **mossback whale** (*Eschrichtius robustus*), the humpback whale, and the **sei** (pronounced "say") **whale** (*Balaenoptera borealis*), can exceed 70 years. Ages can be estimated in baleen whales by counting the layers in the continually growing "plugs" found in the ear canals, and in toothed whales by counting the growth rings in the teeth themselves. Both methods are not precisely accurate since the rings can vary depending upon environmental circumstances.

The color of cetaceans is usually a blend of black and white or various shades of blue or gray. The sulphur-bottom whale's yellow belly gains its color from **diatoms** (*Cocconeis ceticola*) attached to the whale's ventral surface. Many species are countershaded—dark on top and light on the bottom—for camouflage. **Cuvier's beaked whale** (*Ziphius cavirostris*) is dark on the bottom and light on top, for unknown reasons. Individuals of this species are often colored quite differently from each other. Some cetaceans, such as the **pilot whale** or **blackfish** (*Globicephala melaena*), are all black, while the **beluga** or **white whale** (*Delphinapterus leucas*) is completely white when mature. Albinism seems most frequent in **sperm whales** or **cachalots** (*Physeter catodon*), but it is also seen in the **bottle-nosed whale** (*Hyperoodon ampullatus*), the **bowhead** or **Greenland** or **Arctic right whale** (*Balaena mysticetus*), **Dall porpoise** (*Phocoenoides dalli*), gray whale, killer whale, and **southern right whale** (*Balaena glacialis*). There have been "false albino" sightings of countershaded whales swimming on their backs. The **Amazon River dolphin, boutu,** or **pink dolphin** (*Inia geoffrensis*) is a light pink in color and susceptible to sunburn. With continued exposure to sunlight, it can become quite dark and even black, although age is also a factor in its coloration. A color phase of the **Indo-Pacific humpback dolphin** or **Chinese white dolphin** (*Sousa chinensis*) is white with reddish fins and black eyes. The fin whale is said to be the only mammal with asymmetric pigmentation. It is black to the left of its lower jaw and white to the right. This color contrast may be used to herd fish for feeding.

The **common dolphin** (*Delphinus delphis*) and others become more brightly colored during the mating season. In toothed whales, the genitals and nipples are often highlighted with striped markings, as in killer whales, thought to direct conspecifics' attention to these areas during courtship and nursing. Other markings on the killer whale, such as on the head, serve as disruptive coloration, while those on its side are thought to aid in schooling behavior. The **rough-toothed dolphin** (*Steno bredanensis*) is reported to be blackish purple with pink spots, and **spotted dolphins** (*Stenella attenuata* and *Stenella plagiodon*) become more spotted with age. The young of some cetaceans may be colored differently from their parents. Young belugas are born pink, turn brownish or blue-gray, next yellow, and finally white at five to seven years of age. In later life, they become somewhat yellow in color. Infant killer whales are pale orange in the areas that later become white, and newborn Commerson's dolphins are brown, taking on the distinctive black and white pattern with age.

One of the primary characteristics of mammals is the presence of body hair. Many cetaceans, such as the beluga, have no hair except as embryos, losing it by the time of birth. Many species have a small amount of sparse hair, located primarily on the head. Frequently, it is found in the same places as human facial hair, on the chin and top of the head and as eyebrows, moustache, and sideburns. Some species have hair only on their snout when young. In most, the few bristly hairs they do have take on a sensory function. Most freshwater dolphins have many tactile hair bristles on their snout. The humpback's unique warts on its head and snout each sprout two bristly hairs used in a sensory way. It is thought that sensory bristles may aid those cetaceans who have them in detecting water movements set up by currents and prey species, thus aiding both navigation and feeding.

The largest number of teeth found in any mammal is credited to dolphins. The **spinner dolphin** (*Stenella longirostris*) has an average of 224 teeth with a maximum of 260. This is rivaled by a subspecies of the **common dolphin** (*Delphinus delphis tropicalis*) known to have an average of 238 with a maximum of 250. None of the baleen whales have teeth except as embryos, and a few of the toothed whales are essentially toothless. No cetacean produces "milk teeth."

In sperm whales, largest toothed animal in the world, only the lower jaw has teeth. Although there are vestigial teeth in the upper jaw, these rarely erupt, and the upper jaw is mostly studded with sockets into which the lower teeth fit when the mouth is closed. Teeth are not required for feeding in the sperm whale since often the lower ones do not even erupt until the animal is about ten years old. It may be that sperm whale teeth are a male secondary sexual characteristic used in dominance fights.

Dall's porpoise has horny projections on its gums between the teeth to hold slippery prey such as squid. With time, these projections wear down to expose the real teeth, which are the smallest of any cetacean. In beaked whales, the males generally have two teeth at the forward end or middle of their lower jaw, while the females are usually toothless (their teeth usually don't break through the gums). Sometimes, as in the bottle-nosed whale, the males fail to sprout teeth as well. In **Arnoux's beaked whale** (*Berardius arnuxii*) and Baird's beaked whale, the two or occasionally four teeth are embedded in a cartilaginous mechanism that allows the teeth to be extended at will, perhaps for a better bite. In the former species, both sexes have exposed teeth. In **Shepherd's beaked whale** (*Tasmacetus shepherdi*), both sexes also have exposed teeth; this is the only beaked whale to have a full set of functional teeth in the mandible, in addition to the large pair at the end of the jaw. **Gray's beaked whale** (*Mesoplodon grayi*) is the only beaked whale with functional maxillary teeth. In any individual beaked whale, the teeth might not erupt, can be worn down, or even may have fallen out. In the **strap-toothed whale** (*Mesoplodon layardii*), the two teeth grow to a foot long and curve up and over the upper jaw; they may at times hinder complete mouth opening. They may, however, help channel fish down the gullet by preventing them from slipping out the side.

The **narwhale** (*Monodon monoceros*) has been dubbed the "unicorn of the sea." It is generally thought that the oryx began the unicorn myth, but the narwhale has contributed its fair share to the legend. As the narwhale matures, it loses most of its teeth. The left upper canine in males (females usually become toothless but may occasionally sprout a tusk or two) is not dropped; instead it grows out to protrude from the upper lip. It becomes a formidable tusk reaching ten feet in length. In rare instances, males have been found with both canines grown out into tusks. These teeth always spiral to the left and are the only spiralled teeth found in the animal kingdom. The tusk may be used for fighting, breaking through ice (although the melon has been seen to serve this purpose), or stirring up mud to uncover prey. It also might serve as a badge of rank like antlers in deer. There are recent reports of males "dueling" for dominance. At these times, the tusks are crossed both above and below the surface, and some males sport facial scars and broken tips, possibly as a result of such contests. Broken tusks have also been found embedded in the jaws of males. Other less probable theories of tusk function say it is a device to radiate away excess body heat, can attract females, or may be used to direct their vocalizations towards rivals. It is probably safest to say the tusk is a multifunctional tool used primarily in dominance fighting between males. Humans have used the tusk for many purposes ranging from aphrodisiacs to cure-alls. At times, they have been worth their weight in gold. Sperm whale teeth have also found their way into commercial ventures, primarily for jewelry and other items after engraving as scrimshaw.

The cetacean skeleton is very similar to that of other mammals. All four limbs are present in embryos, but the hind legs degenerate and disappear. Occasionally a whale is found that still possesses vestigial hind limbs, small, symmetric, and containing bone and/or cartilage representing normal mammalian leg bones. In one humpback whale, these legs were three feet long. In many species, the seven cervical vertebrae are fused to some degree. In the **right whales** (*Family* Balaenidae) and bottle-nosed whale, all seven are fused, and in the family Delphinidae the first two are usually fused. The result is that neck motion is limited. The beluga, narwhale, and **river dolphins** (Family Platanistidae) do not have fused vertebrae and can turn their heads easily from side to side. Rorquals also have seven separate neck vertebrae. The bones of baleen whales float due to their high fat content. In fact, up to one third of the total oil obtained from a whale comes from the bones.

As a rule, cetacean skin is thin, fragile, and extremely sensitive. It is easily damaged by trauma and is important in swimming as will be described. In large baleen whales, the skin is only about a quarter of an inch thick. There are exceptions, for the skin of belugas, narwhales, and some river dolphins can be tanned into attractive leather. The skin of the penis of large baleen whales (up to nine feet long in the blue whale) is tough and was converted into overalls by early whalers, who cut out appropriate holes. These overalls were generally worn by the mincers who cut up the blubber and required some type of protection from the penetrating grease. Another exception is the sperm whale. It is covered with the thickest skin in the animal kingdom. In some areas of the body, the skin can measure fourteen inches thick.

Cetacean skin is often parasitized by algae, barnacles, copepods, and various species of whale lice. Many of these parasites both are species-specific and specialize in the anatomical sites to which they attach themselves. In humpbacks, the main three species of parasitic barnacle are found primarily at specific locations, depending on the barnacle type. The **stalked barnacle** (*Conchoderma auritum*) grows only on **acorn barnacles** (*Balanus balanoides*) already attached to the skin, or occasionally to the front teeth of sperm whales. Beaked whales' teeth are also adorned with barnacles at times. The **gray whale barnacle** (*Cryptolepas rhachianecti*) is found only on the skin of gray whales and is responsible for its alternate name of mossback whale. It is interesting to note that newborn gray whales are known to pick up lice from their mother's vaginal area during the birth process.

Since a single humpback whale may have 1000 pounds or more of barnacles attached to its skin, the importance of removing some becomes obvious. Humpbacks rid themselves of a few by breaching, rubbing up against rock shelves, and wallowing in shallow water. Lobtailing (tail-slapping) may also help remove barnacles, as do the changes in water temperature and salinity encountered in migration or in traveling up rivers.

The skin of right whales is marked with large callosities that often contain patches of hair. These are strategically located on head areas and thus resemble beards, eyebrows, moustaches, sideburns, and hair. They are used by researchers to easily identify individual whales because of their variation in

157

size, shape, and location. Males occasionally bump each other with these callosities, sometimes referred to as bonnets, when fighting for dominance. Some whale lice (*Cyamus* spp. and others) are actually known to clean algae and other parasites from these bonnets in addition to feeding on some of the whale's tissues. The callosities appear in shades of pink, yellow, orange, and white, depending on the type and number of lice in residence. Some researchers feel that the callosities actually function to keep parasites in these isolated spots and thus free most of the whale's skin from infestation.

Kelp and **hooded gulls** (*Larus* spp.) will sometimes alight on surfaced whales to pick off lice. A few gulls have inadvertently found themselves on top of the blowhole as the whale spouts, causing the birds to be blown skyward with the expelled air. **Gray phalaropes** (*Phalaropus fulicarius*) winter at sea but can obtain some food by feeding on killer whale skin parasites. The skins of some whales, including the rorquals and humpback, often exhibit crater-like depressions and oval, white scars caused by bites from the **cookie-cutter shark** (*Isistius brasiliensis*) or possibly by parasitic worms such as *Penella* spp. The skins of smaller cetaceans are usually free of parasites because of the high turnover rate of their skin cells.

As a general rule, large whales have poor eyesight while the dolphins and porpoises have keen eyesight. In order to visualize objects clearly while underwater, cetaceans (and seals, fish, and other animals) have spherical lenses in their eyes. Terrestrial animals have elliptical lenses because light bends to a greater degree upon entering the eye from a gaseous rather than a liquid medium. Spherical lenses bend the light rays more acutely and enable them to be focused on the retina. One consequence, however, is that most cetaceans are short-sighted when their eyes are above the water's surface. Some dolphins can see well above and below water since they have double slit pupils enabling them to focus in either situation. Some species can move their eyes independently. The **pygmy killer whale** (*Feresa attenuata*) can protrude its eyes outward from the sockets to see behind itself. The killer whale and others also possess a tapetum allowing for vision in dim light. It is said that they can see as well underwater as a cat can on land.

The susu and **Indus River dolphin** (*Platanista indi*) have atrophied, lensless eyes and are essentially blind to all but light and dark. This is not much of a handicap since they usually inhabit muddy river waters. The Chinese lake dolphin also has poorly sighted, atrophied eyes, while the Amazon river dolphin has stereoscopic vision with most of the visual field residing overhead, which explains why it is often reported to swim upside down over river bottoms in visual search of food. They are aided by the presence of sensitive sensory hairs on the snout. When using sonar, this dolphin swims and catches prey in the normal manner. It also has very seal-like front flippers that are reportedly used to pull the animal over shallow sand bars. The Chinese white dolphin and the **Gulf porpoise** (*Phocoena sinus*) are also reported to wallow over mudbanks to get to deeper channels. Most cetaceans secrete oily substances to protect their eyes from the many irritating chemicals found in the sea, and most can "close" their eyes because of special skin folds, even though they lack true eyelids.

Hearing rather than sight is the dominant sense of most cetaceans. Like bats, many find their way about with the help of sonar, although theirs extends to a higher frequency than that of bats. The beluga can produce sounds with frequencies exceeding 300,000 Hz, the highest frequency mammalian call so far detected. Sonar—or echolocation—is the process of transmitting sounds and analyzing their echoes. It is probably used by all toothed whales and possibly some baleen whales (which produce

158

eaf-Chinned Bat (*Mormoops megalophylla*)
Referred to on page 51)

Wrinkle-Faced Bat (*Centurio senex*)
(Referred to on page 47)

Patas Monkey (*Erythrocebus patas*)
(Referred to on page 72)

Black and White Ruffed Ler
(*Varecia variega*
(Referred to on page

Above and right: Gibbon (*Hylobates* spp.)
(referred to on page 74)

Gorilla, male silverback (*Gorilla gorilla*)
(Referred to on page 78)

Orangutan, male (*Pongo pygmaeus*)
(Referred to on page 75)

airie Dog (*Cynomys ludovicianus*) (Referred to on page 110)

lympic Marmot (*Marmota olympus*) (Referred to on page 107)

Naked Mole-Rat (*Heterocephalus glaber*)
(Referred to on page 148)

Giant Squirrel (*Ratufa* sp)
(Referred to on page 10

Killer Whale (*Orcinus orca*)
(Referred to on page 155)

Bobcat (*Felis rufus*)
(Referred to on page 179)

Arctic Fox (*Alopex lagopus*) (Referred to on page 193)

Dingo (*Canis familiaris dingo*) (Referred to on page 194)

Coyote (*Canis latrans*)
(Referred to on page 193)

Ring-Tailed Cat (*Bassariscus astutus*)
(Referred to on page 214)

Polar Bear (*Ursus maritimus*) (Referred to on page 216)

own Bear (*Ursus arctos*) (Referred to on page 217)

arbor Seal (*Phoca vitulina*) (Referred to on page 226)

African Wild Ass (*Equus asinus*)
(Referred to on page 258)

PHOTO: WILLIAM VOELKER

Hippopotam
(*Hippopotamus amphibi*
(Referred to on page 2€

PHOTO: WILLIAM VOELKER

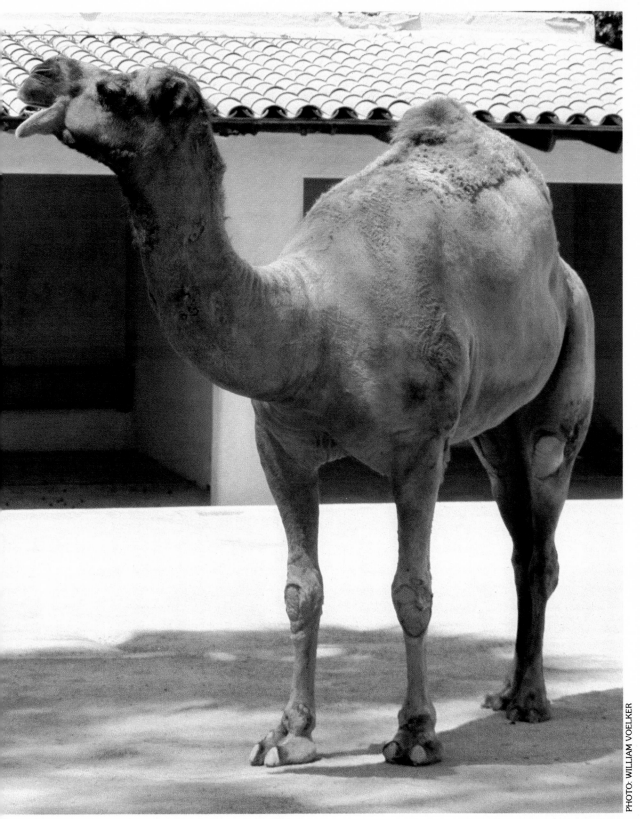

Arabian Camel (*Camelus dromedarius)* (Referred to on page 272)

Pere David's Deer
(*Elaphurus davidianus*)
(Referred to on page 276)

Roan Antel
(*Hippotragus equin*
(Referred to on page 2

ala (*Tragelaphus angasi*)
eferred to on page 286)

Bongo (*Tragelaphus euryceros*)
(Referred to on page 284)

Gerenuk (*Litocranius walleri*)
(Referred to on page 283)

Musk Ox (*Ovibos moscha*
(Referred to on page

suspicious clicks), although this has not been proven. Since cetaceans have no vocal cords, sounds are produced by forcing air over valves and flaps in nasal areas below the blowhole. The air can be internally recycled in the process so that sounds can be produced without losing air or producing bubbles even during the long songs of humpbacks.

Clicks may actually be produced by forcing hard nasal plugs across bony edges in the skull. The sounds are transmitted from the animal's melon (forehead region) and can involve up to 1200 clicks per second, which travel at five times the speed of sound in air. Thus it only takes 2 seconds for a dolphin to detect prey a mile away. Both high- and low-frequency sounds can be produced. High frequencies are better suited for short distances and resolution of detail; low-frequency sounds are better for echolocating at long distances and for communications. It is thought that most baleen whales do not truly echolocate but emit low-frequency sounds for communication, while toothed whales use both high and low frequencies. In one study of dolphin echolocation, three basic groups of sounds were produced. The first group was low-frequency clicks with good penetration used to determine topography. The second group was high-frequency whistles (higher in pitch than the supersonics of other marine animals) used for communication. High-frequency clicks made up the third group and were used primarily during hunting; they too, were at a frequency higher than the hearing apparatus of prey could detect.

Some of the sounds produced have so much energy that if they could be detected by human ears they would be deafening, and there is one episode of a **bottlenose dolphin** (*Tursiops truncatus*) actually imploding a human swimmer's eardrum with a quick sound burst. Sperm whales and dolphins may be capable of stunning or killing prey with very loud bursts of sound. Blue whales can produce sounds that approach 190 decibels and can be detected hundreds or thousands of miles away. Thought to be the loudest of all animal sounds, they must once have made truly long-distance communication possible. Today, however, human noises (mainly shipmotors) undoubtedly restrict the whales' communicative range. Killer whale sounds can be heard up to fifty miles away underwater, and sperm whales produce loud clicking sounds whose energy output is comparable to a jet engine at full throttle twenty feet away. Many cetaceans, including the sperm whale and bottlenose dolphins, produce unique clicks or whistles, referred to as "signature sounds," that identify the individual animal.

Cetacean ears are well developed for complete reliance on sound. They are protected from loud sounds by a thickened skull medial to the sound-producing areas. Some cetaceans do show external ear openings, but none of the openings communicate with the internal ear directly; the outer canals dead-end in the tympanopetromastoid bone that surrounds the inner ear with air-filled cavities. In baleen whales, the auditory canals become blocked with ear plugs up to four feet long. While low-frequency sounds travel directly to the inner ear, high-frequency sounds are picked up by the melon, the sides of the head, and probably most importantly, the jaw, so that echo interpretation is aided by horizontal movements of the head.

The cavities around the inner ear help the cetaceans to determine sound direction by slowing down the sound waves as they hit the air within them. Further directional sensitivity comes because the two inner ears are separated medially by a foamy, insulative substance, and each side of the jaw transmits sounds to the inner ear on the same side. The inner ear air cavities are protected from pressure changes by a special nitrogen absorbing oil/mucus emulsion.

The valuable oil or spermaceti found in sperm whales and bottle-nosed whales may be a part of their transducer system but may serve to absorb nitrogen from nasal passages in deep dives. (The most recent evidence seems to imply that it is most important in the regulation of the whale's specific gravity and hence its buoyancy.) It is interesting to note that a dolphin's melon is capable of transmitting and receiving sound waves simultaneously, and that dolphins are capable of communicating with other dolphins while simultaneously using their sonar, since both whistles and clicks can be produced at the same time. This is possible because the left and right nasal sacs can be used independently at the same time.

The efficacy of the echolocation system is almost beyond belief. It is extremely accurate and so well developed at detecting shape, texture, and density that dolphins can tell the difference between similarly shaped objects of different metals, ball bearings that differ in diameter by a quarter of an inch, and even different species of fish that are the same size and shape. It is even thought, since the system works much like the ultrasound equipment used in diagnostic medicine to produce sonograms, that dolphins can detect their prey's as well as each other's internal workings. They may be capable of detecting a conspecific's emotional states, sexual arousal, and health problems, including tumors, strokes, and heart attacks.

At times, cetacean sonar may get the animals into trouble. Since it is generally directed forward, it may fail to detect gentle, sloping bottoms which reflect sound waves off in the wrong direction. The cetaceans might also run into problems with muddy bottoms which absorb sound waves and some rocky bottoms that can produce confusing echoes. All the preceding can lead to disorientation or panic, causing beach strandings of whole pods (as groups of cetaceans are known) or even collision with rocks, as shown in right whales who injure their jaws by colliding with rocks in shallow water.

Currently, experts believe that illness and/or parasites, especially certain helminths that infect the ear (*Nasitrema* spp. and *Stenurus* spp.), are responsible for some whale strandings by interfering with their orienting ability. Others speculate that weak magnetic fields may be responsible for some strandings. Terminally ill dolphins are also known to ground themselves intentionally, possibly to allow the animals to remain at the surface and breathe with less effort. Stranding usually dooms large whales to death from suffocation, for their respiratory muscles are not strong enough to overcome the weight of their own bodies when out of the weight-supporting water, and their lungs collapse because of their "floating ribs" (see below). The gray whale and many dolphins have survived strandings at low tide, and some believe that they may come into shallow water purposely to sunbathe. Attempts to rescue stranded whales almost always fail because after a whale is dragged into deeper water, it heads right back to shallow water, perhaps in answer to the distress calls emitted by its still-stranded companions. Some experts believe that in these situations the whales may not be answering the cries of their fellows so much as fleeing in such fear of the killer whales that drove them ashore in the first place that they once more seek cover in the shallows and thus become beached again. There is probably no single reason for strandings, which continue to remain a mystery.

Some people think that baleen whales do use not sonar but smell to find krill. They propose that this is why these whales have two blowholes instead of one, as in the toothed whales. (Toothed whales do have two nasal passages below the single blowhole entrance.) Two nostrils would enable the whales to locate their food by odor since they could be guided in the direction of the stronger stimulus. Most cetologists doubt that a good sense of smell exists and state that cetaceans are anosmatic (without a sense of smell). Actually, toothed whales are anosmatic, but baleen whales are at least

160

microsmatic (slight sense of smell).

In toothed cetaceans, the respiratory and gastrointestinal tracts share no common connections; each is complete in itself. The larynx sits just behind the oral cavity, keeping them separate as is also the case in many marsupials. It is thus impossible for them to choke on food or seawater that enters their mouth during feeding. They are also unable to breathe through their mouths. Toothed whales have a tight throat sphincter to prevent entry of seawater to the stomachs. The first stomach pulverizes prey, and many whales will regurgitate bones and other hard materials instead of passing them through to the remaining stomachs.

As a fetus, a whale's nostrils are located in the same position as other mammals, but with development the nostrils migrate to the top of the head. Many whales are equipped with ridges and fairings that act as a barrier to keep water from being channeled to the blowhole when the whale is at the surface. Right whales' callosities serve the same purpose when located forward of the blowhole. The blowholes also have a built-in safety measure in that they automatically shut tight should any water enter the opening. The underlying nasal passages contain a series of chambers and valves so that water that does get by will become trapped in the first chamber, to be blown out with the next expiration.

Cetaceans do not spout water as many suppose. Their visible plumes are simply the condensation from warm, moist air that cools with expansion upon release from the high-pressure respiratory system. For example, in large baleen whales, 2000 liters of air can be exhaled and then inhaled (making a total of 4000 liters per breath) in two seconds. This explains why the spout is formed in the tropics as well. Air speed from the blowhole can approach 900 miles per hour in some species and can be heard up to a mile away in large whales. Other factors that contribute to the visible spout are the addition of some water on top of the blowhole, as well as in the first chamber as described, and mucus from within the respiratory passages.

Due to the variety in blowhole anatomy, various species of whale can be distinguished from their spouting alone. Right whales form two sprays that angle away from each other to form a large V. Sperm whales shoot a spray up to 25 feet high, angled 45 degrees to the left since their blowhole is displaced by the large reservoir of spermaceti. The rorquals spout straight up, forming a large inverted pear-shaped spray. In the blue whale, this spout can reach a height of 50 feet. When frightened or pursued, right whales and others will exhale underwater in order to cut down the time spent at the surface, enabling them to swim faster or to catch prey more efficiently. Susus have fetid bad breath, and it is reported that the combined breath of some whale pods reeks so badly that it can be detected from some distance. Some researchers say they have never encountered this. When fasting, many baleen whales have the odor of acetone on their breath, due to their use of fat for energy.

Many cetaceans can stay submerged for long periods of time. The sperm whale can stay under for over an hour and a half, while a few of the beaked whales (such as the bottle-nosed whale) and the **minke** (pronounced "minkey") or **piked whale** (*Balaenoptera acutorostrata*) can do so for an unbelievable two hours. Several anatomical and physiological adaptations make this possible. First, cetaceans have more "floating ribs" (ribs that are not firmly attached to the sternum or breast bone) than other mammals. In fact, right whales have only one non-floating pair of ribs. This anatomical difference lets cetaceans exchange up to 95 percent of the air in their lungs with each breath, where humans at rest exchange less than 10 percent of the lung volume. Some rorquals can thus exhale and

161

inhale 500 gallons of air in two seconds.

Second, while humans carry about a third of their oxygen reserve in the lungs, whales carry less than a tenth and sometimes none in their lungs. Whale blood carries twice as much hemoglobin (the blood's oxygen-carrying chemical) as human blood, and cetacean muscle has ten times as much myoglobin (an oxygen-carrying protein) as human muscle. This causes the muscles of some species to appear almost black. (In birds, myoglobin concentration accounts for the difference in color between light and dark meat.) This extra myoglobin lets cetaceans store almost ten times as much oxygen as land mammals.

Third, the bodies of whales (as well as beavers, manatees, muskrats, and seals) are less sensitive to the build-up of carbon dioxide, so that they don't feel the need to breathe for longer periods of time. (Humans who practice the dangerous act of hyperventilation before diving can also feel less need to breathe, but in them oxygen can fall to a dangerously low level, until they pass out and drown.) When cetaceans dive, their bodies conserve oxygen by shutting down all processes not needed for swimming or sustaining life. Blood is shunted from unneeded areas and pooled in special blood vessel networks, the retia mirabilia, located at the base of the brain and along the spine. This reservoir both holds a tappable source of extra oxygen reserves and prevents sudden blood pressure changes to vital areas during deep dives. That is, it pumps blood to the brain when the animal's pulse drops too low to ensure cerebral perfusion.

Fourth, lactic acid produced by the muscles does not enter the circulation because the blood flow is cut off until the cetacean surfaces. Since abundant new supplies of oxygen are taken in at this time, the large amount of lactic acid released can be quickly metabolized. During deep dives, the pulse rate drops from half to a quarter its normal value, and metabolism falls to half its regular rate. In Dall's porpoise, the pulse will fall from 125 to as low as 15 beats per minute. These same changes are seen to some extent in all air-breathing animals that dive underwater and are collectively known as the "seal reflex" (as mentioned in the order Monotremata). They are developed only slightly in humans, but a few extreme examples include some penguins, in which heart rate falls from 200 to 20, and the marine iguana, whose heart is even known to stop beating for several minutes when it dives.

One question that had long puzzled biologists was how whales could dive to great depths for such long periods of time without contracting the "bends" or decompression sickness. The answer came when it was found that as the water pressure increases with depth, the air inside the lungs' alveoli (air sacs where gasses are exchanged with the blood) is forced into the bronchi and trachea, which are further connected to large air sacs near the underside of the skull. These storage sacs are not connected to numerous vessels for gaseous exchange. Some species can prevent air from re-entering the alveoli with special muscles that clamp the bronchioles shut. Thus, the lungs literally collapse to as little as one tenth their normal size, preventing air from getting into the bloodstream while under high pressure. Should absorption occur, the air would expand upon resurfacing to form air bubbles in the blood and tissues, causing the "bends." The whole process is made possible by the fact that very little air is inhaled prior to diving, so that there is not much in the lungs to begin with. The air sacs used for storage become filled with a foam produced by an emulsion composed of air, water, and a special oil thought to absorb nitrogen. Some feel that this foam contributes in part to the visible spray seen in a whale's spout, particularly in the tropics. Whales pant after long dives, and as a rule they must take one breath for every minute they are submerged. Interestingly, whales have nitrogen-absorbing fats in

162

their bloodstream which probably protect them against nitrogen narcosis as well.

In both cetaceans and pinnipeds, the nostrils are normally closed when relaxed. In some, such as the beluga and various dolphins, breathing is initiated through conscious effort only. Thus, should a cetacean be knocked unconscious or anesthetized, it would die from respiratory failure. This also explains why some species cannot afford to go into a deep or complete sleep. It has been found that "sleeping" or resting dolphins remain in a small group as they slowly swim in circles. Each dolphin keeps its outward-facing eye open as if to watch for danger. This can continue for several hours, after which the group reverses direction and its members close their other eyes for several hours. Both eyes are closed simultaneously for less than five minutes per 24 hour period. These facts, along with other information, have led investigators to believe that some cetaceans probably only let half of their brain sleep at a time. The susu rarely sleeps, while others, such as the Amazon river dolphin, lie on their backs on river bottoms and go into deep sleep with periodic rises for breathing. Sperm, fin, and right whales are known surface sleepers as evidenced by their many collisions with boats, and killer whales and humpbacks are said to snore loudly, with the latter known to sleep for two to three hours at a time. On the other hand, migrating gray whales are thought not to sleep at all.

Cetacean tails are horizontal, not vertical like a fish's. This gives them primarily vertical mobility, a necessity for a marine, air-breathing mammal for quick ascents and descents. Cetacean lungs are also somewhat dorsally located which, accompanied by their blubber distribution, oily bones, and flipper arrangement, helps to keep them upright. Cetaceans' tail flukes are their main means of propulsion, at speeds that vary greatly from species to species. Power is generated by the upward movement of the flukes. The slowest whale is the southern right whale, with a cruising speed of two knots (about four km/h or a little over two mph), although it can "sprint" at up to five knots. The fastest are the sei whale, Dall's porpoise, and the killer whale, with speeds clocked at or near 30 knots (57 km/h or 35 mph). There are unsubstantiated claims of 40 knots for killer whales and even 60 knots for the common dolphin, but most cetologists consider these fantasy. A 90-foot blue whale can travel at 20 knots for ten minutes or more, requiring the tail flukes to be generating 520 horsepower. Although several species, including the killer and bottle-nosed whales, are known to dive deeper than 3,000 feet, the champion cetacean diver is the sperm whale, known to dive over 10,000 feet in search of food. The pressure at this depth is about 4,450 pounds per square inch, or in excess of two tons per square inch.

Dolphins can swim faster than theory and models said was hydrodynamically possible. This is due to the fact that turbulence is kept to a bare minimum. The outer skin and blubber overlay another, pleated skin layer. Four fifths of the water content of the true outer layers of skin can be channeled into the longitudinal folds of this pleated layer beneath. This helps to eliminate turbulence at its point of origin to preserve smoother, laminar flow and allow greater speed. (The fur padding of pinnipeds seems to function in the same manner.) Other factors also aid in obtaining higher speeds. To increase streamlining and reduce drag, the teats, genitalia, and anal openings are all housed in folds of skin. Antiturbulent "oils" (ethylene oxide polymers) are secreted by skin cells in some species to increase laminar flow. Another way dolphins can increase their swimming speed is by clearing the water's surface when rising for a fresh breath of air. If only part of their body breaks through the surface, an increase in drag will result from an increase in turbulence, even though dorsal fins and/or humps will force away vortices of water to decrease some turbulence. Some species including Dall's porpoise and the **white-beaked dolphin** (*Lagenorhynchus albirostris*) produce "rooster tails" when swimming at speed. This behavior has earned the Dall's porpoise the alternate name of spray porpoise.

Cetaceans use many interesting swimming techniques. Some, such as the beluga, are agile swimmers and can even swim backwards. Humpbacks have also been reported to swim backwards with the help of their long flippers. The Amazon river dolphin occasionally swims on its back when searching for prey, while the Ganges river dolphin often swims using a sidestroke technique with its right side down. The leading edge of its flippers is very sensitive and can be used to "feel out" prey for consumption. Some of the large baleen whales, such as the fin whale, are occasionally observed swimming on their backs and have caused false albino sightings. The narwhale, too, often swims belly up.

The spinner dolphin is named for its spectacular spins as it leaps from the water to heights of fifteen feet in the air. Although they usually spin on their longitudinal axis about two and a half times before splashing back, they have been known to spin up to seven full times in a single leap. **Hourglass dolphins** (*Lagenorhynchus cruciger*) and others are also known to spin with some jumps. **Dusky dolphins** (*Lagenorhynchus obscurus*) and **Pacific white-sided dolphins** (*Lagenorhynchus obliquidens*) occasionally somersault when they leap from the water in the wild. **Cameroon** or **Atlantic humpbacked dolphins** (*Sousa teuszii*) occasionally do backward somersaults. The killer whales hold the cetacean high jump record with leaps reaching forty feet above the surface.

Pilot whales are so named because they are thought to have a leader or "pilot," usually a large male, as they travel about in large pods, at times numbering into the thousands. In fact, whole pods were once easily taken by beaching the lead whale since the others would blindly follow. Gray whales exhibit a technique known as "evasive swimming." While so engaged, the whale exposes only its nostrils upon surfacing and then slowly and quietly exhales so that no noise or spout is produced. It then sinks down and swims on. The procedure is continued until the whale is out of danger, whereupon it resumes its normal pattern of surfacing and spouting. Gray whales and others also perform a maneuver known as "spy-hopping" or "pitch-poling" in which they rise vertically headfirst out of the water and remain in such a position for a minute or more. This has long been attributed to their looking around to orient themselves visually when migrating, but some observers feel it aids in swallowing, particularly shellfish scooped up from the muddy bottom. Some spy-hopping whales are known merely to be standing upright on their tails in shallow water.

Whales and other animals, such as the manta rays, are known to make great leaps from the sea in displays known as breaching. The reason for this is not completely clear, but theories abound. It may aid digestion, rid the animal of external parasites such as barnacles, be a form of play, take place when feeding to stun or panic prey, aid the animal in visual orientation, allow immediate change in swimming course, or be a form of communication or threat. A large killer whale's breaching can be heard five miles away.

Right whales will actually use their large tail flukes as sails. The whales sit vertically in the water with their flukes exposed to the wind at right angles allowing themselves to be pushed along by the breeze. They often start from a favorite point to which they return after sailing awhile to begin over. They may engage in such activity for three to four hours at a time. Whale-watchers often see tracks of "whale prints," smooth water slicks where the whale dives down. These result partially from body oils but primarily from water vortices produced by the flukes.

Temperature control is of utmost importance to any ocean-living mammal. A cetacean's blubber is such efficient insulation that a whale removed from the water for an extended period can expire from heat stroke. Dead whales left aground for periods of two days have been known to get so hot that their internal flesh is literally cooked. The throat pleats of **rorquals** (*Balaenoptera* spp.) (baleen whales with dorsal fins, including blue, Bryde's, fin, humpback, minke, and sei whales) may fully expand from bacterial gas production within twelve hours of death, and some dead whales have literally exploded, throwing entrails for hundreds of yards.

To keep cool, cetaceans and pinnipeds can increase blood flow through their fins and flukes. These are covered with little blubber so that excess heat can be lost quickly to the cold ocean water. The large, pleated throats of rorquals may aid in temperature control. When the pleats are opened they expose many small blood vessels to dissipate heat rapidly. It is thought that these folds serve other functions, primarily to increase mouth capacity when feeding, but also to stabilize motion, improve streamlining, and possibly to act as a braking mechanism for quicker stops. Baleen whales with pleats are sometimes referred to as **balaenopterids**; those without are called **balaenids**. Blubber, in addition to being their main means of insulation, is also the major factor contributing to buoyancy. It can be up to 28 inches thick in the bowhead and can account for as much as half the body weight.

Cetaceans, like desert animals, must conserve water since they cannot drink seawater in large amounts and survive. This is because their kidneys cannot produce a urine more concentrated than the salt water of the oceans. Thus, like humans, should they drink seawater, they would actually lose body water in order to excrete the excess salt. Cetaceans do not sweat and are capable of concentrating their urine to a greater degree than humans and most other mammals. The kidneys of some of the larger whales contain up to 3000 lobules enabling them to do so. Cetaceans depend mostly on the ability of their prey to filter out excess salt from seawater, thus obtaining their "fresh" water from the body fluids of their catch. They also obtain fresh water from the metabolic breakdown of fat (blubber). In fact, baby whales get their fresh water to a large extent by this method, showing an advantage to the high fat content in the females' milk.

Cetaceans are very intelligent animals, although the exact level of their intelligence remains unknown. A few researchers have actually claimed some species are smarter than humans. Others say they are about as intelligent as dogs. The largest brain of any living animal is the twenty pound "computer" found in the sperm whale. The brain of the bottlenose dolphin, made famous by "Flipper," is about the same size as a human brain. The brain-to-body weight ratio is fairly close for the two species. Actually, not counting the dolphin's blubber, its brain makes up 2 percent of its weight, while the human brain makes up about 2 percent of the human body, the highest of all mammals. In comparison the smallest mammal brain/body weight percentage is 0.027% as found in the water vole.

Within the brains of both species, there is a region known as the limbic system. It is the center for the emotional aspects of behavior, is related to survival of both the individual and the species, and contributes to the mechanisms of memory. It is primitive in function, however, when compared to the neocortex or "gray matter," where reside the higher functions of perception, thought, self-control, objectivity, humor, creativity, and so on. Although the cetacean cortex has only six cell layers, compared to the seven of the human cortex, it is interesting to note that cetaceans have a higher neocortical-to-limbic ratio than humans. Many captive dolphins have exhibited both humor and self-control when even patient humans would find themselves hard pressed. In fact, dolphins have rarely attacked or been harmfully aggressive to humans, despite occasional abuse. Some whales have three times as many brain neurons as humans.

Cetacean insight and intelligence are demonstrated in their "play," during which they often use extraneous objects, pushing them, throwing them into the air, and sometimes even catching them. In aquariums, these playful captives throw objects from their tanks, tease other resident animals such as fish by pulling at their fins, and sometimes squirt spectators. In one instance, two dolphins were playing "keep-away" with a reluctant eel when the latter found the game not to its liking and hid in a small pipe. One of the dolphins gently grasped a small venomous fish nearby and pushed it into one end of the pipe. The eel fled, and the game resumed.

Captive **false killer whales** (*Pseudorca crassidens*) have learned tricks simply by observing performing dolphins. In one of the more humorous ancedotes involving mimicry, a young bottlenose dolphin observed a human blowing out a cloud of cigarette smoke. She returned to the spot after taking in a mouthful of milk from her mother and proceeded to engulf her own face in a "cloud" of milk resembling smoke. There are even reports of released dolphins allowing themselves to be easily recaptured or swimming close by in apparent hope of being recaptured.

In the wild, cetaceans have been seen to take part in various forms of play. Dolphins often surf in the wakes of boats or even large whales. For example, **rightwhale dolphins** (*Lissodelphis borealis*) will ride the bow waves of gray whales. Some species will surf in waves, at times even apparently awaiting the arrival of large swells as do human surfers. Even adult gray whales are known to surf, and sei whales have been observed riding large storm waves at sea. Humpbacks often engage in frolicsome backward somersaults, and right whales have taken up sailing, as already described. Humpbacks have recently been discovered playing sound games. They vary their sound production to listen to the rebounding echoes from underwater canyons. Some of these behaviors, of course, may stem from innate drives and could occur for other reasons than interpreted.

Cetaceans communicate through both wide-ranging clicks and whistles and additional behaviors. For instance, jaw clapping signals intimidation, barking and mewing signal feeding, and a whimpering has been detected during mating. Both belugas and humpbacks communicate through long songs. (The only mammals known to be true singers include the bearded seal, beluga, humpback whale, humans, and the white-lined bat.) The beluga has been dubbed the "sea canary" for its frequent vocalizations, easily detected above the surface. Male humpback whales produce complex and eerie (but soothing) songs lasting half an hour or more. The actual sound pattern lasts about eight minutes but is continually repeated. All males within a single group sing identical songs, and yearly variations lead to the singing of completely new songs about every four years. The songs can be heard hundreds of miles away underwater, and it is thought that, prior to acoustic cluttering of the ocean by human vessels, they may well have been perceptible thousands of miles away.

The function of the songs may be to draw whales together for breeding. They may also play a role in courtship, and as the sound waves travel through the bodies of the females, they may stimulate mating behavior. Others postulate that songs serve to establish male dominance hierarchies. Cetaceans transmit warnings to others of their kind by slapping their flukes repeatedly against the water's surface, a behavior known as lobtailing. It is an effective measure for distances of several miles and is the communicative method of choice when high winds disturb the water's surface and interfere with low-frequency communication. Breaching may serve the same purpose under certain conditions.

Some feel that lobtailing and finning (the slapping of water with fins) may be used by humpbacks and others to stun or herd prey. Fluke wriggling may serve as a threat, and curiosity is shown by the release of a single, large bubble underwater. It is said that the most alarming signal given by dolphins is silently sinking down into the water. Over two thousand different whistles, with at least thirty linked to various messages, have been distinguished in dolphins, and for a while there were many popular claims that dolphins were capable of imitating human speech. At best, the sounds are garbled, but some faintly resemble human words.

Paradoxically, the largest animals in the world feed on some of the smallest. The gigantic baleen whales feed on plankton and krill (a catch-all term for swarms of various crustaceans and molluscs). Up to 90 percent of the zooplankton in the Antarctic consists of the **red prawn** (*Euphasia superba*), often referred to as "the" krill, while the primary animal in Arctic zooplankton is the **sea butterfly** (*Clione* spp.). The Southern Hemisphere whales are usually larger than their northern counterparts because of the greater food supply of Antarctic waters. Many of the largest whales are anatomically limited to such small prey because of their narrow throats. On occasion some, such as the humpback, have been known to choke to death by accidentally swallowing a bird such as a cormorant that happened to be feeding in the same location.

All baleen whales use the whalebone plates for which they are named in feeding. Baleen plates are not really bone, but keratin, like human fingernails and hair. The plates range in size from the eight inch slabs in the minke whale to the fifteen foot plates of the bowhead. Those of the blue whale may reach 2½ feet in width. The blue, fin, humpback, and sei whales may all grow 400 plates or more on each side of the jaw. The fin whale holds the record, with up to 480 on each side. At birth, the baleen plates are soft and pliable. They harden and become functional at about the time of weaning but grow throughout life.

Baleen whales use two general methods of feeding. The first is known as "gulping." In this method, the whale takes in a large mouthful of water and then forces it out through the baleen plates, which act as a strainer. After the excess water is forced out, the whale swallows the food trapped by the plates. Whales using this method of feeding include the blue, Bryde's, fin, humpback, minke, and sei whales. Some ingenious styles have developed in gulping. For example, humpbacks often obtain their mouthful of krill by "herding" the prey. This is done by flicking the tail and therefore krill towards its mouth or by swimming in a circle while blowing bubbles, followed by rising vertically and engulfing whatever occupies the center of the entrapping curtains of bubbles. At times, two or more humpbacks will work together to form a bubble net measuring 100 feet in diameter. These whales may even vary the bubble size, producing small bubbles for small prey and larger bubbles for bigger prey. Humpbacks also perform what is known as "flick-feeding," in splashing water forward over their heads with the flukes. This startles the krill directly ahead, causing them to pause just long enough for the whale to take in a mouthful. At other times, the long flippers herd fish towards the mouth. An 80 ton blue whale can engulf up to 1,000 tons of water, with a single mouthful giving it the appearance of a gigantic tadpole. A whale this size requires about 3,000,000 kilo calories per day. This in turn is equal to four tons of krill or about 40,000,000 red prawns. Larger blue whales can eat up to eight tons of food daily, although we should remember that they fast for seven to eight months of the year.

Some of the rorquals that "gulp," such as the Bryde's, fin, and sei whales, feed primarily on fish such as sardines and small **herrings** (*Osmerus* spp.), but also on krill. Bryde's whale has fairly large gaps between its baleen plates so that many of the smaller organisms slip through, and it must depend more on fish and squid. Sei whales will herd fish prey. They circle schools of herring while swimming sideways. Their highly reflective, light underside causes the school of fish to pack more tightly together, whereupon the whales charge through and obtain more fish per mouthful. One Bryde's whale was found with fourteen large penguins in its stomach. The minke whale (and some toothed whales) is sometimes found with rocks in its "first stomach" or forestomach (all cetaceans have at least three stomach chambers, and a few toothed whales have more). The rocks are thought to aid in digestion by helping to break down chunks of food, but some experts feel that they are only accidentally swallowed, perhaps while occasionally feeding on the bottom. Sperm whales have been known to swallow rocks, sand, clams, sponges, and even a boot.

The second common method of feeding is known as "skimming" or "sieving" and is more efficient than gulping. Sieving whales travel with their mouths open while the water is strained by the longer baleen plates found in these species. Such large plates often produce a clacking sound when the whale is feeding. After having collected a substantial amount, the mouth is shut and the food is swallowed. This method is used by the bowhead, fin, gray, right, and sei whales (fin and sei whales use both methods). The really long baleen plates found in the right whales must bend back to let the mouth close.

A third feeding method known as "grubbing" has also been described. Gray whales use tongue movements to suck in water and then force it out through the straining baleen. These whales sometimes feed by sucking up amphipods and other crustaceans from the ocean bottom. When using such methods in Arctic waters, they stir up large amounts of mud along with crustaceans, worms, and gastropods. These are then devoured at the surface by many species of sea birds. The 10,000 gray whales in the Bering Sea can supply such food to hundreds of thousands of birds in this manner. This method of feeding also reveals that gray whales are right- or left-lipped, as witnessed by worn baleen plates and the absence of barnacles on the preferred feeding side.

Gray whales make one of the longest known migrations of any mammal, up to 12,000 miles round trip, in order to feed on the rich, polar krill blooms found during the spring. The longest known mammalian migration is carried out by humpbacks, one group of which is known to migrate from the North Pacific to the Indian Ocean and back, a total distance of 20,000 miles. Such migratory whales spend the winter breeding and birthing in the tropics. The food supply during this time is quite scarce, forcing the whales to fast for as much as eight months of the year. The gray whales and others are known to do some feeding even though they can make do with the large amounts of blubber put on during springtime feasting when the krill is abundant. A pregnant female gray whale can lose eight tons after leaving the Arctic. Much of this is from fasting, but a good share is due to the delivery of a 1.5 ton baby and subsequent milk production.

The smaller toothed whales feed on larger prey than do their huge cousins. Most prey on various species of fish and squid. Very few species of whale are capable of swallowing a human whole. One species that is quite able to do so is the sperm whale, but despite nineteenth-century whaling yarns, some historians doubt that it happened very often or at all.

Sperm whales dive to ten thousand feet in search of giant squid and other prey, including octopus and occasional sharks. One sperm whale had almost 30,000 squid beaks in its stomach, while another was found to have swallowed whole a ten-foot shark. The contrast between a sperm whale's white gums and purple tongue may act as a lure to squid near the surface, but in the inky abyss it would be invisible. Some sperm whales carry bioluminescent organisms on their teeth, which may attract prey such as squid even in deep darkness. Large "sucker marks" up to eight inches in diameter, from what must be huge giant squid, scar the skins of some sperm whales. When consumed, a squid's beak may possibly act as an irritant to stimulate the production of ambergris from the sperm whale's stomach and intestines (which can reach lengths of 500 feet). Large chunks of ambergris weighing up to 1,003 pounds have been recovered. It is thought that a special bacterium may be responsible for its production. In the past, it was an important perfume fixative, and in times gone by it was literally worth its weight in gold (up to $50 per ounce). Sperm whales are known to gather near sand banks during new and full moons. This is currently believed to be related to squid behavior.

At one time, killer whales were alleged to be the most ferocious and fearsome of all animals on Earth. They can grow to lengths of 30 feet, but 20 feet seems to be the average for both males and females; males are not generally twice as long as females, as was once thought. The male's triangular dorsal fin reaches six feet high; females can be distinguished by their smaller, sickle-shaped dorsal fins. Despite their gruesome reputations, killer whales (actually large dolphins) have been thought to cause at most two human deaths. Off British Columbia in 1956, one reputedly turned over a boat carrying a lumberjack who had earlier been sliding logs down at the killers, injuring one in the process. The man was never seen again. Another unconfirmed report can be found of a fisherman killed off Baja California in 1977. There is a recorded attack on a surfer off California in 1972 who had his leg bitten and released. There are no "proven" fatalities.

Killer whales are the largest predators of warm-blooded prey. A four-ton killer whale eats about 5 percent of its body weight per day, and a blue whale about 1%. In contrast, a one-ounce shrew eats up to 150 percent its body weight daily. In another comparison, a four-pound rabbit burns 22 kilo-calories/pound/day, while a large whale burns about one kilo-calorie/pound/day. Killer whales have been known to break apart ice floes three feet thick or to tip them to one side to force penguins or basking seals into the water for easy capture. (Belugas, narwhales, and bowheads also use their heads to break up ice in order to keep open breathing holes or "savssats" where they are known to congregate.) These same techniques were used on early explorers in the Antarctic, but some biologists believe the whales mistook the men for large penguins (emperor penguins can grow to almost four feet high and 100 pounds in weight). Should the killer whales fail to break up a floe to get at their prey, they will sometimes rush it in unison. At the last moment, they will dive while flipping their flukes to form a wave that will occasionally wash the victim off the ice into the water.

Killer whales can stun and even kill sea lions by smacking them with their powerful tail flukes. The resulting blow has been known to send sea lions and porpoises weighing several hundred pounds 20–30 feet into the air. They have been known to toss sea lions and other prey back and forth to each other like a ball in play. This may actually represent a training session for younger whales. Orcas are also known to chase seals right up onto the beach, from which they must wriggle themselves back into the water.

Orca packs occasionally round up prey cooperatively for easier feeding. In one such episode involving dolphins, several killers rushed in to make kills or cripple the dolphins by biting across their tails. One often quoted story has a killer whale being found with thirteen porpoises and fourteen seals in its stomach. However, the whale actually had only remnants such as bone fragments and teeth from this many animals, not whole bodies.

Killer whales are the only cetaceans to feed regularly on warm-blooded prey. The pygmy killer whale, false killer whale, and the **melon-headed whale** (*Peponocephala electra*) are known occasionally to kill and feed on dolphins. Killer whales have captured ducks at the water's surface and even have jumped into the air to catch them as they attempted to fly away. They have been reported to attack other whales only to consume the lips, skin, or tongue, leaving the rest intact, but this is rare behavior. During such attacks on larger whales, it is not uncommon for one of the killer whales to cover over the prey's blowhole in an attempt to suffocate it. Their teeth are designed to interlock giving them a good grip on slippery prey such as squid and fish (including sharks), which comprise about seventy percent of their diet. Other mammals, including sperm whales and other killer whales by some reports, form only a small part of their prey. Interestingly, dolphins and other whales have been filmed swimming within packs of killer whales. Some experts feel that the killers only attack other cetaceans when the prey is weak, wounded, or giving birth.

Each year thousands of dolphins are killed by the tuna industry. In 1966, there were 250,000 dolphin deaths, but this number fell to 17,000 in 1979 as a result of better methods and fewer dolphins. The deaths result when the dolphins drown when tangled in the tuna nets. Certain species of dolphin, including spotted dolphins, spinning dolphins, common dolphins, and about five other species to a lesser degree, have the curious behavior of swimming just above large schools of tuna. The tuna fishermen locate the tuna by spotting the dolphins. They then proceed to encircle the pod with their large purse seine nets to trap the tuna, but they invariably entangle and drown many of the dolphins. It is not known why the tuna and dolphins associate together, but they may do so to feed on the same prey.

Some of the cetaceans possess unusual feeding characteristics. As mentioned, the Amazon river dolphin has binocular vision only above its head since large cheek pads block it below. Thus it often feeds on the bottoms of rivers while turned upside down. The "sightless" susu hunts fish in as little as eight inches of water by swimming on its side. The Cameroon dolphin feeds on fish and is not a strict vegetarian as was first thought. In fact, there are no vegetarian cetaceans, although the Chinese lake dolphin is reported to feed on a few eel-like herbs. The gray whale, which may purposely eat kelp at times, is the only baleen whale to eat any vegetation, although many experts think it only mouths kelp in play or ingests it incidentally. Grays have been observed to defecate kelp floats. The finless porpoise has been reported with rice, grain, and other vegetation in its stomach. The boutu makes a habit of feeding on the deadly piranha and is actually known to masticate some of its food (armored fish) before swallowing; all other cetaceans swallow their prey whole. Belugas prefer bottom-living fish, such as halibut and flounder, but will also gorge themselves on jellyfish. Bottlenose dolphins off Georgia have been filmed forcing fish ashore with small waves. They then swim partially onto the beach to eat them before sliding back into the water to repeat the performance. Bottlenose and common dolphins are known to catch bonito and flying fish in mid-air. Dusky and common dolphins cooperatively herd fish for efficient feeding.

Some cetaceans live out their lives in solitude, but most are fairly gregarious and roam about in groups known variously as pods, gams, herds, packs, or schools. Pods may have from five members to as many as 1,500 in white-sided dolphins, 2,000 in narwhales, 2,500 in spotted dolphins, 3,000 in **striped dolphins** (*Stenella coeruleoalba*), 4,000 in sperm whales, 10,000 in belugas, and a staggering 300,000 in the common dolphin. Humpbacks tend to be solitary, and they (and other baleen whales) probably group together for feeding more often than for social reasons. Generally, females are dominant to males. Single-sex pods are common in some species, including belugas, **common porpoises** (*Phocoena phocoena*), and sperm whales. In sperm whales, the old bulls are known to travel singly. They are the only species of large whale known to be polygamous. A dominant bull will often lead a harem of cows and their calves, numbering up to thirty total. At times, two large males will fight each other over the possession of such harems. In these duels, the jaws are used to tear at each other's flesh or occasionally interlocked leading to broken teeth and sometimes broken jaws. Fin, gray, killer, blue, sei, and other baleen whales may mate for life and live in family groups without promiscuity. In some social whales, the dominant members have been reported to mark subordinate and group members with special secretions, but this has not been substantiated.

Many female whales become sexually mature after four or even three years, which has aided some species in survival against the onslaught of the whaling industry. In smaller cetaceans sexual maturity occurs between 5 and 16 years of age. Whales court by singing, clicking, grunting, moaning, rolling, caressing, biting, and jumping together. They may court for hours on end, although actual mating requires only a few seconds. Male dolphins often display their swimming skills during courtship. They will even charge the female, turning away only at the last possible moment. **Shortfin pilot whale** (*Globicephala sieboldi*) and **longfin pilot whale** (*Globicephala melaena*) pairs butt in head-on collisions during courtship. Courting humpback whales caress and slap each other with their long flippers. The resounding claps can be heard miles away. They will even use their long appendages to embrace each other. Another ploy used by males during courtship is to expose their genitals to the females. The penis of a blue whale can be over nine feet long, while the right whale's testes—largest in the world of mammals—weigh a ton a piece. The markings of some, such as the Baird's dolphin, become more pronounced during the mating season. These markings, like those in the killer whale, seem to point out the genitals in the males and the teats in the females (the latter probably for the benefit of the young). Male bottlenose dolphins are known to masturbate by rubbing up against other males or even other animals such as sharks and sea turtles.

Copulation usually takes 5–20 seconds. Belugas, harbor porpoises, humpbacks, killer whales, pilot whales, right whales, rorquals, and sperm whales mate ventral-ventrally (belly-to-belly) while perpendicular to the water's surface. Others do it side by side, with the male twisting his tail end underneath the female. Sexual encounters among some of the large whales, including the gray whale, call for the presence of two males. The first one handles the actual mating while the second lies across the mating couple during intercourse to keep them in proper water position.

Females can employ several maneuvers to discourage mating. They can hold the lower part of their body vertically out of the water or can roll onto their back. This is not completely effective since as soon as she rights herself for a breath of air a waiting male can swim under her and quickly mate. In fact it is not unknown for several males to wait nearby for up to 25 minutes to make a mad dash with much pushing and shoving in order to get into position first. Females can circumvent even this problem by carrying out these maneuvers in water that is too shallow to let the males get below her.

The longest average gestation period for a cetacean is 17 months in Baird's beaked whale. The sperm whale averages 16 months, and the killer whale about 15. The blue whale, largest of all animals, has a gestation of only about 11 months. Suckling periods may last a year or so, with reported instances of whales up to thirteen years old nursing when the opportunity arises. With gestation and nursing lasting two or more years, the females can only give birth once every three to four years. In smaller species, birthing intervals range from one to four years, and even some large whales have been known to give birth to calves only a year apart. The incidence of twins is about one in every hundred births (about the same as for humans), and there are recorded instances of triplet and quadruplet fetuses in female fin whales.

The calves are born underwater and are generally not helped to the surface by either the mother or an "aunt," another female, possibly an older sister, that assists the mother in delivery and sometimes in raising the young. It has been found that aiding a newborn to the surface is generally reserved for those in trouble or stillborn. Most healthy cetacean young get and need no help to the surface.

Cetaceans and some bats are among the few mammals born tail first. Dolphins and porpoises have been filmed giving birth in this way, and there are records of baleen whales and killer whales giving birth both head first and tail first. The only two beluga births observed were head first.

The calf is stimulated to take its first breath by the exposure of its skin to the surface air. Humpbacks and others will purposely break the cord prior to expulsion of the placenta by turning somersaults in the air. Should the umbilical cord not break away, the newborn is in danger of pulling out the placenta, which can cause it to sink and drown. Female cetaceans do not eat their placentas. Like many other animal females, whales will usually abandon deformed or sickly offspring, although females have been seen to remain with their infants for some time after death.

Cetacean breasts are located on each side of the genital orifice. The whale calf need not suck on the teat to obtain milk. In fact, it cannot, for cetaceans lack lips with which to suck. Fortunately, the glands contract on their own, automatically squirting milk into the calf's mouth. Baleen whales have a middle gap between their baleen plates in which the mother's teat fits. Their tongues can be partially rolled, like straws to direct the flow of milk. Baby sperm whales and some other toothed whales are known to take the teat into the side of their mouth. During nursing, the mothers will often roll onto their side so that the young can breathe while nursing. Female gray whales often "lean" on another female or on the bottom in shallow water to counteract the calf's push and prevent rolling.

Whale milk is very rich, containing 40–50 percent fat rather than the 4 percent of cow milk and 3 percent of human milk. It therefore mixes very poorly with seawater. Its higher percentage of protein, 12 percent, about twice that of land mammals, is essential to support an incredible growth rate. A blue whale grows from a 0.000035 ounce egg to about 29 tons in two years, a 30 billion-fold increase. This is the most rapid growth shown by any living organism. A newborn 25 foot, three ton blue whale can grow at a rate of almost ten pounds per hour with an increase of 1.5–2 inches in length per day. This is supported by up to 150 gallons of milk daily in about 40 feedings. This equals about 375,000 kilocalories per day. It is no wonder the females can lose up to 50 tons of weight through birthing and nursing.

Mother whales generally watch over their youngsters carefully and won't allow them to stray very far. Frisky right whale calves frolic about their mothers, occasionally giving her playful blows. When her patience finally runs thin, she will give the calf a firm "hug" with her flipper causing the calf to settle down, at least for awhile. Bottlenose dolphins discipline calves by holding them underwater until they become frantic for air. This is also thought to be a method of teaching young to increase their diving time.

172

If a dolphin calf ignores its mother's call, she can retrieve it by swimming on her back underneath the calf and then literally lifting it out of the water to take it where she wants. Finless porpoise females actually carry their young on their backs. The dorsal area used for this is covered with warty tubercles to help hold the calf in place. Larger whale females, such as the gray whale, can exploit the eddy currents set up by their huge bodies to piggy-back their children. Female gray whales are also known to play with their young by swimming underneath and spouting large volumes of bubbles that cause the baby to spin and whirl. Female sperm whales are known to transport their young, at times, in their mouths. Many young whales keep up with the pod by riding bow waves or the currents set up by adults. Female dolphins occasionally bite heads off fish before presenting them to the young to prevent injury from skull bones. "Babysitting" has been observed in some, including bottlenose dolphins.

Like most other animals (with the prominent exception of humans), cetaceans die soon after they lose their reproductive abilities. Cetaceans have few natural enemies. They are preyed on by killer whales and humans. Occasionally rorquals, such as blue, sei, and minke whales, are found with the beaks of swordfish or marlin buried in their flesh, but these wounds are not fatal. A few cetaceans may be lost to giant squids and large sharks. A lone dolphin stands less than an even chance against a large shark, although a pod of dolphins can easily dispatch a menacing shark. However, dolphins and sharks are not natural enemies and are known to tolerate each other's presence peacefully. The **pygmy sperm whale** (*Kogia breviceps*) resembles a shark anteriorly with its underslung jaw, sharp, pointed teeth, triangular head, and false gills on its neck. It reportedly basks frequently at the surface and if frightened will produce a cloud of reddish-brown feces before quickly diving to safety.

When alarmed, some spotted dolphins "play dead" by sinking straight down tailfirst. A gam of sperm whales will sometimes assume a defensive formation known as a "Marguerite Flower" by forming a circle, raise their flukes vertically, and then perform lobtailing en masse. Calves and/or injured whales are located in the middle of the circle and are effectively protected against marauding killer whales. Whalers knew that this might occur if they first killed a group's leader. The others in the group would then remain in position while the whalers easily picked them off, one by one.

Since many cetaceans, such as the bottle-nosed whale, rarely desert an injured companion, whalers could often destroy many whales at once. (When an animal aids or rescues a companion, it is known as epimeletic behavior.) The whalers also knew that many whale species exhibit chivalrous behavior. Male blue, gray, and humpback whales quickly come to the aid of injured or troubled females; females (with the exception of humpbacks) generally abandon males in similar circumstances. Thus whalers would first kill females in order to increase their catch. Despite romanticized sea stories, most sperm whales were dispatched with little trouble or effort.

Some whales do seem to keep an eye out for danger. Some experts believe spyhopping in gray and sperm whales is a way to watch for approaching trouble. Resting gams of sperm whales have been observed to post apparent sentries to watch for approaching danger. These are usually the less dominant animals, as in many other species, so that the dominant members are preserved for propagation of the species. Some cetaceans are known to jam the ultrasonic detection devices used by modern whalers by blowing and fleeing behind a screen of bubbles. When gray whales are approached by a pack of killer whales, they can, some say, occasionally become so terrified that they either roll over on their backs to await their doom or charge blindly toward the beach and strand themselves on shore. Although such stories are probably bunk, gray whales are known to take cover in kelp beds,

173

where the air-filled bladders of the plants impede the sonar of the killer whales. During these times, the gray whales exhale underwater to conceal their tell-tale spouts.

Whaling is a classic example of human greed and stupidity in the exploitation of a limited resource. Whale products have now all been replaced with other materials, but at one time or another they included candles, cosmetics, crayons, fertilizer, food for pets, gelatin, glue, hormones, ink, lamp oil, lard, lubricating oil, margarine, paints, perfume fixatives, shoe polish, soap, varnishes, vitamins, and more. At one time, the baleen plates were the most valuable parts, being used for busks in women's corsets, hoops in crinoline skirts, umbrella ribbing, and even whips for schoolmasters. Disobedient children would get a "whaling" with these instruments.

The right whale was so named because in early whaling days it was the "right" whale to harpoon. This was because it was so slow, had a lot of whalebone, and didn't sink once killed, thanks to its large amount of blubber which could be converted to large quantities of oil. Since sperm whales also didn't sink when killed, they were also a favorite target, with the added bonus of their spermaceti and possibly some ambergris. A sperm whale may hold over 500 gallons of spermaceti in its head.

Whale products have been used in some interesting ways. Fin whale oil was once used as an insecticide to keep locusts out of Japanese rice fields. The blubber from Cuvier's and Baird's beaked whales, used as a laxative by the natives of Kamchatka, was sometimes prepared and served to offensive guests. Narwhale skin is consumed by Eskimos for its high concentration of Vitamin C.

The human-cetacean relationship through the centuries has been an ambivalent one. Even though humans have been their worst enemy, cetaceans have on many occasions aided and even saved the lives of humans. Around Burma, the Irrawaddy dolphin is known to herd fish into the nets of local fishermen, who alert the dolphins to the presence of the fish by beating on the water. The dolphins come to feed and at the same time force the fish into the nets of the natives. The boutu does the same thing in South America, and the Chinese river dolphin does so in China. Killer whales have betrayed their relatives by leading whalers to baleen whales and then preventing the latter from sounding to escape. On some occasions, the killers even drove pods of whales to the shallow bays of the whalers' home ports. The killers were repaid with some of the leftover scraps. There is also the well known account of Pelorous Jack, the famous **Risso's dolphin** (*Grampus griseus*) that guided ships past treacherous New Zealand reefs for almost 25 years. Finally dolphins have been known to rescue people in the ocean since the times of the ancients. During World War II, six downed American airmen in a raft were pushed to an island by a dolphin. Even within the past 20 years, bathers off Florida have been rescued from dangerous undertows by wild dolphins. These acts may stem from their well known epimeletic behavior as mentioned. In one recorded case, a dolphin even "helped" a shark it had killed by lifting it to the surface to "breathe" for over a week's time. Dolphins have been observed to nudge along various objects on the water's surface including boxes, dead and live fish, turtles, balls, and mattresses. Possibly the saving of humans stems from this natural pushing behavior.

12

ORDER CARNIVORA
(238 SPECIES)

Carnivores are generally defined as flesh-eating mammals. Many mammals outside the order Carnivora are certainly carnivorous, and some members of the order, such as the giant panda, are primarily herbivorous. But as a rule, this order contains the ultimate mammalian predators, which live by feeding on other animals. The smaller carnivores are important pest-controllers, keeping many rodent species' and other pests' numbers in check. The larger carnivores keep the populations of herbivorous mammals healthy by weeding out excess numbers, the old, and the sick. In so doing, they help to prevent needless suffering and even mass starvation in many prey species.

Family Felidae
(cats—37 species)

The cats range in size from the three pound **rusty-spotted cat** (*Felis rubiginosus*) to the **Siberian** or **Manchurian tiger** (*Panthera tigris altaica*), whose adult males average almost 600 pounds. The largest individual felid on record is an 857 pound **Indian tiger** (*Panthera tigris tigris*).

The classification of felids is, as usual, controversial. One simple method is to divide them into two

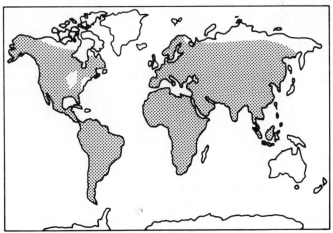

Family Felidae

Jaguar (*Panthera onca*)

Tiger (*Panthera tigris*)

major groups (genera), the **big cats** (*Panthera* spp.) and the **small cats** (*Felis* spp.). As a rule, the big cats have an incompletely ossified hyoid bone, enabling them to roar, have circular pupils, and crouch with their front feet extended. The small cats have elliptical pupils, crouch with their front feet tucked underneath, and have a completely ossified hyoid bone so that they cannot roar. They can purr during both inhalation and exhalation, while the big cats can purr only during exhalation. The many exceptions include the **snow leopard** (*Panthera uncia*), which doesn't roar and can purr during inhalation and exhalation; and the **black-footed cat** (*Felis nigripes*), which roars like a tiger except one octave higher. Some of the small cats, such as the **clouded leopard** (*Felis nebulosa*) and the **ocelot** (*Felis pardalis*), crouch with their front legs extended, while others such as the **jaguarundi** (*Felis yagouaroundi*), the **lynxes** (*Felis* spp.), the **pallas cat** (*Felis manul*), and the **puma** (*Felis concolor*) have spherical rather than elliptical pupils. The **cheetah** (*Acinonyx jubatus*) is accorded its own genus because of its many dog-like characteristics.

Felid fur coloring is quite variable, and many species are adorned with beautifully striped or spotted coats. This has in many cases led to their demise at the hands of furriers. The **lion** (*Panthera leo*) is the only cat to exhibit sexual dimorphism. The males carry thick manes for protection against the bites and blows of rivals. The mane generally appears at about 18 months of age, although some males never develop one. (In some lion subspecies, the manes are almost nonexistent in all the males.) On rare occasions, a female lion will develop a small mane. In all other felids, the sexes are identical (except for size in a few). When there is a size discrepancy, it is usually the male that grows larger.

It is interesting to note that the stripes of tigers are unique to the individual, much as fingerprints are in humans. Also, the two sides of the same animal are not symmetrically identical; and the stripe pattern may change somewhat as the tiger ages, often by reducing the number of stripes. Each **leopard** (*Panthera pardus*) also possesses a unique spot pattern. The spotted coats of the **jaguar** (*Panthera onca*) can be distinguished from those of the leopard by the fact that at least a few of the jaguar's rosettes possess a central black spot. It is more difficult to differentiate the **margay** (*Felis wiedii*) from the ocelot. The ocelot's spots are generally aligned more in bands while the margay's are more randomly placed.

Probably the best known example of melanism in animals is that of the **black panther**, a melanistic leopard. This phase is more common in Asia but rare in Africa. Actually, the terms leopard and panther are equivalent and interchangeable in all circumstances. If one looks carefully, he can see the normal spotted pattern in the black variety. Jaguars, too, will frequently and pumas will occasionally appear in black form, while tigers more often will tend to a "pseudo-albinism." Lions rarely exhibit either melanism or albinism.

Cats keep their coats clean by licking them. This both removes dirt and debris and rids the coat of loose hairs. Their saliva is thought to kill bacteria and fungus on the coat. Licking also helps to cement social bonds. Surprisingly, jaguars are known to lie quietly in the water to allow small fish to pick their coats clean of ectoparasites.

Unusual genetic characteristics can be found in the more than 30 breeds of **domestic cat** (*Felis catus*). Some breeds, such as the **calico** and **tortoiseshell Persians**, are always female since the characteristic color pattern is sex-linked. Male **blue-cream** and tortoiseshell Persians are always sterile, and pink-eyed albino cats and blue-eyed **white Persians** are usually deaf. Should the cat have one blue eye and one brown eye, it will probably be deaf only on the blue-eyed side. It may be of interest to note that one "cat study" showed female **Siamese** cats to be the most "neurotic." A Siamese cat's beautiful blue eyes are produced by light-scattering proteins in the iris rather than blue pigments.

The tails of cats come in varying lengths. In fact, one breed of domestic cat, the **manx**, is tailless as a result of genetic mutation. True manxes have no tail and are called rumpies; those adorned with small vestigial tails are known as stumpies. The short tails of the **bobcat** (*Felis rufus*) and the **Canadian lynx** (*Felis lynx*) can be used to differentiate the two species. The bobcat's is black on top with white underneath; the lynx's has an all black tip. The lynx generally has longer ear tufts and "sideburns" as well. The snow leopard will wrap its long, bushy tail up over its nose to keep warm during storms and cold spells. The long, vertically flattened tail of the cheetah increases its effectiveness as a rudder and balance organ, a function of most cats' tails. Lions are the only cats to have a tuft of fur at the end of the tail. Lions use their tails as "fly swatters," to signal emotion, to stroke each other, to make their location known, to balance, and even as a toy for the young to help sharpen their hunting skills early. The tuft occasionally conceals a small, thorn-like spine at the tip. This is actually an exposed portion of keratin not connected to the distal tail vertebrae. The white underside at the tip of a leopard's tail is thought to be used by the female to signal her young to follow her. Baby pumas obtain their first lessons in hunting skills by stalking and pouncing on the sinuously waving, black-tipped tail of their mother. Many felids twitch their tails while hunting, a behavior that probably serves to release nervous energy.

The feet of some cats have several unusual features. Almost all can retract their claws into protective sheaths. The only exceptions are the **flat-headed cat** (*Felis planiceps*), **fishing cat** (*Felis viverrinus*), and the cheetah. Cheetahs can retract their claws fully for about the first four months of life; later the sheaths degenerate and they lose the ability to retract them fully. Retraction is valuable in that it keeps the claws sharp by preventing them from scraping on the ground as the animal travels about; it also enables the cat to approach prey more quietly. The claws are kept shortened by frequent scratching upon hard surfaces to remove the outer "shells." In the cheetah, the non-retractile claws serve as "cleats" to increase traction and maneuverability when pursuing prey.

The Canadian lynx, Pallas cat, snow leopard, black-footed cat, **desert cat** (*Felis bieti*), and **sand cat** (*Felis margarita*) all have haired soles. In the first three species, this serves to protect the feet from the cold and increases their surface area in the manner of snowshoes to give added traction in the snow. In the second three species, the hair protects the feet from excess heat and gives added traction in sand. Large feet also increase the swimming ability of many.

Felids often hunt at night and possess eyes adapted for night vision. They are equipped with a special structure, known as the tapetum lucidum, designed to shift wavelengths to more sensitive ranges of color and to reflect light in such a way that it literally bounces back through the retina a second time, enabling cats to see in very dim light. In fact, cats require only a sixth as much light as humans need; no cat can see in total darkness. The tapetum is also what is responsible for their "glowing eyes" when a light is shined on them. This results from a reflecting back of the light rays that penetrate their eyes. A few species are known to have color vision.

Cats use their whiskers as feelers in the dark. With them, a cat can tell instantly whether it will be able to fit through a small opening. The whiskers of many cats were once ground up and used as poison by natives, but this property is based on superstition only. The growth pattern of felid whiskers are unique to the individual and can be used for identification.

Sight and hearing are the senses most used by felids. Some, such as the tiger, hunt primarily by sound since smell and vision are poor in comparison. The sand cat has the largest middle ear cavity in proportion to size of all cats. Its hearing is so sensitive that it can detect the faint scratchings of the small rodents upon which it feeds. The large ears of the **serval** (*Felis serval*) are thought to be sensitive enough to hear the underground diggings of the mole-rats which are one of its major food sources. The ears of lynxes and bobcats are tufted in order to improve their sound-collecting properties. Their hearing acuity and ability to localize sound can be significantly decreased experimentally by cutting off these tufts. It is thought that many of the smaller felids, including domestic cats, are capable of hearing ultrasonic sounds produced by mice and other rodents, which explains why cats proverbially crouch at mouse holes. Their high-frequency hearing probably also explains why cats are more receptive to the voices of women than to those of men.

Felid tongues are quite rough due to the presence of many coarse papillae. The lion's tongue is so rough that it can reportedly take some skin right off the back of a human's hand. This is a great asset in self-cleaning to remove parasites and dead hair, in scouring the meat from the bones of its prey, in debriding wounds to promote healing and, for females, in splitting the amniotic sacs enclosing the newborn kittens. When the tongue is used for lapping fluids, it is curved chinwards rather than nosewards as most people generally assume.

One of the main features of carnivores is their teeth. Most have fang-like canines for tearing into prey. Those of the clouded leopard are about two inches long. Felids and other carnivores slice meat away from the carcass with their tightly fitting carnassial or side teeth, the last premolars in the upper jaw and the first molars in the lower.

The big cats generally roar, while the small cats purr and meow. The big cats can purr, but only during exhalation. The mechanism of purring has never been fully elucidated; some researchers feel it results from the circulatory rather than the respiratory system being produced by aortic vibrations. The cheetah, domestic cat, puma, snow leopard, and others are known to purr when content. This is sometimes accompanied by a forward protrusion of the ears and whiskers. The purr uttered by some cats when sick or hurt is generally a coarser type of purring. Purring may also serve to signal feeding in kittens. Other animals known to purr include bear cubs, civets, genets, female hyenas, and mongooses.

Roaring is best performed by the lion, whose roars can be heard up to five miles away. Cats can, of course, produce many other types of sounds. The puma has long been known for its blood-curdling screams which, though apparently rare, are responsible for at least some of the many names given to this animal—American lion, brown tiger, catamount, cougar, deer killer, deer tiger, king cat, long-tailed cat, mountain devil, mountain lion, mountain screamer, painted devil, painter, pampas lion, panther, purple panther, red tiger, screamer, silver lion, sneak cat, and others. Pumas also produce a high whistling sound similar to certain bird calls and thought to be used for intraspecific communication. The cheetah produces chirps that were long mistaken for those of nearby birds. Tigers occasionally produce a bell-like chatter known as titting, and the Bengal tiger is reported to imitate the bark of deer species upon which it preys, although some experts relegate this to myth. Lions, leopards, and others are known to "cough" when angered or about to charge. Anger is also frequently accompanied by laying the ears back flat against the head. The arched-back posture of fear is familiar to almost everyone.

Most people associate balance, speed, stealth, and strength with cats. Most are aware that falling cats usually (but not always) land on their feet. In stalking, cats place their hind feet in the tracks of their forefeet. This ensures a quiet stalk since it avoids twig snapping or leaf crackling by the rear feet. The cat need pay attention only to the placement of its forefeet.

The clouded leopard is the most nimble of all cats when in the treetops, where it catches birds, monkeys, and squirrels. It can swing suspended by a single hind paw (as can margays) and can jump directly onto its prey's back from a tree limb. In fact, only a few cats, such as the clouded leopard and the margay, will jump from trees directly onto their prey. Others that ambush from trees generally jump to the ground first and then attack. Pumas and other cats will spread their legs to help slow their fall when jumping from heights. The puma has jumped from as high as 80 feet without apparent harm.

Lions can run up to 40 miles per hour and have been known to make leaps 12 feet high and some 36 feet long. They can break the neck of a zebra with a single blow from their powerful forepaws, and they have a front leg pulling power equal to that of ten men. Although most of the lion's prey can run faster, the cat can accelerate faster; thus, if it starts close enough, it can catch the prey. Tigers have been recorded to make vertical jumps of 18 feet and have pulled adult gaurs in excess of 30 feet which thirteen men could not even budge. One leaped 13 feet high while carrying a carcass weighing 150 pounds. The puma is the champion jumper of cats. It has been known to jump vertically almost 20 feet and horizontally 40 feet. It is rivaled by the snow leopard, which on occasion may be able to leap 45 feet and has even been seen to turn somersaults in mid-air. In some instances, pumas have been known to carry prey distances of 20 feet in their initial attacking leap. They can bring down animals that outweigh them more than eight-fold. One of the most shy and wary of cats is the leopard. To eat its meal in peace, the leopard will often drag a carcass, which can outweigh it two-fold, up a tree.

The speed and agility of the small cats is aptly illustrated by the **desert lynx** or **caracal** (*Felis caracal*). After quietly stalking a flock of birds, they have been known to strike down ten birds in less than two seconds. These small, 15 pound cats can jump up to six feet in the air while swatting down the fleeing birds. The serval has jumped ten feet to snag birds.

181

The fastest of all felids, and indeed all mammals, is the cheetah. It can accelerate from zero to 45 miles per hour in two seconds flat, and it has been timed at speeds of 60 miles per hour over short distances. One reportedly covered 700 yards in 20 seconds (71.6 mph), but this is often discounted because of the excess distance. In most chases, cheetahs average more like 40 mph. Because they overheat and tire quickly, they can put on tremendous bursts of speed for only several hundred yards. In fact, in a 400-yard sprint a cheetah's temperature rises to 105°F. If the cat pushed further, its temperature could rise to brain-damaging levels. After a sprint, cheetahs rest and pant for 15–20 minutes to bring their temperatures down before eating or wandering off to make another attempt (four out of five sprints end in failure). While in such weakened states, gazelles that have just escaped have been known to return and actually attack the exhausted cheetah with their horns. At full speed, the cheetah makes bounds of 25 feet or more. Their shoulder blades are quite mobile and serve as "shock absorbers." There are two sets of muscles that control leg motion, serving as a low-speed "first gear" and a high-speed "second gear." Occasionally, cheetahs lose their balance or trip in full sprint, producing spectacular falls.

It is said that cats dislike water, but what they really dislike is very cold water. Cats that are frequent swimmers include the fishing cat, **Geoffroy's cat** (*Felis geoffroyi*), jaguar, jaguarundi, **leopard cat** (*Felis bengalensis*), lynxes, ocelot, serval, tiger, and others. Tigers are in fact superb swimmers and will take to the water to cool off for hours on end. The jaguar will actually hunt prey in the water, as well as submerge most of its body to escape stinging insects. The fishing cat inhabits swampy areas and is adapted with webbed feet for water life. It readily dives for prey.

The cats, often referred to as the great hunters, take a wide variety of prey, ranging from insects such as grasshoppers and lady bugs to even an adult elephant, observed to be downed by two large tigers. Lions can kill rhinos three quarters grown but are also known to feed on insects. Lynxes have killed weakened moose on rare occasions.

Cats will also feed on other felids. For example, tigers have been known to attack and kill leopards, lions, and even other tigers during the breeding season. Leopards and lions have been known to eat baby cheetahs and conspecifics. Jaguars are also known to be cannibalistic at times. The puma will kill and eat smaller cats such as the lynx and bobcat, and the lynx will in turn feed on smaller felids or other lynxes it can overpower.

Lynxes also kill and eat foxes. The puma and bobcat are known to take porcupines deftly, sometimes quilling their mouths, which can lead to death by starvation. Their scat often contains quills that have passed through safely. Although leopards are known to feed on many animals ranging from rats and frogs to humans and dogs, it seems that individual leopards (and many other carnivores) tend to specialize in one or several species. In other words, one leopard may feed primarily on impala, while another feeds mostly on bushbuck, despite the ready availability of other prey. Leopards often develop a liking for dogs or—rarely—humans. Leopards can overcome lone baboons, but groups of the latter have often treed and even killed unwary leopards. It is occasionally stated that leopards have a particular liking for male antelope since male prey outnumbers females by two to one or more. This actually can be attributed to the fact that males without territory do more wandering into areas of likely ambush; territorial males and the females of their harems stay in the more desirable open territories, within and from which they can easily spot lurking predators. Man-eating has been documented in lions, tigers, leopards, and jaguars. One tiger thought to hold the record was credited with 438 human deaths in eight years. Often, the man-eater is an old, sick, or injured cat, but some apparently develop a preference for human flesh without obvious reason. Sometimes the cats attack and kill but do not eat the body. One theory of why such attacks occur holds that the animal sees the human as an intruder into its territory. This is supported by the fact that attacking cats often preface their charge with aggressive gestures such as laying back the ears, drawing back the lips, and snarling, all of which are used to warn off others of their kind but not in attacks on prey.

Most cats are not completely carnivorous, although a total vegetarian diet would lead to blindness as it would be deficient in the amino acid taurine. Humans and most other mammals can synthesize it from other proteins. The **African wild cat** (*Felis silvestris libyca*), jaguar, jaguarundi, and others will feed on fruit such as wild figs. The flat-headed cat has been known to take fruit in preference to mice. Bobcats occasionally eat grapes and the fruit of the prickly pear cactus. Lions and leopards obtain water by eating melons, and the sand cat can survive without any free drinking water, getting by on the body liquids from its prey. Lions, tigers, and domestic cats are frequently seen to eat grass. This may be because they like the taste, but some experts feel that the grass may serve to help coat sharp objects such as bone fragments in the stomach to protect the esophagus when these are regurgitated, or that the grass may help the cats rid themselves of hair balls in the stomach by stimulating emesis. Indian tigers occasionally eat soil.

Felids use many techniques in capturing their prey. Most ambush it directly. Jaguars, leopards, pumas, lynxes, and others leap from the branches of trees or sneak up as close as possible, using any available cover, and then pull down the victim after a short sprint. All felids but lions are usually solitary hunters. Cheetahs, leopards, lynxes, ocelots, pumas, **wild cats** (*Felis silvestris*), and others will occasionally hunt in pairs during bad weather or at mating time, occasionally even cooperating in the hunt. Tiger pairs are known to hunt cooperatively, but when the breeding season ends the two go their separate ways. A female lynx and her kittens will sometimes form a hunting line to flush prey.

Unlike other cats, the cheetah runs down its prey but gives up if it doesn't overtake it quickly. Cheetahs will frequently not carry through an attack if its victim stands its ground. In fact, flight acts as a stimulus to most predators and will trigger the chase response. Once a cheetah overtakes the animal, it will bring it down by tripping it with one of its forepaws.

Although the cats do not often make a swift kill, and the prey appears to struggle a lot, it is thought that there is not much suffering. The prey animals often go into shock with the first bite of the long canines. Humans who have lived through attacks state that they felt no pain or even fright during the attack itself. Many of the cats are known to snap back their victim's head, sometimes breaking the neck in the process. Many bite into the neck, and some, like the lion will suffocate their prey by gripping it over the nose and mouth.

Bobcats have lured fish to the surface by tapping the water with their paw; they flick the fish to the bank when it comes near. The jaguar (like some coyotes, domestic cats, foxes, and raccoons) will lure fish to itself by dangling its tail in the water and quickly scooping out with one paw any fish that ventures near to investigate. Some jaguars reportedly attract fish by allowing saliva to drip into the water. They will not hesitate to follow prey into the water; while there, they are known to take caimans and turtles. Leopards have taken animals through the technique known as charming. They writhe on the ground to lure in curious deer. After the latter have approached within striking distance, they jump up and attack (see canids for more on charming).

Although the cats are good hunters, their efficiency is usually overrated. For example, only 10–30 percent of cheetah attacks end in a kill. Statistics for lions vary from 5 to 30 percent. For this reason, most cats put away a lot of meat per feeding and, with the exception of the cheetah and bobcat, they are not averse to eating carrion. A 360-pound male lion can eat up to 90 pounds of meat at a sitting, after which it can easily fast for a week. Right after such a feast, the animal appears uncomfortable with movement. There are reports of large tigers consuming more than 100 pounds of food at once, although this is exceptional. Large felids can swallow chunks of meat as large as a football.

Most cats will skin or defeather their prey to some extent before eating. Ocelots remove almost all feathers before eating birds. Again, the exception is the cheetah, which is known to eat its victims hair and all. After eating their fill, most cats will store the excess for later use. Most of them, including the bobcat, puma, and tiger, will bury their leftovers and return later to feed; when the meat spoils, it is left for scavengers such as hyenas and vultures. Cougars gut prey before burying it. This is known to delay the onset of spoilage. A few tigers have been reported to relish maggot-infested, spoiled meat. Tigers will sometimes store leftovers in water to keep it from scavengers. Leopards hoist their excess into the forks of trees, where it is safe from hyenas, jackals, and—especially—lions (the practice is not observed where there are leopards but no lions). In many areas, lions actually do more scavenging than do spotted hyenas. In one study, it was shown that lions scavenged up to 80 percent of their food from the kills of spotted hyenas; on the average, they scavenge about a third of their food, finding it by watching vultures or tuning in to the laughter of spotted hyenas, which is known to signal feeding. Cheetahs eat only what they kill and dine only once on a kill; they store no leftovers. In some studies, more than 10 percent of cheetahs kills are stolen by other predators, including hyenas, wild dogs, and even vultures. After eating, cats such as the lion will use their dew claws like toothpicks to dislodge meat caught between their teeth.

Most cats engage in long hours of sleep or rest. The average lion spends about 20 hours of the day either sleeping or staring into the distance. Frequent yawns indicate not tiredness but threat; they are also used to complement panting to reduce body heat.

The lion is the only truly social cat, although tigers and cheetahs are more sociable then was once thought. Lions gather in 3–40 member groups known as prides. A pride is usually a group of related lionesses and a few unrelated (although at times related) males. The males rarely number more than three; occasionally, there are as many as seven. The male population is not stable, and new males may wander in and fight the present ones, sometimes to the death, for control of the pride. In such fights males with the thickest manes may have a slight advantage since the manes better protect them from the blows of conspecifics. A pride's males change every three to four years on average. In one study, less than 20 percent of single males had control of a pride, about 60 percent of paired males, and over 95 percent of male groups numbering three or more were head of a pride. Male groups hold prides longer and do not exhibit a dominance hierarchy. Old males that get ousted often become the prey of hyenas.

Female offspring usually stay to become part of the pride. Male offspring usually wander away after maturing to find their own. Females usually do most of the hunting, while the large males provide the pride with security. The males are generally poorer hunters due to their conspicuous manes but will sometimes work as "beaters" to drive prey into an ambush of waiting lionesses. Actually, the males do more hunting than they are usually given credit for. In one study, more than 10 percent of primary kills were made by males. After a kill, the males eat first, followed by the females and then the cubs, which explains why the young are the first to starve in times of famine. Although even the weakest male is dominant to any of the females, there have been occasions where females have defended their kills from old, weakened males.

Felids, like most mammals, lay claim to personal territory. In the case of the Siberian tiger, this can measure up to 7,500 or more square miles. Most mark their areas with feces, secretions from anal glands, scratchings, urine, or a combination of these, often right after a kill. Cat urine is quite strongly scented as a result of its concentration. Felids can concentrate their urine five times more than humans, an adaptation to rid themselves of excessive urea, a major by-product of a carnivorous diet. Many can urinate straight backwards for marking. The lynx, puma, and others are known to bury their wastes except when near the outlying boundaries of their territory, where the wastes are left exposed as "no trespassing" signs. They often defecate on rocks, stumps, piles of dirt and/or vegetation ("scrapes") they scrape together, or in small mounds of wastes to make their boundary lines more conspicuous. The wild cat marks its territory with secretions from special glands between the pads of its feet.

When a certain area of ground is common to more than one cat, the "age" or freshness of the scent becomes important. If a cat comes across a fresh mark, it will frequently head in another direction to avoid confrontation or competition. Some cats, such as the fishing cat and the leopard cat, eliminate most of their wastes in the water, possibly to help keep their presence a secret from other predators. Bobcats, cheetahs, leopards, lynxes, pumas, tigers, wild cats, and others will also mark out their areas with prominant tree scratchings. The familiar head rubbing exhibited by domestic cats actually serves in marking territory with scent. Lions, in addition to scent, verbally mark out their territory with loud roars audible up to five miles away. The forceful expulsion of air during a roar often kicks up an appreciable amount of dust in front of the lion's face. Bobcats, jaguars, leopards, and tigers may also use sounds to help signal ownership of territory. The careful stalking of humans without attack reported in pumas, jaguars, and tigers is probably a result of the cats "escorting" intruders from their territory. It may also reflect their intense sense of curiosity.

Many of the smaller felids take over the dens of other animals or will form their own in rock crevices, hollow trees, or dense shrubbery. A few, such as the sand cat, will occasionally dig out their own. Well-constructed nests are usually produced by the females only when about to give birth. The clouded leopard builds sleeping nests in trees, the **Spanish** or **pardel lynx** (*Felis pardina*) occasionally takes up residence in storks' nests as much as 45 feet off the ground, the black-footed cat is known to reside in termite mounds, and the wild cat has been noted to sunbathe while lying in the nests of birds of prey. Lions, too, may take to trees on occasion to avoid the greater concentration of bothersome insects nearer the ground. A few cats, including the **marbled cat** (*Felis marmorata*), **red cat** (*Felis badia*), ocelot, and margay, are arboreal and spend most of their time in trees.

Most cats live solitary lives and come together once a year for the purpose of mating. The few exceptions include the social lions as well as the jaguar and ocelot, in which an occasional pair may stay together for extended periods and even raise the cubs cooperatively.

When mating time arrives, both sexes usually become excited by the hormones released in the urine of the opposite sex. When coming into season, the females do much licking, sniffing, and rolling. Catnip is thought to give off an odor similar to mating hormones and thus triggers such premating behaviors in many cats. Males often fight at such times for dominance and/or territory; by winning, they often gain mating rights to the female. In lions, the female will sometimes mate with the loser. In some of the big cats, such as the lion and tiger, the males have been known to fight to the death. On rare occasions, the winner has even eaten from the carcass of the loser.

When the sexes finally come together, the pair often engage in a sexually stimulating premating tussle. During copulation, male felids frequently grasp the females by the scruff of the neck. Although this is generally quite innocuous, female lions and others have been accidentally killed by overzealous males. This neck bite may cause the females to become passive for mating, just as neck-holding makes young kittens easily carried. In some species, the act of mating stimulates the female to ovulate immediately, thanks to the physical effect of "barbs" on the male's penis with the female's reproductive tract. In the social lions, the dominant female has first choice of mates. Some male lions may copulate up to 50 times per day. Due to high cub mortality and the frequency of male matings, it was estimated by one researcher that a male lion must copulate 3,000 times to produce a single cub that will live through adulthood.

Felids generally produce 1–6 kittens at birth, although cheetah females have had up to nine offspring at a time. Bobcats, clouded leopards, leopards, and others often give birth in tree hollows. Snow leopards, ocelots, and pumas often use caves, and many use thickets, rocky areas, or other similar cover for birthing. The females sometimes line the den with grasses and some of their own fur. Lionesses leave the pride to give birth and may occasionally be accompanied by another lioness that acts as a midwife. Most females consume the afterbirth. It is thought that this stimulates increased milk production.

Newborn cats are generally blind for the first week or two of life. For their first few days of life, baby felids must be stimulated to urinate and defecate by their mother, who does so by licking their anus. Females generally drive the males away because of the threat they pose to the young. This is not true, however, in the case of cheetahs, golden cats, jaguars, leopards, lions, ocelots, and Siamese cats, whose males are good fathers most of the time. Male jaguars bring food to their mates with young, and adult cheetahs will sometimes prechew food for their cubs. Unmated female cougars and wild cats occasionally help train the young of another female. Young lions are not introduced to the pride until 4–6 weeks of age.

Only about a fifth of lion cubs live to maturity, and 70 percent of cheetahs die before three months of age. Maternal care in lions often leaves a lot to be desired, for the females will rarely share any food until they themselves are satiated, and it is not unknown for females to kill their cubs accidentally by rolling over on top of them. Males occasionally kill the young, and the sickly or deformed are generally destroyed regardless of numbers. Lions do raise their cubs communally, so that the females will nurse or play with any of the cubs, regardless of whose it is. It is not unknown for females to abandon their whole brood. Female felids often teach their young hunting skills by letting them stalk their tail tips and later by presenting them with live but crippled prey. The coats of young cheetahs mimic that of the very aggressive ratel or honey badger and may thus give them some protection. Female cheetahs often skin prey for their young, which are incapable of cutting through the hide on their own.

Population control in lions, tigers, and others is sometimes accomplished through cannibalism. Cannibalism (found in other mammals, birds, reptiles, amphibians, fish, and insects) is usually a result of overcrowding, a feeding frenzy, or unusual psychological stress. Cub killing is said to occur

186

routinely only between lions of different prides. It is also observed when a newly risen male takes over the pride. This is thought to serve several purposes. It increases female sexual receptivity towards him and hastens the arrival of his own offspring. It also eliminates the offspring of the previous dominant male who may later, if they are allowed to mature, prove able to overthrow the new "king." This same behavior of killing the young in a newly taken-over group is also seen in hippos, wolves, and some primates, such as the langurs.

Though the main enemy of cats is humans, other animals do cause problems. Dholes have killed and eaten tigers and leopards, and cape hunting dogs and hyenas have taken big cats or their young. At times, prey animals such as giraffes and some antelope have inflicted mortal wounds on attacking cats.

Felids respond to fear by spitting and raising their hair. The latter behavior causes the jaguarundi to change its apparent color, since its hair is lighter at the roots which are then exposed. While most of the small cats flee when threatened, some of the big cats will respond with an attack. Lions, leopards, and tigers frequently "cough" when angry or about to charge. When pursued, the bobcat, lynx, ocelot, and others are known to backtrack. For example, one bobcat being chased by hounds was observed to climb a tree, go back down, backtrack about fifty yards, and then head off at a 90 degree angle just before the dogs came into view. The dogs followed the scent to the tree where they stayed, sure they had treed their prey.

Remarkably, humans have exploited the hunting prowess of cats despite their independent nature. The African wild cat was once trained to retrieve game, while caracals, cheetahs, cougars, and lynxes have been used directly for hunting. Cats have been trained to retrieve objects, and there is even a "seeing eye" cat on record.

Family Hyaenidae (hyenas and aardwolf—4 species)

While superficially resembling dogs, hyenas are more closely related to the civets and mongooses. The **brown hyena** (*Hyaena brunnea*), **striped hyena** (*Hyaena hyaena*), and **aardwolf** (*Proteles cristatus*) all sport an erectile mane along their backs. When frightened or alarmed, the manes are raised. Hyena jaws are quite powerful, and the **spotted** or **laughing hyena** (*Crocuta crocuta*), reputed to have the strongest jaws proportionate to size of any mammal, can crack open large elephant bones to get at their nutritious marrow. While feeding in this way, they often consume many bone fragments. Hyenas are often quite hard to locate when heard, since they are capable of throwing their voices with considerable skill. Their sense of smell is well developed, enabling them to locate a rotting carcass from a distance of

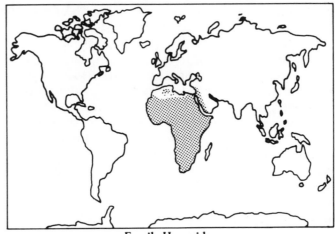

Family Hyaenidae

187

Striped Hyena (*Hyaena hyaena*)

several miles with a good wind.

The word hyena is still often associated with the word scavenger. This stereotype fits best the striped and brown hyenas, which do scavenge much of their food although they are also known to hunt down and kill prey such as small mammals, birds, reptiles, and insects. The striped hyena frequently seems to take a liking to snakes, and both species feed on fruit. During the dry season, up to 50 percent of the brown hyena's diet may consist of melons and other fruits. Brown hyenas consume hooves, horns, skin, bones, and even the dung of other predators and ungulates. Spotted hyenas are also known to eat dung, possibly for its vitamin content. Brown hyenas will steal prey from cheetahs, jackals, leopards, and servals. They are also known to dig up graves to feed on the decomposing remains of humans. The gastrointestinal tracts of hyenas, vultures, and other scavengers are mysteriously resistant to putrefying microbes and their by-products, allowing them to feed on rotting flesh that is harmful to many other creatures. Brown hyenas are known to store food in brush and holes, while spotted hyenas sometimes cache excess food in shallow water. Spotted hyenas have been known to dive underwater to retrieve carcasses of drowned animals.

The term "scavenger" is not aptly applied to the spotted hyena, for it kills up to 80 percent of its food. This aggressive species, able to run 40 mph, has been known to attack and overcome lions, baby elephants, humans, other hyenas, and even small rhinoceroses. Its neck muscles are so strong that a single hyena can pull a fully grown zebra carcass about at will. They are not averse to scavenging and will eat or attempt to eat almost anything, including canned goods, car tires, fruit, garbage, pots, and shoes; they will even lick blood from the ground. On the African savanna, this animal is the most important and numerous predator involved in the regulation of prey species. Adult hyenas can pack away 30 pounds, about a fifth of their weight, at a single sitting. After making a large kill, a pack or clan may participate in a feeding frenzy. This can occasionally cost some participants toes and parts of the snout. On rare occasions, it has been reported that a hyena has been killed in the turmoil and actually become part of the meal. A large wildebeest can be completely devoured in less than fifteen minutes.

Spotted hyenas are active primarily at night. At daybreak, their kill is often taken over by lions. The hyenas must then wait at the periphery for leftover scraps from their own kill. This explains why it was originally thought that these animals were primarily scavengers. Now it is known that lions are the thieves far more often than the victims. The lions often home in on the spotted hyena's laughter, a frequent signal of feeding but also emitted when chased or attacked. In contrast, the brown hyena's bark resembles the hoot of an owl. All hyenas, as well as lions and jackals, are known to be vulture-watchers, while vultures are known to be hyena-watchers, each thus using the other to find food. Spotted hyenas mark excess food with their scent. This is usually enough to keep other hyena clans away from their larder and territory.

The aardwolf feeds on small animals, particularly insects such as termites, with its foot-long, sticky tongue. One was found to have 40,000 termites in its stomach, and over 200,000 can be eaten in a single night. Much dirt is consumed in the process. Aardwolves bury their wastes and, after feeding, thoroughly wash themselves. The long tongue can be rolled back on itself to clean out the oral cavity; it can also reach up to wash off the eyes.

Only the spotted hyena consistently lives in groups known as clans. A clan may number as many as one hundred animals. During the rainy season, brown hyenas are known to form small groups of 10–12 animals. Clans mark out their territories with their feces and glandular secretions. These glands are found in the anal area and between toes. Since a large portion of their fecal matter is composed of calcium powder from ingested bone, the markings resemble small piles of lime. Neighboring clans generally respect each other's property rights and have been observed to give up the chasing of prey when the chase leads them into the territory of their neighbors. Should they intrude into the foreign land, they would most likely be attacked and driven out. In fact, clans often send out members to patrol their borders against invasion.

Male spotted hyenas fight to determine a hierarchy, at times even fighting to the death. Brown hyenas fight for dominance through ritualized neck biting, with the dominant animal throwing the subordinate to the ground while continuing to bite it for 15 minutes or more. Afterwards they may get up and feed side by side. In spotted hyenas the females are generally larger and more dominant than males, possibly to help protect the pups. The males are larger in the other hyenas.

In spotted hyenas, the external genitalia of the two sexes are so very similar in appearance that it is impossible to tell the two apart. The female's clitoris is as large as the male's penis, and it is accompanied by a scrotum-like fusion of the labia behind it. In addition, it can erect to the same degree during confrontation ceremonies, although it does not do so during the mating process. These similarities led to the early false belief that the animals were hermaphroditic. The two sexes do have about equal amounts of testosterone in their bloodstreams.

At birth all species are unmarked. Only later do they develop their characteristic markings. In spotted hyenas, neither sex brings food to the pups, which are exclusively nursed for their first 18 months of life. The female has been known to abandon her pups for so long while hunting that by the time she returns they have become emaciated or even died of starvation. In contrast, brown hyenas and aardwolves are known to raise their pups communally so that a female will suckle any pup until they are about 14 months old. In addition, it is not unknown for young, male brown hyenas to participate in raising the pups, even though the actual fathers are nomadic and long gone.

When alarmed, hyenas will grumble, paw the ground and, with the exception of the spotted hyenas, erect their manes to appear larger. When cornered by dogs, striped hyenas will sometimes sham death. The instant the dogs lose interest, the hyena jumps up and runs off at full sprint. When the aardwolf is threatened, it can eject a foul-smelling fluid from its anal glands. The hyenas' worst enemy is the lion, which kills and eats them. Lions account for about half of all hyena deaths. Humans are their second worst enemy. Other hyenas rank third on the list of mortal threats. Both spotted and striped hyenas have fatally attacked humans.

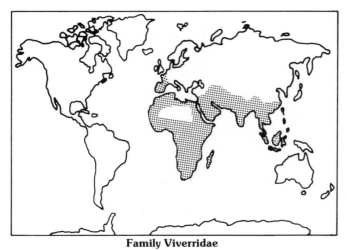
Family Viverridae

Family Viverridae (civets, genets, and mongooses—70 species)

Although there are many species in the Viverridae, we really know very little about this family, the civets, genets, mongooses (not mongeese), and their relatives. These animals have adapted in various ways to living in the trees, on the ground, under the ground, and even in the water. The **binturong** or **bear cat** (*Arctictis binturong*) has roughened foot pads to increase traction on tree branches, and also a prehensile tail. The **African palm civet** (*Nandinia binotata*) increases its foot traction with special glands that keep its pads moist. The **true mongooses** (*Herpestes* spp.) can close their ear canals to prevent the entry of dirt, sand, or water. Many, such as the **fossa** (*Cryptoprocta ferox*), **genets** (*Genetta* spp.), and **linsangs** (*Prionodon* spp.), have retractile claws, and while most are capable swimmers, a few, such as the **water mongoose** (*Atilax paludinosus*), **water civet** (*Osbornictis piscivora*), and **otter civet** (*Cynogale bennettii*), are aquatic. While the latter two have at least partially webbed feet, the water mongoose is paradoxically the only mongoose without any sort of webbing between its toes. The ears and nostrils of the otter civet can be closed when diving.

The various viverrid species show many colors and patterns in their fur. The faces and bodies of many genets are patterned with white spots which stand out at night and help the animals easily spot and recognize each other. The **Indian palm civet** or **toddy cat** (*Paradoxurus hermaphroditus*) and the **masked palm civet** (*Paguma larvata*) also display white facial markings. These may serve the same purpose and/or serve as warning coloration in the manner of skunks since they maintain a similar defense.

Most of the viverrids are quite long and slender with short legs. Many, such as the lithe linsangs and genets, move with their bellies so close to the ground and with such sinuous movements that they are often mistaken for snakes at first glance. The **banded mongoose** (*Mungos mungo*) further enhances this effect by slithering about in groups lined up in single file. The true mongooses can run straight backwards for short distances, up to ten feet.

Most viverrids live solitary lives, although some species are known to hunt in pairs or small groups. **Dwarf mongooses** (*Helogale parvula*), the smallest viverrid, form unique associations. The dominant leader is a female. She and the dominant male produce all the offspring, preventing subordinates from mating. These subordinates have their own dominance hierarchy, with younger group members dominant to older. A very few species, including the **suricate** or **meerkat** (*Suricata suricatta*) and the **yellow mongoose** (*Cynictis penicillata*), are known to live in large colonies numbering as many as 50 animals. Although they will occasionally dig their own burrow systems, they more often take over and modify existing structures built by termites, rabbits, or squirrels. With squirrels, the squatters often share the underground dens, and it is not unheard of for squirrels, suricates, and yellow mongooses all to live together. The animals are quite clean and often designate certain areas for waste material. The suricates are habitual sunbathers. They come out at

Dwarf Mongoose (*Helogale parvula*)

dawn, stand on their hind feet, and face the sun in such a manner that it is often called "sun worshipping." At night they sleep piled atop each other, as many as five deep, to conserve heat. The African palm civet is arboreal and sleeps in trees.

Although primarily carnivorous, most viverrids are omnivorous. Some, such as the **banded palm civet** (*Hemigalus derbyanus*), are wholly carnivorous, while others, such as the members of the subfamly Parradoxurinae (which includes the binturong and most of the palm civets), feed mainly on fruits, occasionally on types known to be harmful to humans. The Indian palm civet (or toddy cat) has a strong yen for fermented sap collected by humans to produce the liquor toddy. The vast majority of the species in this family will attack almost any small animal they are capable of overpowering. In many species it is movement of the prey that stimulates the attack. Although the majority hunt alone, a few such as the **brown mongoose** (*Herpestes fuscus*) hunt in pairs. In this species, one animal will occasionally herd a shoal of fish into shallow water while the other waits nearby to pounce on the fish. The dwarf mongoose and the **cusimanse** (*Crossarchus obscurus*) are also known to hunt in small groups. Banded mongooses have been known to roam in packs numbering up to 50. These large groups have overcome prey as large as young antelope, as well as large predators. Individuals of this same species are also known to search through elephant droppings for insect meals. Some mongooses, including the banded mongoose, will mark their excess food and each other with their scent glands. Genets when satiated will play with their prey in the manner of cats. At the height of such play, they are known to hop up and down on their victims while on their hind legs.

The **common mongoose** (*Herpestes auropunctatus*) and other mongooses are probably best known for eating snakes, particularly cobras. They attack snakes instinctively, but they do not deliberately seek them out. They are protected by their extreme quickness; bristly, thick fur; the short fangs and slowness of the cobras; and a partial immunity to the cobra's venom, being eight times more resistant than a rabbit. They will occasionally join forces to dispatch a cobra. Despite all these advantages, the little mongoose will on rare occasion come out the loser, at times even after beginning to devour the snake. It has been found that after being swallowed, the snake's fangs had penetrated the gastrointestinal tract and injected venom directly into the bloodstream to cause death. Snake-lovers may be soothed to hear that large pythons and the quicker pit vipers usually have no trouble making a meal out of the relatively defenseless mongooses.

The snake-killing reputation of mongooses has led to their introduction into many areas in hopes of lowering snake and rodent populations. Such projects begin with good intentions, but because mongooses are voracious predators, once introduced they have caused much destruction to the endemic wildlife, even causing the complete demise of some desirable species. Part of the problem stems from the mongoose's well known appetite for eggs. After obtaining an egg, banded and dwarf mongooses crack it open by throwing it through their hindlegs against a solid object such as a rock, much as a center hikes a football to the quarterback. The water mongoose, cusimanse, and **crab-eating mongoose** (*Herpestes urva*) break open shellfish in the same way, and the dwarf mongoose so treats snails.

Mongooses use additional techniques to break open crabs, eggs, nuts, pillbugs, shellfish, and snails. Some grasp the object in their forepaws and then smash it against the ground. If it is too large to be lifted the mongoose may throw rocks at it. The **narrow-striped mongoose** (*Mungotictis decemlineata*) lies on its side or back so that all four paws can be used to hurl the egg against a rock. The dwarf mongoose drinks by dipping its paws in water and then licking them dry.

Most viverrids use scent glands to mark their territory. Genets occasionally do so from handstand positions. The scent of many viverrids is strong and putrid although some are pleasing to human smell. For example, the binturong's odor is similar to the smell of cooked popcorn. The **African civet** (*Civettictis civetta*) and others mark territory with piles of droppings. When the **Indian civet** (*Viverricula indica*) and the African palm civet encounter a strong-smelling plant chemical, such as tincture of valerian, they rub their entire bodies over it to impart the scent to themselves. In groups of banded and dwarf mongooses, where the oldest female is the most dominant, the group members will mark each other with their glandular secretions. Presumably, this is done so that members of a single group can quickly recognize each other through a common scent and thus easily spot an outsider. Female dwarf mongooses mark their young with anal glands by hiking them like eggs beneath their bodies, marking them as they go by. Most species in this family, however, are solitary.

Most species mate in spring or summer. Not much is known about the techniques, but male suricates often won't take no for an answer. If the female tries to reject one, he grabs her by the neck gently to show who is in control. In the fossa, copulation never lasts less than an hour in encounters so far observed. The male's long penis is held so firmly that the pair often climb about without separation.

Viverrid females give birth to 1–6 young. Female genets consume their offsprings' wastes and vigorously defend them from predators, including the males. At about the time the young genets start on solid food, they occasionally lick saliva from the corner of their mother's mouth, perhaps obtaining antibodies or bacterial flora for their digestive systems. In some mongooses, such as the dwarf mongoose and the suricate, all members including the males help raise the young. In banded mongooses, the young will suckle from any available female, and males babysit while females forage.

Defense techniques vary among the members of this family. When alarmed, genets raise their hair, particularly that on the tail, to form the bottle-brush look, and they have been heard to purr so loudly that witnesses claim it sounds like a boiling kettle of water. Almost all the viverrids (with some exceptions, including the **fanaloka** [*Fossa fossa*], linsangs, and some palm civets) have scent glands that can produce foul-smelling substances, and many can, like skunks, spray enemies with an obnoxious secretion from their anal glands. Some of the true mongooses react to danger by rolling into

192

a ball. The suricate and binturong defend themselves by first raising their hair and spitting at the intruder. If this fails, they attack with vicious bites. If possible, suricates dive into their burrows; if trapped away from a burrow, they may resort to displacement behavior by beginning to dig a hole. The water mongoose escapes predators by immersing itself in water except for the nostrils, which remain above the surface for breathing. Banded mongooses and suricates have attacked predators such as eagles en masse to rescue captured conspecifics.

As is often the case, humans have exploited the viverrids. The secretions known as civet are milked from the Indian civet, **oriental civets** (*Viverra* spp.), and the African civet to make expensive perfumes. The African palm civet is often kept as a pet to keep houses clear of rodents and insects, and the masked palm civet is said to make a good ratter since it is not known to go after the owner's poultry as well. The fossa has even been trained to help people hunt water hogs.

Family Canidae (dogs and relatives—36 species)

Excluding **domestic dogs** (*Canis familiaris*), the canids range in size from the 3-pound **fennec** (*Fennecus zerda*) to the 120-pound **wolf** (*Canis lupus*). One large wolf reportedly weighed 210 pounds. This range broadens with the inclusion of domestic dogs, of which there are about 400 breeds weighing from 2 pounds (the **chihuahua**) to 295 pounds (**the St. Bernard**). Male canids are generally larger than females. The wolf has the largest natural range of all terrestrial mammals excluding humans and introduced mammals (including domesticated species and some rodents).

Domestic dogs come in a great variety of shape and form. There are even breeds that are hairless. Humans have bred and trained them for numerous activities, including guarding, hunting, truffling, herding, police work, rescue, sight for the blind, sled pulling, and tracking. Most commonly they take on the role of family pet, but abandoned pets can become feral and form packs that are destructive both to wildlife and to other domestic animals. Their ravages have often been wrongly blamed on wolves or **coyotes,** also known as **prairie wolves** (*Canis latrans*).

The difference in appearance between the **Arctic fox** (*Alopex lagopus*) and the fennec illustrates several biological principles. The first, Allen's Rule, states that the colder the environment an animal inhabits, the smaller its extremities, including ears and tail, become. Another, Bergmann's Rule, says that the colder the habitat, the rounder the body shape. Such adaptations lower the surface to volume ratio and enable the animals to conserve heat more efficiently. Of course, the reverse also holds true; animals inhabiting warmer areas will more readily dissipate heat with their greater surface area. A third principle, Gloger's

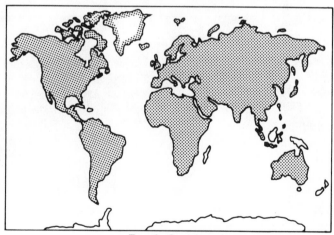

Family Canidae

193

Rule, says that animals are generally less pigmented in cool or dry areas and darker in warm or humid habitats.

The Arctic fox is well adapted to its habitat and does illustrate these principles. Its rounded, densely furred body helps it survive temperatures of -100°F without the aid of hibernation. In fact, it has been observed hunting in temperatures of -50°F. Its feet are densely furred to increase traction on ice and snow as well as to help maintain body heat. Its long, bushy tail can be used to keep its nose warm. The shorter day lengths of winter stimulate it not only to grow denser fur but also to change its color to white, and less often blue, from the summer gray. It and the **Corsac fox** (*Vulpes corsac*) are the only canids to exhibit seasonal dichromatism.

The warm-climate foxes, such as the fennec and **bat-eared fox** (*Otocyon megalotis*), have large ears that both help to dissipate heat and enhance their hearing. The bat-eared fox can even hear insects below the ground, often digging up and eating them. Both of these foxes' ears grow many fine hairs to prevent the entry of sand or dust. The fennec also has hairy-soled feet for better traction in sand.

As a rule, the canid sense of smell is well developed. The noses of some species have 225,000 sensory cells, compared to 5,000 in humans. They are said to be able to sense the "smell of fear," and males can detect about 50 different aromas in the female's urine. They can thus determine the exact stage of her estrous cycle as well as the contents of recent meals, sex, age, possibly status, and reportedly her "emotions." Tracking dogs can differentiate the scents of identical twins and even that of human touch three weeks old. Wolves can sense strong smells at distances up to 1.5 miles. Dogs also have small heat receptors in their noses, which is probably how St. Bernards can locate living but not dead humans as much as seven feet beneath the snow.

Most canids are capable of barking and howling. Several species of dog are said to be barkless, or to bark only rarely. These include the **basenji** breed of domestic dog, **raccoon dog** (*Nyctereutes procyonoides*), **dingo** (*Canis familiaris dingo*), and **dhole** or **red dog** (*Cuon alpinus*). Generally these dogs limit their vocalization to howls or whines. Canid howls may serve to announce the presence of food, ownership of territory, organizing of a pack, or possibly the pleasure of "singing." The most recent evidence indicates that wolf howling primarily functions as inter-pack communication to keep groups separated. Wolf howls can be heard up to ten miles away, and **black-backed jackals** (*Canis mesomelas*) often produce strange howls when rain is imminent. Dholes and bat-eared foxes use whistling sounds to keep in contact.

Canids, like most mammals, pant to rid themselves of excess heat. In panting, the air flows in through the nostrils and out the mouth. This one-way air flow allows

Bat-eared Fox (*Otocyon megalotis*)

194

greater heat loss per cycle of inspiration and expiration. The air passages contain special "sweat glands" that saturate the air with water. This increases heat loss with the aid of evaporation. Generally, canids pant at a rate up to 400 times per minute, felids pant only at rates up to 250 breaths per minute. Many canids can also sweat away excess heat from the soles of their feet. With the exception of the raccoon dog, canids are quite tolerant of the cold. The raccoon dog occasionally spends the winter in "hibernation" although, as in bears, it is not a true hibernation. It may increase its weight by 50 percent prior to its winter sleep.

Before lying down, many canids, including domestic dogs, will run in tight circles. This helps to form a bed, makes sure the area is safe, and enables the animal to determine wind direction so that it can take an advantageous position facing upwind.

Most canids do not climb trees. The exception is the **gray** or **tree fox** (*Urocyon cinereoargenteus*), which actually takes up residence in hollow trees and abandoned bird nests. The corsac fox will also take shelter in trees, but most others keep to the ground.

Most canids live in burrows that they either dig themselves or take over from other animals such as aardvarks, armadillos, badgers, squirrels, and termites. Wolves are even known to den in beaver lodges. When excavating their own dens, most species dig with their feet. The **swift fox** (*Vulpes velox*) uses all four legs to spread excavated dirt widely and rid the entrance of tell-tale dirt piles. The **maned wolf** (*Chrysocyon brachyurus*) digs its burrows with its teeth. Arctic foxes' burrows may have as many as 100 entrance and exit holes. The desert canids often stay underground in their dens during the hottest part of the day to keep cool. Wolves, foxes, and others will curl up and allow themselves to be covered with an insulating layer of snow during storms.

Canids are often considered intelligent and crafty. In folklore, **foxes** (*Vulpes* spp.) are almost universally associated with cunning and cleverness. Although most of this lore is pure fancy, many of the stories have been shown to be based in fact. Trained biologists have seen foxes using folk-tale methods to rid themselves of fleas and other pests. The fox takes hold of a stick or piece of wool in its mouth and then slowly submerges. After several minutes, when the free-loaders have supposedly worked their way to the still dry object, it is released to float away. Although this procedure has been witnessed, it is probable that there is a different motive, as it is doubtful that any parasites are actually gotten rid of in this manner.

Many experts believe predators are more intelligent than their prey, but the predators' intellectual advantage may exist only because the predators get more practice in refining old techniques and learning new ones. Prey animals must be able to avoid predators from the outset, since their mistakes in learning are likely to be fatal. Their "intelligence" is necessarily genetic, built in by natural selection. Predators appear smarter because of their greater opportunity to learn and their genetic ability to take advantage of the opportunity. Certainly, persecution seems to make some canids, such as the coyote, which has been the target of mass extinction campaigns for well over a century, more wiley. They have learned to shy away from traps and poisons, and they have proved able to expand their ranges despite all efforts to stop them.

Many predator species have been shown to possess an "oddity preference." That is, they attack odd-appearing animals in preference to others. This serves the function of eliminating mutant or crippled prey animals. These are usually less able to escape, a fact that probably is responsible for the observed behavior. Whether through awareness of their vulnerability or through built-in instinct, many prey animals will do their best to hide an injury or defect, often making a veterinarian's job that much more difficult. As predators, the canids ensure that the animals they prey upon will not suffer from deformity, disease, injury, or senility. This fact has been adequately shown in studies of caribou, moose, and musk oxen that are preyed upon by wolves. It has also been demonstrated that predators help to prevent epidemics of contagious diseases within their prey species.

With the exception of the dhole and the **Cape hunting dog** or **wild African dog** (*Lycaon pictus*), the canids are not strict carnivores. Actually omnivores, they feed on both meat and vegetation. The maned wolf and **jackals** (*Canis* spp.) have even become pests in certain areas by feeding on sugar cane. Jackals also feed on coffee beans, figs, pineapples, and mushroom species that apparently cause them to get high and may lead to irrational behavior. Coyotes are known to eat grass, prickly pears, and watermelon in addition to animal prey including skunks, rattlesnakes and gray foxes. Gray foxes can subsist soley on plant food, including manzanita berries and even bitter juniper berries during hard times or certain seasons. In fact, some gray foxes are known to eat as much vegetable substance as flesh, and as a species they eat more plant food than any other canid.

Some substances may be ingested for their therapeutic value. Many canids may eat grass for its laxative effect. Black-backed jackals often eat the fruit *Balanites aegyptiaca* which acts as a natural dewormer. Maned wolves eat wolf's fruit (*Solanum lycocarpum*) which is thought to rid them of giant kidney worms (*Dioctophyma renale*). Red foxes and others occasionally eat clay, most likely for its mineral content.

The canids are generally opportunistic hunters, feeding on whatever they can catch. The gray fox eats almost anything, including garbage, tinfoil, leather, centipedes, and scorpions. It and other foxes often take eggs, which they break open by smashing against rocks. With the exception of the Cape hunting dog and a few others, most will readily take carrion. In extremely hard times, Arctic foxes, coyotes, wolves, and others are known to become cannibalistic, killing and eating injured or sickly members of their own kind. More frequently, canids will take other family species. For example, wolves readily eat coyotes, jackals, or foxes; and coyotes will eat gray and swift foxes. Dogs frequently take nibbles of their own excrement, but the Arctic fox quite frequently will feed on the feces of polar bears, reindeer, and musk oxen. Coyotes are known to have fed on bison and cattle feces.

Canids use various methods to obtain food. Although most are solitary hunters some, such as the web-footed **bush dogs** (*Speothos venaticus*), Cape hunting dogs, dholes, and wolves, are known to form packs. When small game is plentiful, wolves generally move about in small family groups that subsist primarily on rodents. When hard times hit, these small groups often come together to form large packs of as many as 36 wolves to tackle big game. The pack frequently performs "yawning" and/or howling choruses, apparently to synchronize the group's mood for hunting. Wolves can hear each other's howls five or more miles away. The howls are thought to serve a number of functions, as mentioned previously, but none have been definitely proven. The pack may travel up to 125 miles in a single day while hunting for food.

As has been found in some of the felids, a wolf will not alarm prey species by its presence alone. It must first exhibit hunting intentions, shown by a lowered head and ruffled, raised shoulders. Usually wolves will not attack a large prey animal that stands its ground, and many healthy prey species can easily outdistance pursuing wolves, which give up the chase within a few minutes if they are not gaining. The prey's speed may signal the wolves that they are after a healthy and possibly dangerous opponent. Many attempts are abandoned after about 1000 yards if the prey is not caught. Many a full-grown moose has stood its own against attacking wolves. Not surprisingly, wolf packs are only successful about ten percent of the time. Wolves can sprint up to 45 miles per hour, and a single wolf can consume as much as 20 pounds at a time, allowing it to fast for two weeks or more without harm.

The Cape hunting dog forms packs of up to 90. Dhole packs number up to 40. Both species are often credited with hunting down prey using relays. It is usually stated that one or a few hunters approach the prey animals from downwind and chase them until relieved by fresh pack members. This continues until the prey is worn down and caught. Although the dogs are known to take turns at chasing the prey, closer observation has shown that these "relays" come about more by accident than planning. As the chased animal switches direction or circles about, it allows dogs further back to cut in at an angle and take over pursuit. Eventually the dogs' stamina allows them to overcome the animal.

The prey is literally eaten alive when caught, but it has been found that more than 90 percent die within two minutes after the attack begins. Cape hunting dogs can entirely consume an impala in ten minutes. Usually, the larger pack members attack and kill the prey, but sometimes this chore is left to be finished by the younger, inexperienced animals to help them learn. Although dominant animals have first rights, the younger animals are often allowed to eat first.

Cape hunting dogs are very efficient and known to be successful in up to a staggering 70 percent of their chases. They successfully overcome animals such as wildebeests and lions. Dholes often hunt in packs using their sense of smell and can overcome prey as large as bears and tigers. When being pursued by Cape hunting dogs, prey such as antelope, gazelle, and wildebeest have been known to run into ponds and lakes and have even accidentally drowned in their attempts to flee. Some have successfully escaped by running up to the sides of hunters or shepherds, into huts, or into farmyards. Bush dogs often feed on pacas, which will frequently attempt to escape by diving into the river. Unfortunately for pacas, bush dogs are excellent swimmers, even able to swim underwater, and have a ploy of their own. One or several of the dogs hide on the bank while the others dive into the water and actually drive the paca(s) to the shore and inevitable ambush.

The rest of the canids hunt singly, in pairs, or in small family groups. Coyotes will frequently team up with another coyote or even with some other species of animal. Occasionally, a coyote and badger will cooperate to take prairie dogs. The badger begins to dig them out, and the coyote waits to capture any animals flushed to the surface. Neither animal is known to share a kill with the other despite some writings to the contrary. Coyotes are also known to get food through distraction and ambush techniques. While one coyote puts on a show for the prey by running, dancing, yelping, or waving its tail, the other will sneak up on the unsuspecting prey from behind. Another technique used by coyotes is to entice prey such as a crow to it by arousing its curiosity. This is done by playing dead. When the animal ventures near enough, the coyote jumps up and quickly dispatches it. **Golden jackals** (*Canis aureus*) are known to use this same technique to capture birds in Africa. It has been reported that coyotes will also deliberately drive prey into an ambush of other coyotes waiting in hiding.

Foxes also exhibit ingenious hunting behaviors. They are credited with pushing rolled up hedgehogs into water to force them open, and may even accomplish the same result by urinating on them. Foxes (and coyotes) exploit fires by hunting just ahead of the flames to catch fleeing rabbits and rodents. Coyotes also take prey flushed by cattle, elk, and bulldozers. Foxes often catch their prey by jumping high into the air and pouncing on the game with all fours. Sometimes the prey is grabbed and quickly thrown into the air and caught several times in succession. They have at times become pests by raiding chicken houses and killing more than they can eat. Fennecs often play with their food before eating. They, too, will throw their prey but usually up over their backs. They then will frequently restalk it, pounce, and repeat the performance over and over before consuming the carcass. Bat-eared foxes are known to gang up on falcons to steal their freshly caught prey. By far the most fascinating technique used by foxes (as well as by some coyotes, grizzly bears, martens, mongooses, stoats, weasels, and others) is that of "charming." The fox begins by rolling around on the ground, chasing its tail, or turning somersaults. Stoats do it in pairs enabling them also to "box," play "leapfrog," or take part in some other antics. This behavior soon draws spectators or captivates the attention of prey already nearby. The predator slowly "plays" its way toward the victim until it gets near enough to jump up suddenly and make an easy kill. **Red foxes** (*Vulpes vulpes*) have made lunging leaps of up to 15 feet in their attacks on small prey and may charm by playing dead.

Many canids, such as the jackals and foxes, will drag fresh kills to cover, often to their den, before eating. The majority of predators, however, will do no hunting in their nesting area. This inhibition may help prevent adults from attacking and eating their own offspring. It is also responsible for reliable reports of predators playing with but not harming prey animals near their den areas. For instance, foxes have been seen to play with bunnies and pheasants, without ever attacking. It may also explain why it is possible to find an unusual group of animals such as foxes, badgers, polecats, rabbits, and ducks all inhabiting common areas in what appears to be peaceful coexistence.

Some of the canids obtain food by following large carnivores to feed on their leftovers. This feeding pattern is common in the scavenging jackals, which frequently follow lions. The black-backed and golden jackals often feed on the afterbirths of wildebeests. Golden jackals can recognize females that haven't expelled the placenta and will follow them until they do. Jackals do hunt and kill their own prey at times, and it has been found that when they pair up they are about four times as successful in getting prey as when they hunt singly. Arctic foxes will travel more than 1,000 miles in their searches for food. They frequently follow polar bears to feed on leftover scraps or even droppings. When the foxes are bothersome, the bears are known to throw pieces of meat towards them to keep them quiet and away. Should the foxes make the mistake of getting too close, the bears have been known to kill and eat them. Arctic foxes have been reported to nibble on sleeping humans and are known to have eaten the corpses of natives in burial caves.

Most of the canids hide or bury excess food so that it can be eaten at a later date. Jackals wallow in their food to put their scent in it and keep other jackals away. Red foxes often urinate on inedible items so time will not be wasted in reinvestigation next time they pass by. They will sometimes dig up stored food only to rebury it in the same place, as if checking to be sure it is still there. The Arctic fox stores prey in the snow, but many other canids build special storage rooms in their dens. One Arctic fox cache contained fifty lemmings and forty little auks all neatly aligned. The **Pampas** fox (*Pseudalopex gymnocercus*) is the "pack rat" of the dog world. It stores not only excess food, but cloth and leather goods as well.

All canids can swim, and many are quite adept at it. The Portuguese water dog has been trained to retrieve fish and fishing gear that has fallen overboard. Some canids obtain their food by diving. The raccoon dog dives for fish, and the **crab-eating fox** (*Cerdocyon thous*) occasionally catches and eats crabs, although it feeds primarily on rodents and other small animals. Coyotes in desert areas will at times dig "wells" to get at underground water, and fennecs can survive without free water.

Social canids, like other social animals, arrange themselves in definite hierarchies. Wolves and some others show two hierarchies, one for males and one for females. The most dominant animal is generally the alpha female. In times of adequate or excess population only the dominant pair of animals do any mating. In fact, the head female in wolves and Cape hunting dogs will prevent other females from mating; should the latter succeed in becoming pregnant, it is not unknown for the dominant female to kill the resulting pups. There is usually one member of the pack that borders on being an outcast and serves as a "scapegoat." In some studies, the scapegoat's presence seems essential to maintaining social order within the pack by allowing members to work out latent aggressions.

Although posturing and bluff are often all that is needed to determine position, fights are not uncommon. When one canid has had enough, it signals submission by exposing its neck to the victor. The latter usually proceeds to mark the ground with urine before allowing the loser to wander off. This submissive posturing inhibits the attacker from further assault, lending credence to the Biblical injunction to "turn the other cheek," but it does not work with trained domestic dogs, which have been known to tear submitting coyotes apart. Wild canids have been known to fight to the death, but this usually happens only when the foes are from different packs.

Subordinates acknowledge superiors by nuzzling, licking, or nipping at the mouth; exposing their necks; or by lying down and rolling onto their sides. In wolves and others a subordinate will urinate in front of a dominant animal to show submissiveness, particularly if feeling very threatened. The dominant animal will then often sniff and even taste the urine. Domestic dogs are known to show some of these same postures to their dominant human owners. It is sometimes said that Cape hunting dogs will even suckle each other in the excitement of greeting. Actually, teat licking, jaw nudging, and creeping under the belly of another dog are all behaviors aimed at suppressing aggression.

Most carnivores mark territories, and canids are no exception. They use urine, feces, and special glandular secretions from sebaceous glands located around the anal opening. These are what newly acquainted dogs are checking when they meet. From the odor of these glands, a dog can tell the sex of the other, and since these secretions usually make their way to the feces a dog can tell the sex of a dog by sniffing at its wastes. Foxes produce their "foxy odor" from glands located at the dorsal base of their tail. Maned wolves mark with fecal piles averaging 16 inches high.

Canids also mark territories verbally. Wolves and coyotes howl, and foxes yelp; these sounds may also serve other purposes. In maned wolves and jackals, both sexes hold and freely mark territory. Often pairs do so together. A large wolf pack may hold up to 5,000 square miles of territory. In many canids, the fringes of territories overlap. Markings in these areas can prevent encounters which occasionally may end in a fatality. A fresh marking signals "keep out," a moderately old one "caution," and a very old one "proceed freely."

Some canids, including coyotes, dholes, foxes, some jackals, and wolves, may occasionally mate for life. In social species, only the dominant pair may mate. Although males are usually the aggressor, it is the female wolf that courts the male. When two coyotes fight for dominance, it has been noted, the female will just as often choose the loser as the winner for mating. The female jackal greets the male by touching his shoulder with her nose and then rolls onto her back to await acceptance. After he acknowledges her, she carefully grooms his entire body. Male Cape hunting dogs show their preference for a particular female by urinating in the same locations she uses. They sometimes do this simultaneously, but the male must often stand on his forelegs to carry it out. Generally it is only the dominant animals (male and female) that cock their leg to urinate. Female bush dogs let males know they are ready to mate in a unique way. She climbs up a tree backwards; once she is in a handstand position, she urinates and applies scent markings. She then carefully slides down.

A feature common to Canidae, Procyonidae, some rodents, pinnipeds, some primates, many bats, some insectivores, and Ursidae is the "mating tie." The males have a baculum or penis bone which makes up more of the erect penis than does the erectile tissue. In mating, the engorged vulva of the female clamps down upon the penis with sufficient force to prevent uncoupling. Although this usually lasts only 10–20 minutes, it can be as short as one minute in the Cape hunting dog and as long as an hour in the jackals.

In most species of canids, inbreeding is prevented because adolescent males leave to form or join up with other packs. In the Cape hunting dog, it is the females that emigrate, a rare behavior seen also in a few other species, such as the chimpanzee, gorilla, hamadryas baboon, and red colobus monkey.

In preparation for birth, female canids construct a litter nest. Female foxes line theirs with some of their fur, particularly that around the teats where it is easily pulled out at this time to expose them for the young. Bush dog males actually deliver the female's pups and eat some of the afterbirth. All other canid females give birth without help from the male. The females usually eat the placenta, which may help to stimulate lactation.

Pairs of female red foxes, Arctic foxes, and others will occasionally raise their young communally. Female coyotes will help each other, too, but usually only one will have pups at a time. In black-backed jackals, older siblings often aid in raising new litters. In Cape hunting dogs, all the pack members share in raising the young, but the females play the major role. In this species, the females often steal and occasionally kill the pups of subordinate females. If disturbed, female canids may move their young or divide them between more than one den site.

When a canid pack goes on a hunt, some of the adults remain behind to guard the pups. The hunters, whether they be wolves, Cape hunting dogs, or others, fill up with an extra amount of food. This is brought back home and delivered "fresh" to the pups, the guard dogs, and other dogs that may be too old, sick, or lame and thus unable to go on the hunt. To get delivery, all these pack members must do is nudge the side of the hunter's mouth or lick the returning dog's lips while giving a submissive display. Food will be regurgitated three or four times upon stimulation, and occasionally as many as eleven times. The dogs usually regurgitate 1–2 pounds of food. The occasional unaccounted vomiting of domestic dogs is attributed to being a vestige of this behavior.

In foxes, the male or reynard often does the hunting for his vixen and her pups. Arctic fox pairs have been known to hunt for up to 19 hours per day to keep their pups fed. The litter may require 2400 lemmings (60/day) to be successfully reared. Young pups are nursed from a prone position, and older siblings from a standing position. Female coyotes wean their pups by nipping at them to keep them away.

Canid mortality is often high. In one study of gray foxes, it was found that 50 percent of the litters die during their first summer and 90 percent before the first winter is complete. Thereafter, there is an annual 50 percent mortality. In the social canids, should the parents get killed, another pair, often pupless, will take charge of the young. In many foxes, if the female is killed, the male will take over raising the pups; he may even be partially successful if they are near or beyond weaning.

Babysitting is common in the social species. Baby coyotes are allowed only a certain distance from the den. After this is learned, it can be amusing to see them run about but come to a direct halt when they reach this limit, just as if there were an invisible wall in front of them. In addition to bringing the females and young food, male coyotes, Arctic foxes, and others will lure enemies from the den area. Coyotes, foxes, and others teach their young to kill by giving them crippled prey upon which they can practice.

Since canids are predators, they have few enemies except larger predators and humans. They generally react to danger by fleeing. If having just eaten, red foxes and others may vomit so they can escape at a faster pace. The pampas fox responds by playing dead and refuses to move even when touched. The fennec's main defense is to bury itself quickly in the sand. The bat-eared fox protects itself from predators by frequently backtracking as it wanders in search of food. The infamous red fox will not only backtrack and then hop to the side when being chased, but will also run through water to hide the scent of its trail. It can make six-foot vertical jumps without problem. Some are even reported to give their presence away in order to be chased by domestic dogs. The maned wolf erects its mane when excited, and most canids exhibit typical posturing to advertise their intent. For example, domestic dogs show fear by laying back their ears while drawing their mouth corners down and back. To signal aggression, they open the mouth and hold up both the corners of the mouth and the ears.

In closing, we should mention that despite all the stories and legends, wolves have never been known to take the life of a human in North America—unless it was rabid at the time. By contrast, domestic dogs are implicated in 7,000 reported attacks on humans annually in the United States alone.

Family Mustelidae (weasels, badgers, skunks, otters—64 species)

Weasels (33 species)

The world's smallest carnivore weighs less than an ounce. It is the **American least** or **dwarf weasel** (*Mustela nivalis rixosa*), a subspecies of the **common** or **European weasel** (*Mustela nivalis*). This small bundle of energy is small enough to pass its body through a large wedding ring.

Many mustelids, such as the **polecat** (*Mustela putorius*), are covered with thick fur overlying an abundant amount of fibroelastic tissue. This usually possesses enough "give" to protect the owner from

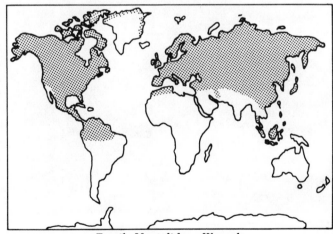

Family Mustelidae—Weasels

serious injury when bitten by an enemy. The fur of the **wolverine, mountain devil,** or **glutton** (*Gulo gulo*) is resistant to fungal growth and will not allow ice crystals to adhere to individual hairs because of its oil contents. Since this allows snow and ice to be easily shaken free, the fur has become valuable as a lining for hoods on snow jackets. The wolverine is further adapted to snowy conditions with large, furred feet that can function like snowshoes.

The fur of a few weasels, such as the common weasel, **long-tailed weasel** (*Mustela frenata*), and **short-tailed weasel** or **stoat** (*Mustela erminea*), turns wholly or partially white during winter. Individuals inhabiting warm southern areas sometimes remain brown year-round. In the most northerly habitats, they are occasionally always white. This is thought to be genetically inherited since brown weasels, if moved north, remain brown in winter, and white weasels, if moved south, continue to turn white. The color change is brought on by hormonal secretions triggered primarily by shorter days, although low temperatures also exert an influence. The short-tailed weasel becomes completely white except for its tailtip which remains black. When in this dress, it is generally referred to as an **ermine.** The black tailtip often serves as the focal point of attack from predators such as hawks and thus diverts injury from the body. The white hairs, but not the dark ones, are air-filled to increase their insulating quality. Since it is the presence of air that gives the fur its white appearance, some biologists believe that the reason many mammals are white or turn white in cold habitats is an adaptation to preserve body heat rather than for camouflage. The **American marten** (*Martes americana*) and others grow fur on their soles during winter to increase traction. It and many others travel under the snow during winter.

Weasels often take over the burrows of other small mammals such as mice and squirrels, usually after having killed and eaten the original owner. In fact, it is reported that the female least weasel will frequently kill its victim, take over its home as her own, and line the nesting area with the victim's fur. **Black-footed ferrets** (*Mustela nigripes*) often take over prairie dog burrows. The prairie dogs may try to seal the predators in by plugging up such tunnels, but the ferrets can easily dig themselves free; they are quite capable of digging out 150 pounds of dirt in a single night. **Mink** (*Mustela vison*) usually steal the nests of beavers or muskrats but will occasionally build

Marten (*Martes americana*)

202

their own. The only access is usually through an underwater entrance, although air shafts are often present. They line their dens with moss, grass, hair, and feathers. Other weasels are known to include bones and meat. The **European pine marten** (*Martes martes*) will sometimes house its young in the nests of large birds of prey.

Weasels feed on everything from insects and snakes to rodents and deer. The wolverine has even been known to kill moose and elk, although these victims were undoubtedly old, sick, or deformed. The partially web-footed mink can swim at two miles per hour and is capable of overtaking fish in a pond. They sometimes attack ducks and at times do not deliver the fatal bite until airborne, which leads to crash landings. Martens will roll slugs between their paws and the ground to remove much of the slime before eating them, and the pine marten will snatch bumblebees from the air to eat. Polecats will eat toads but are known to roll them repeatedly on the ground to remove much of their poison gland secretions. **African striped weasels** (*Poecilogale albinucha*) kill venomous snakes much as do mongooses. Mink take many birds, usually plucking them before eating.

Although rarely catching fish on its own, the **fisher** (*Martes pennanti*) is named after the word "fiche" used for the European polecat and its fur or for its habit of "fishing" bait (sometimes fish) from traps and stealing fish from other predators. It is the fastest tree-traveling mustelid and can make leaps of 16 feet. Fishers are also one of the few animals to prey readily on porcupines. They first attack the relatively unprotected head and then may flip the porcupine over to get at the defenseless belly. With snow on the ground, the fisher may actually burrow underneath to make a surprise attack on the belly right off. They frequently consume many of the quills, which can be found in their feces. They seem to be more resistant to the quills than other mammals, and they have been successfully used in Vermont to control porcupine populations. Fishers are also known to attack, kill, and eat beaver, deer, raccoons, and American martens. The mink occasionally becomes cannibalistic. Various species, including the American marten, **sable** (*Martes zibellina*), **stone marten** (*Martes foina*), **tayra** (*Eira barbara*), and wolverine, supplement their meat diets with berries, fruit, nuts, and honey. In fact, tayras are at times significant pests on banana plantations, and the marten may sport purple-stained lips as a result of its berry eating.

Most weasels hunt singly, or occasionally in pairs. A few, such as the tayra, are known to hunt in groups and will use "charming" (see Canidae). Mink may stalk prey by slithering on their bellies. Polecats, stoats, and weasels generally hunt by scent. Some readily attack animals many times larger than themselves, usually emerging the victor; however, when an overzealous individual takes on more than it can handle, it can end up paying with its life. Weasels have been known to overpower animals that outweigh them a hundred-fold. Some can drag prey that weigh more than ten times their own weight. Long-tailed weasels have even attacked humans who have come between them and their prey.

Weasels are frequently dubbed the "fiercest mammals for their size" or the "most bloodthirsty animals in existence." It is a fact that many, including the common weasel, mink, short-tailed weasel, stone marten, and **zorille** (*Ictonyx striatus*), will often kill much more than they can eat. These surplus killings occasionally take the form of mass slaughter; in one instance, a common weasel killed almost 50 chickens in a single night. Some experts have stated that weasels may kill for the sheer "joy" of it. More probably, the massacres happen because weasels instinctively kill everything at hand, eat their fill, and store the rest for later, *and* because humans hold prey in artificially small, inescapable areas. They may also relate to the different levels of motivation required for stalking, catching, killing, and eating since most stalks do not end in a kill (see Felidae). The depredation of domestic fowl is often overexaggerated and is well made up for by the fact that an adult pair of weasels can kill more than 2,000 rodents within a year. In New York state, it is estimated, weasels kill 60,000,000 rats and mice yearly.

It is often stated that a weasel may merely suck out one victim's blood before proceeding to the next. Most mustelids kill their prey with a bite to the back of the neck. This is an instinctive behavior. Many will then lap up some of the resulting blood before going after another prey animal in the same vicinity, before eating, or before removing the carcass to a place of concealment, but none are true bloodsuckers. Occasionally, the prey may carry an attacking weasel some distance before being killed.

Weasels are voracious feeders, eating up to 40 percent of their weight daily. The **grison** or **huron** (*Galictis vittata*) is probably the biggest glutton in the group. Captives have allegedly eaten themselves to death. Other species have been known to gorge themselves into unconsciousness.

The largest of the weasels is the 40-pound wolverine, one of the most powerful of animals for its size. It often scavenges its food. In so doing, it is known to become impatient and drive off an original predator such as a cougar, wolves, or a bear. They are probably successful in this displacement because of their scent glands, which can produce an obnoxious spray. Some experts believe the larger carnivore is probably somewhat satiated in such instances, since a determined adult grizzly or cougar cannot be driven from a wanted kill by a wolverine. On rare occasions, wolverines are known to hide in trees and ambush prey by leaping on to their backs. Wolverine eyesight is not very good, and they have frequently been observed to shield their eyes from bright light with the aid of an upraised paw. Their teeth and jaws are powerful enough to crack almost any bone and can snap through tree branches as much as two inches in diameter. They have dragged carcasses three times their own weight for distances in excess of a mile. Wolverines sometimes protect uneaten portions of their prey from other predators by placing it in a tree fork or by spraying it with their obnoxious anal gland secretions. They occasionally urinate on fresh kills. (Fishers are also reported to urinate and rub anal scent on carcasses.)

One very frustrating behavior of wolverines is their tendency to kleptomania. Many trappers, hunters, and campers have lost equipment such as axes, blankets, clothes, eating utensils, knives, and traps to the wily beasts. These are generally carried off to be buried, although they sometimes are destroyed on the spot. Trappers swear by the wolverine's cunning. Countless are the stories relating how these animals have taken the bait yet avoided the trap. Nevertheless, though they hold vicious reputations, wolverines have never been known to attack a human.

The marten, mink, wolverine, zorille, and most others will generally store excess food, while a few, such as the common weasel and stoat, will more often than not leave it to rot. Martens store prey in tree forks, bird nests, and even squirrel dreys. Wolverines generally bury theirs, and caches of 100 grouse have been found. After burial, the wolverine will at times camouflage the site with vegetation. Mink store food caches that can last them for as long as a month, and polecats are known to store live frogs for later eating. The frogs are paralyzed since the neck bite effectively piths the frog by destroying the spinal cord at the base of the neck. Some will take their numerous kills down into their burrow; they can then retire for a week or more without having to hunt. A fisher once actually made a home inside the carcass of a frozen moose and continued to feed until it ate itself out of a home.

Most weasels hold territories that they mark with urine, feces, and scent gland secretions. A mink's territory may be up to 16 miles in diameter, and wolverines may stake out as much as 500 square miles or more. Martens have special belly glands for territory marking, and male minks have fought to the death in territorial disputes.

Mating time varies with the species. Female American martens cluck to show their readiness to mate. In many species, including martens and stoats, males drag reluctant females by the scruff of the neck until they become more cooperative. Some have been dragged about by the male for an hour or more. Mink and sable have the longest coitus of any mammal, with the act lasting up to eighteen hours. Mink pairs are said to purr during the process. In all mustelids, mating induces ovulation.

Most mustelids (and armadillos, bats, most bears, kangaroos, lesser pandas, pinnipeds except walruses, roe deer, and wallabies), with the exception of the common weasel, polecat, and least weasel, use delayed implantation, in which a fertilized egg lies dormant in the uterus. Only at a proper time, controlled by various factors such as hormones and environmental conditions, including the amount of sunlight, does the egg become embedded in the uterine wall to continue development. This allows the young to be born when conditions are best for their survival. It has been found that there is often a longer delay in animals inhabiting northern areas; there is little or no delay in the same species found in warmer climates. The fisher has exhibited the longest delay, up to ten months.

Some animals (certain bats, sharks, snakes, surf perch, turtles, and others) exhibit another adaptive measure known as delayed fertilization, in which the female stores the male's sperm until needed. One act of mating can thus be effective for multiple pregnancies, even over a span of several years.

Young mustelids are cared for almost exclusively by the females, although male long-tailed weasels may occasionally help their mates. Female polecats will sometimes cool their babies by wetting their own bellies and then curling their bodies around the young. Many females, as in stoats and polecats, consume all of their youngs' waste products. Mink mothers frequently move their offspring from nest to nest to foil predators. They and most others carry their babies in their jaws, although when in the water the female mink carries them piggy-back. On occasion, some mink broods have been known to kill and eat their mother. Female wolverines will pre-chew their young's first solids. In stoats and long-tailed weasels the males are known to mate with female nestlings prior to the latter leaving the nest.

Some mustelids give warnings when alarmed. The fisher sometimes hangs by its hindfeet and drums on the tree trunk with its forepaws. If on the ground, it may beat the earth with its front legs. If threatened, many of the weasels can defend themselves by releasing foul-smelling fluids from special anal glands. Many other mustelids also have such glands, as do antelope, deer, dogs, and cats; they are generally used to mark territory, flag food burial sites, and aid in bringing the sexes together. Mustelids that can use them to deter enemies include the polecats, some martens, skunks, striped weasels, wolverines, and the zorille. Polecats can project their spray up to 20 inches, while the wolverine has a range of up to ten feet. The zorille not only looks like a skunk but acts like one, too. Its fluid is said by many to be worse than that of skunks, and at close range it is capable of causing temporary blindness. One zorille was known to chase away and to keep at bay nine full-grown lions simply by raising its tail while it leisurely nibbled away at their zebra kill. If a zorille's spray fails to work, it will sham death. Several species, including the common weasel, stoat, and others, have attacked humans when threatened, cornered, or captured.

Like other mustelids, weasels are prone to play. Martens occasionally do so by repeatedly jumping from trees into snowbanks. Mink, like otters, make slides in snow and mud banks.

Humans have greatly exploited the weasels, mostly by trapping and hunting them for their pelts. Occasionally fishers and others will chew off a leg to escape a trap. Some species have been used to hunt down other creatures. The **ferret** (*Mustela putorius furo*), developed from the polecat, is used for ratting and to hunt down birds and rabbits. The ferret is sent down a rabbit hole while the hunter waits above with his gun. Until the practice was made illegal, the grison was used to hunt down chinchillas. Some people have even kept the zorille as a pet because it is such a good mouser, but they must be careful not to frighten or upset it. Stoats and **yellow-bellied weasels** (*Mustela kathiah*) have also been kept as mousers.

Honey badger (1 species)

The **honey badger** or **ratel** (*Mellivora capensis*) is best known for its commensal association with a small bird known as the **honeyguide** (*Indicator indicator*). When the honeyguide discovers a bees' nest, it often searches out a honey badger and leads it to the site. The ratel tears open the nest and feeds on the sweet honeycomb while the bird feeds on the bee larvae and pupae. The bird also feeds readily on the hard-to-digest wax; its intestinal tract houses

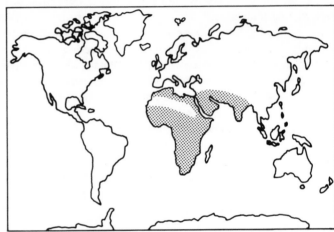
Family Mustelidae—Honey Badgers

special bacteria for just this purpose. Among the honey badger's arsenal of defenses are nauseous secretions from anal glands, strong claws, and a loose, thick skin to prevent injury despite venomous snake bites, insect stings, and even porcupine quills. The skin is so loose that the ratel can actually turn to bite an animal that has a hold on its neck. Like its weasel relatives, the honey badger is very aggressive. It has been known to attack horses and cattle, and one even held its own against a pack of dogs. There are documented cases of their having attacked and killed a ten foot python and even a full-grown Cape buffalo. When threatened, honey badgers will dig their way to safety; if cornered, they will put up a fight and even squirt out their anal gland secretions.

Badgers (8 species)

Badgers are strictly nocturnal animals. The smallest species, **ferret badgers** (*Melogale* spp.), climb and even sleep in trees. The **common badger** (*Meles meles*) lives in small groups or clans composed of several generations inside a large underground complex known as a sett. These living quarters are passed on from generation to generation for as long as centuries. In large setts, up to 25

206

tons of soil have been excavated. The **American badger** (*Taxidea taxus*), by comparison, is solitary. Although most mammals are characteristically clean, the badgers are well known for this attribute. They build latrines separate from the main living quarters, often as much as 50 feet away. Their soft bedding of straw, ferns, or moss is frequently aired out and returned or simply replaced. To fetch bedding, a badger places a load of vegetation between its chin and forelegs and then walks backwards as it returns to the burrow with its bundle, making up to 30 trips in one night.

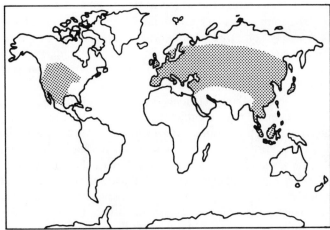

Family Mustelidae—Badgers

Setts are often equipped with air vents, a remarkable feat when one considers that the burrows may be as much as 30 feet deep. Badgers (and some weasels) are known to bury the dead of their own kind.

The main staple of badgers is earthworms. A single badger may catch and eat 200 earthworms in a single day. They also feed on rodents, insects including bees and wasps, and rattlesnakes, which they usually dispatch without being bitten even though they are protected by a thick fur and skin. Fatalities have resulted from snakebites on the nose. Rabbits and hedgehogs are skinned while being eaten. Badgers will also search through cow dung and other feces to obtain beetles. The common badger feeds on large amounts of vegetation, up to three quarters of its diet at times. The American badger is more carnivorous and often takes ground squirrels by plugging up all its exit holes except one. It then digs down through this single opening. When prey is caught above ground, a badger will frequently dig

Badger (*Taxidea taxus*)

itself into the earth or head for its den before eating.

The common badger hibernates for up to seven months in its most northerly haunts. Like bears, it is probably not a true hibernator. It uses its musk for territory marking but will release some when frightened, excited, or while playing with other badgers. Common badger clans mark each other with scent. When out at night, groups of badgers have been observed to perform playful antics described as "leapfrog" and "king of the hill."

Badgers are thought to mate for life. Female common badgers give out a loud scream when in heat and frequently run in circles before actually mating. Implantation is delayed by 2–12 months, depending on the climate. At weaning, female badgers will feed the young regurgitated food.

When threatened, badgers emit a snorting bark and raise their hair on end in order to make themselves appear much larger than they really are, a ploy common throughout the animal kingdom. They are capable of digging themselves to safety within a matter of seconds and can run backwards as quickly as they can forwards. Like many of their relatives, the **Malayan stink badger** (*Mydaus javanensis*), the common badger, and others are capable of resorting to the release of noxious fluids. The stink badger can squirt its stream up to five feet, and it may cause temporary blindness if it strikes the eyes. The **Philippines stink badger** (*Mydaus marchei*) is known to feign death.

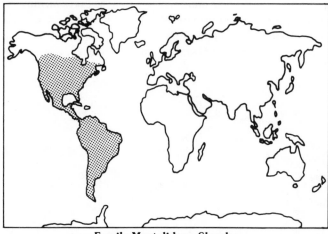

Family Mustelidae—Skunks

Skunks (9 species)

Skunks are primarily nocturnal. They usually inhabit dens dug out by other animals but on occasion will dig their own. Raccoons and opossums have been found in the same dens with skunks. The **spotted skunk** (*Spilogale putorius*) is the only regular tree climber and will sometimes even nest in trees. The coat pattern on the spotted skunk is unique to each animal. All skunks can swim.

Although much maligned, skunks are extremely valuable in that they are one of the best insect extermina-

tors around. Despite the many stories, most skunks rarely feed on poultry or game birds' eggs. In one study of 1,700 skunks, not a trace of either was found in their diet. Spotted skunks feed more on eggs than do other skunks since they are the only ones to climb trees regularly. They are known to open eggs by throwing them back between their hind legs, mongoose-style. Spotted skunks and many of the **hog-nosed skunks** (*Conepatus* spp.) occasionally feed on snakes and are partially immune to the venom of dangerous species. They are 10–15 times more resistant to the venom of rattlesnakes than are rabbits of similar size. In fact, their odor has been shown to alarm rattlesnakes. Skunks raid beehives and are seemingly not bothered by the stings. They roll fuzzy caterpillars before ingestion to remove many of the venomous bristles. Bears, cougars, and lynxes have all been known to abandon a fresh kill to the approach of a skunk. Skunks readily eat carrion. They will also take vegetation, and hog-nosed skunks will feed on the fruit of the prickly pear cactus.

During the mating season, **striped skunks** (*Mephitis mephitis*) participate in communal dances, while male spotted skunks go beserk in what has been described as a "mating frenzy." During this period, the males take part in all sorts of strange activities, leading many who have observed them to believe they are rabid. During this time, they have been known to attack wolves or to run about indiscriminately and without provocation spraying large animals such as cattle or even humans. While so excited, spotted skunks occasionally run around on their front feet. In fights for dominance, they will rarely resort to squirting each other. They are known to be careful about getting any of their secretions on their own fur. Female striped skunks are one of the few mustelids that do not exhibit delayed implantation.

Surprisingly, skunk odor is described as pleasant when very dilute, and secretions from the spotted skunk have been used as a fixative in the manufacture of expensive perfumes. Generally, skunks are hesitant to use their chemical weapon and give ample warning before bringing it into play. Such warning usually consists of thumping the ground with their front feet, hissing, raising their fur, and waving their tails. Striped skunks usually spray from a

Striped Skunk (*Mephitis mephitis*)

back-turned position, while spotted skunks occasionally do it from a handstand posture. In that posture, spotted skunks are known to retreat or even to back towards the enemy if he doesn't take the initial hint. They usually squirt with all four feet on the ground bending their bodies to face their target with both ends.

It is not true that a skunk can't spray if held by the tail, and it need not face its target with its rear end in order to give it a dousing. The twin ejection ducts can protrude to be aimed in various directions. The two sprays join about one foot from their source. Some skunks can send the spray up to 15 feet, although it generally goes less than 10. With a good wind, the spray can be effective 15 or more yards away, and the odor is detectable for a mile and a half. The animals frequently aim for their victim's eyes. If their aim is successful, the repugnant fluid can cause temporary blindness for up to 20 minutes. The odor may linger on victims for as long as two weeks. In the dark, the spray of some is reported to be phosphorescent. The noxious substance common to all skunks is N-butlymercaptan. It is often accompanied by sulfuric acid. Each skunk can store up to three teaspoons of fluid, good for 5–8 successive sprays. It takes a skunk about a week to regenerate two teaspoons' worth of fluid. Young can first spray at approximately six weeks of age. The odor is best removed with tomato juice or ammonia.

Skunks are readily taken by great horned owls, which are not deterred by the scent even though they can usually attack and kill before the skunk brings it into play. Nevertheless, many great horned owls do reek of skunk scent. Skunks are also occasionally eaten by badgers, bobcats, pumas, and gray foxes, but most animals, including rattlesnakes, give them a wide berth. When threatened, many skunks stand their ground, which makes the automobile their major enemy. They can be successfully barred from lawns, gardens, and crawl spaces by putting out mothballs. Unfortunately they are an important vector in the transmission of rabies.

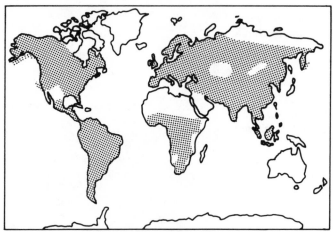

Family Mustelidae—Otters

Otters (13 species)

Otters range in size from the **small-clawed otter** (*Aonyx cinerea*), which runs about ten pounds, to the five-foot, 90 pound **sea otter** (*Enhydra lutris*), the smallest marine mammal in existence today but also the largest mustelid. Most species are aquatic to some degree and are so adapted. Their ears and nostrils can be closed off when underwater, and their fur is waterproof. Foot webbing varies. The forepaws of the **clawless otter** (*Aonyx capensis*) have none, but the **saro** or **giant otter** (*Pteronura brasilensis*) has thick webbing so extensive it handicaps movement on land. The sea otter is the only mustelid with retractile claws. The **Northern River otter** (*Lutra canadensis*) has calloused protuberances on its rear foot pads to increase traction on slippery surfaces.

Otters can see equally well in water and air. This is possible because the eye lens is elastic enough to become round and can be modified anteriorly by the muscular iris. Sea otters shade their eyes with an upraised paw on sunny days and on bright moonlit nights when they are trying to sleep.

The waterproof fur of the sea otter is the most luxuriant of furs. It contains up to 650,000 hairs per square inch. This is almost a billion per animal, making the sea otter the most thickly furred animal in existence. In comparison, there are about 100,000 hairs on the average human's head. The dense fur is necessary because the sea otter lacks the thick, insulating layer of blubber that protects all other sea mammals from the cold.

Air trapped in the fur also gives them the buoyancy they lack without blubber. Since each hair is thin at the base and thicker in the middle, air can be trapped at the bottom to keep water from ever touching the otter's skin. The otter traps air within its fur by frequent somersaulting and rolling. It blows bubbles while holding its fur apart with its paws, and then rolling to seal the air with a coating of water.

The sea otter must take excellent care of its fur for survival. At least 10 and as much as 50 percent

210

of each day is spent grooming the fur. If dirty or contaminated with oil, the fur loses its capacity to hold air and the animal will die from exposure. The fur on the paws does not trap air, and in cold situations they will hold their paws out of the water to decrease heat loss. They may also do this to increase sunlight exposure to enhance the manufacture of Vitamin D. In really cold situations they, like seals, dogs, ground squirrels, and others, will roll up into a ball to decrease surface area and therefore radiate away less heat. Their skin is quite loose and can be rolled to form a belly pouch capable of holding rocks, shells, and sea urchins.

Most otters have very sensitive whiskers. **River otters** (*Lutra* spp.) pick up fish vibrations through their vibrissae. They can also use their whiskers to "feel" out prey such as amphibians, crustaceans, and mollusks on muddy river bottoms. In addition to whiskers, the sea otter uses sensory hairs on its front feet to detect the presence of swimming fish. It is the only sea mammal to catch prey, including fish, with its hands.

Most otters are excellent swimmers. River otters can swim up to seven miles per hour, dive to 60 feet, stay submerged for 6–8 minutes, and can go up to half a mile underwater with one breath. In winter they have been reported to breathe air trapped under ice to increase their down time. They can outswim and outmaneuver many fish species. They are capable of swimming in tight circles to create vortices that suck up mud-hiding creatures or even to pull sheltered fish from beneath underwater ledges. The sea otter can dive to 300 feet and stay submerged for as long as five minutes. They rarely swim belly down when at the surface and reportedly swim fastest on their backs. California sea otters rarely venture onto shore, but Alaskan sea otters do so quite frequently during storms, to give birth, and for other purposes.

Otters are well known for their never-ending indulgence in "play." They frequently perform various types of dives, rolls, and flips into the water. They often spend time skimming down mud or ice slides up to 25 feet long. Such slides may even be cleared of rocks or other obstacles. They may repeatedly fetch rocks they toss into the water, playfully tug on a beaver's tail then swim away, balance a rock or mussel on their head while swimming underwater, or swim in circles chasing their own tails. While swimming at full speed, otters may suddenly flex into ball form resulting in numerous somersaulting rolls. Some play "hide and seek" or "tag" or even shoot fast-flowing rapids. Sea otters are even known

River Otter (*Lutra canadensis*)

to "surf," and their young seem to enjoy pouncing on sleeping adults.

Sea otters use tools more frequently than any other mammals except chimpanzees and humans. They carry rocks and shells to be used as hammers or anvils in breaking open prey such as shellfish and sea urchins. These objects are also used as ballast to increase the speed of descents. These tools can be carried in pouch-like folds of skin, but they are also tucked under the armpit, usually the left. The small-clawed otter holds objects with one forepaw against its side while it runs about on three legs, and the clawless otter will use hard objects to break open mussels.

Sea otters also make liberal use of the **giant kelp** (*Macrocystis* spp.) off the California coast. This kelp can grow to 200 feet in length and is the fastest growing plant in the world, growing up to two feet per day under optimum conditions. A single plant can support more than half a million small animals. Sea otters use the kelp in a number of ways. When sleeping, they frequently wrap themselves in it to prevent currents or tides from taking them out to sea; if disturbed in their kelp-wrapped slumber by raucous gulls, they are known to splash the pesky birds with seawater. They also use bladder-bearing seaweeds as floats and construct small rafts from them for their young. After feeding, the otters have been reported to use sea grass and kelp as dental floss to clean their teeth.

In some areas, the kelp depends on the sea otter for its existence, since it is destroyed by sea urchins which consume its holdfasts. In areas where there are no otters, the urchins have become over-abundant and have wiped out the kelp. Since urchins are a staple of sea otters' diets, the latter help preserve the kelp beds. In fact, sea otters are the only mammals to feed primarily on echinoderms. An adult may consume up to 6,500 small urchins a day, and some otter teeth and bones become purple because of their heavy urchin diet. On rare occasions sea otters have been killed by an urchin spine penetrating the stomach.

In addition to urchins, sea otters eat fish and mollusks. They will even nibble the arms of starfish, which then grow new arm tips for future meals. Being warm-blooded and living in such a cold environment, sea otters must consume up to one fourth their weight in food each day. If deprived of food for a single day, they can lose up to ten pounds; if deprived for three days, they can die from heat loss since they are then unable to maintain their high rate of metabolism. Captive river otters, which also require a high daily intake, eat more food than a lion and cost more to feed than an elephant. Dominant sea otters obtain some of their food by stealing it from subordinate animals (which don't appear to mind the loss).

Otters often dine while floating on their backs, using their bellies as tables. They quickly clear away the resulting mess by performing a few rolls in the water. They can drink some sea water without harm, and they consume about two quarts of it daily to supply a fourth of their water needs. The other 75 percent is derived from their prey's body fluids.

Most of the otters hunt singly, usually seeking their prey in the water. River otters churn up muddy bottoms with their forepaws to flush prey or use their sensitive whiskers to feel them out. Otters occasionally will pull a swimming bird from the surface after an underwater stalk. River otters will persistently chase a single fish until it tires and can be easily caught. They seem to prefer "trash" fish, but this may be because these species are slower and easier to catch. They eat many small animals including venomous snakes, wasps, and occasionally some vegetable matter such as blueberries. Some, such as the Northern river otter and the giant otter, will hunt in small groups (probably families) to drive fish schools into shallow water. Occasionally, otters will cache excess fish meat in vegetation for later use, but most won't eat carrion.

River otters hold territories up to nine miles in diameter. They are marked with spraints which consist of urine, feces, and/or anal gland secretions. All mustelids except the sea otter possess functional anal glands. River otter dens or holts generally have entrances that are accessible only by openings placed about 20 inches below the water's surface. The dens usually have air shafts leading above ground. Should a predator dig in toward the den the otter can quickly exit through the underwater entrance.

Giant otters, smooth-coated otters, and short-clawed otters live in family groups while most other species are solitary. Giant otters are thought to mate for life. Male sea otters signal their amorous intentions by approaching the females while swimming on their bellies rather than on their backs. During courtship, the males hold the females by the nose, often with enough force to cause bleeding. This is the reason why many older female sea otters have scarred noses. Female river otters "caterwaul" during copulation. At such times the males hold them by biting the scruff of the neck. Otters, with a few exceptions such as the Eurasian river otter, exhibit delayed implantation.

River otters give birth in their underground dens. Sea otters used to give birth primarily on land but now do so more often in the sea. The females sometimes use seaweed cribs for their newborn young, which are born at a more advanced stage than any other mustelid. Newborn sea otters are known as woollies, and a group of sea otters is generally referred to as a raft. A raft will usually number between two and sixteen, but congregations of up to 170 have been reported. The females nurse while swimming on their backs, groom the young often, and will blow insulating bubbles into their baby's fur. They have been observed to throw their young up into the air and catch them time after time. They are even known to take some young down on dives for food. Young sea otters are overly buoyant and have a hard time diving while in this stage. In contrast, very young river otters are not buoyant enough until their waterproof underfur grows in; prior to that time they tend to sink. Surprisingly, many of the young are quite fearful of deep water. The mother otter teaches them to swim, but she must often force them in or resort to bribery with food items. She will also teach them to hunt by releasing prey animals in front of the young.

Mother otters will often lead their young in a processional line. Some think that this behavior has led to many sightings of lake monsters such as the Canadian "ogo-pogo." In 1911, Teddy Roosevelt was hunting in Kenya when the Lake Naivasha monster was sighted. When Teddy promptly shot at the middle hump from his boat, all the other humps immediately disappeared. The remaining hump proved to be an otter. A monster so formed could appear to be quite large since the giant otter can run eight feet long, including a 3.5 foot tail, and can weigh 75 pounds. Female sea otters with young will dive with their offspring when threatened. When the baby is very young, this has been known to cause death from drowning since the infant cannot hold its breath as long as its mother.

When frightened, some otters emit a foul-smelling fluid from their anal glands, but most quickly head for the safety of water. The **smooth-coated otter** (*Lutra perspicillata*), when alarmed, emits an explosive snort that has at times frightened away an attacker. If cornered, this otter will attack fiercely while screaming. Sea otters are said to post sentries to watch for danger. They hide in kelp beds at the approach of sharks or killer whales. They are remarkably free of parasites and disease as no wild sea otter has ever been found with ectoparasites.

Several species, including the small-clawed otter and the **European river otter** (*Lutra lutra*), have been tamed by humans and used for fishing. They will both retrieve individual fish and drive schools into nets. The Northern river otter has been trained to retrieve wildfowl for hunters.

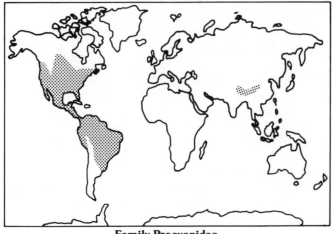

Family Procyonidae

Family Procyonidae (raccoons and relatives — 19 species)

The Procyonidae are a family of medium-sized carnivores. As a rule, they are quite secretive in the wild. Many exhibit anatomical features adapting them to their habitats. The **ring-tailed cat** (*Bassariscus astutus*), also known as the **band-tailed cat, cacomistle, civet cat, coon cat, miner's cat, mountain cat, raccoon fox,** and **squirrel cat**, has hair-covered soles and partially retractile claws that help it keep its footing on the smooth rocks of its home range. The **lesser panda, cat bear, fire fox,** or **red panda** (*Ailurus fulgens*) is similarly adapted for traction on smooth branches and ice; its hairy soles also help conserve body heat. Like its larger namesakes, the lesser panda also has a small accessory foot pad that can be used as an extra "finger" to oppose its true digits and grasp readily the bamboo stems, leaves, and fruit upon which it feeds.

The **kinkajou** (*Potos flavus*), sometimes known as the **honeybear**, and the **binturong** (*Arctictis binturong*) of the family Viverridae are the only carnivores with prehensile tails. The tails of binturongs are only partially prehensile. The kinkajou will sometimes use its tail to grasp fruit hanging beyond the reach of its legs. It is one of the very few mammals capable of climbing up its own tail. Many of the other species in this family, including the lesser panda, **raccoon** (*Procyon lotor*), and the ring-tailed cat have long, bushy tails which can be wrapped around their bodies when sleeping to aid in retaining heat.

Many of these animals are quite agile. The ring-tailed cats are well known as escape artists, and they are agile enough to sneak up on sleeping birds. They can jump ten feet from branch to branch, ricochet off smooth surfaces, and climb up crevices by wedging their backs against one surface and their legs against the other, just as human mountaineers climb rock "chimneys." Raccoons can jump down 40 feet from trees without apparent harm. The **coatis** (*Nasua* spp.) can stand on their hind legs and bat insects from the air with their forepaws.

Procyonids are opportunistic boarders, taking up residence in any convenient shelter, such as a hollow log or empty den or nest. Ring-tailed cats frequently inhabit abandoned squirrel dreys. Raccoons occasionally shelter in muskrat houses, often after consuming the previous occupants. In colder climes, raccoons "hibernate" (not a true hibernation) through the winter and may lose up to half their body weight.

214

The members of this family are omnivorous feeders. Ringtails eat juniper berries, cactus fruits, and other vegetation. Vegetable matter composes about 25 percent of their diet; the rest is small animals, including bats, spiders, and scorpions. Raccoons will feed on almost anything, including bees and hornets, whose nests they readily attack while their thick fur protects them from the stings. They will skin tiger salamanders of their poisonous skins before eating and are known to obtain food by chasing other small predators, such as gray foxes, from their kills. The diet of raccoons is about 50 percent vegetarian. Coatis eat prickly pears, ants, spiders, scorpions, centipedes, and will rub wasps and bees on the ground to remove their stingers before eating them. Many procyonids store excess food as fat in the tail region. The lesser panda and kinkajou are primarily vegetarians. The latter has a particular liking for honey, which is said to give captive specimens a craving for alcoholic beverages. When inebriated, the little beasts have been known to go quite beserk and even become vicious, attacking anything in sight.

It is often stated that raccoons are very clean animals since they wash their food before eating it. Recent studies have shown that this behavior rarely if ever takes place in the wild although it often occurs in captive individuals. For a while there were claims that the raccoon had to moisten its food because its salivary output was low, but this has definitely been shown to be untrue. Some experts say that the animals like the feel of their food in water, this being why some scrub their delicacies until they are no longer recognizable. Others claim the act of wetting their forepaws increases their tactile acuity so they can separate out inedible items by touch.

Raccoons in the wild are known to hunt by feel in small streams, and this is probably the best explanation for observed "washings." The raccoon is even known to attract fish by lightly tickling the water's surface with its long fingers. The real reason

Red Panda (*Ailurus fulgens*)

215

captive raccoons "wash" their food may thus lie with an instinctive pattern of behavior that requires that food first be found, and then be eaten. The raccoon may put its food in the water only so it can discover it and then release the behavior of feeding. We can see a similar behavioral pattern when we feed dead animals to cats, which often go through the ritualized motions of hunting and killing we call "playing with their food."

Some species mark territories but do not defend them as vigorously as most other mammalian families. To show ownership, raccoons and ringtails pile feces on rocks, tree limbs, or the base of tree trunks. Red pandas also pile feces but additionally mark with anal gland secretions. Procyonids are generally nocturnal and solitary. Coatis are diurnal, and the females form social "bands" numbering up to 20. During the breeding season, each such band becomes the "harem" of a male, although he remains subordinate to them. At other times, the males are not tolerated and will be driven away, for they may kill and eat the young.

All members of this family are thought to be promiscuous breeders. In female raccoons, copulation probably triggers ovulation. Delayed implantation can be found in the lesser panda. Male procyonids take no part in raising the young, and it is not unknown for male raccoons to kill and eat the babies. Female coatis and raccoons will adopt and care for orphaned young. Offspring in this family are born with their eyes and ears closed and are sparsely haired. Female ringtails eat their babies' waste products to keep the nest clean, and mother raccoons tear away loose bark from the den tree, supposedly to keep her young from falling.

Most procyonids flee when threatened. The ringtails can release noxious fluids from their anal glands when frightened. Raccoons will lead pursuing dogs to water, where they have been observed to turn on their attackers and hold them underwater until they drown. At other times, they have been observed to run part way up a tree, jump down, and take off, leaving the dogs thinking they have treed their prey.

Family Ursidae
(bears—8 species)

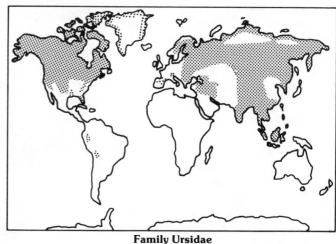

Family Ursidae

Although the single largest bear specimen recorded was a **polar bear** (*Ursus maritimus*) that weighed 2,210 pounds and stood over 11 feet tall, the title of largest terrestrial carnivore goes to the **Kodiak bear** (*Ursus arctos middendorffi*). Adult males in this species average about eight feet in length and weigh about 1,100 pounds, while adult male polar bears average only about 900 pounds.

The recent use of sophisticated serum protein techniques has shown that the **giant panda** (*Ailuropoda melanoleuca*) is in fact a true bear, although many persist in placing it in its traditional place among the Procyonidae. Like other bears, it resembles humans anatomically to an endearing degree. Like dogs, bears have a "wet" nose or rhinarium to increase their sense of smell, their keenest sensory modality.

Although bears are often named by color, a person cannot identify a species through color alone. For example, the **black bear** (*Ursus americanus*) has white (**Kermodes** or **island white bear**; *U.a. kermodei*), brown (**cinnamon bear**; *U.a. cinnamomum*), blue (**glacier** or **blue bear**; *U.a. emmonsi*), and of course black varieties. **Brown bears** (*Ursus arctos*) also show many color phases. As a rule, black bears are tallest in the rump, while brown bears are tallest at their shoulder hump. Although the fur of polar bears is white (it is often tinted yellow as a result of oxidation by the sun or the seal oil in their diet), the underlying skin is black. The white coloration, as in other polar animals, has recently been shown to serve the function of heat retention rather than—or in addition to—camouflage. Although it is true that dark coats absorb more heat, in windy areas light coats dissipate less heat. In addition, since the white is due to air in the hairs, white fur is a better insulator. The hairs also reflect back body heat, their hollows channel ultraviolet light down to the skin for increased heat absorption, and they increase buoyancy when swimming. A few specimens have developed greenish coats as a result of algae growing inside the hollow hairs. The panda's skin is black underneath its black fur and pink below the white areas.

Polar bears show many other anatomical adaptations to life in the harsh Arctic region. They have elongated eyes and nictitating membranes to protect them from snow blindness and the cold polar water, respectively. Their underfur is waterproof and overlies an insulating layer of fat up to four inches thick. There are also skin muscles to control the size of surface vessels that aid in temperature regulation. Their extra wide paws serve as snowshoes on loosely packed snow and help keep them from falling through thin ice. On sunny days the paws are occasionally used to shade the eyes when napping. All four feet are thickly haired to increase traction on ice, conserve heat, and to decrease noise when stalking. Hairy, nonskid soles are present in many other animals as well, including the Arctic fox, Arctic hare, lynxes, wolverines, and some weasels. It can be worn away by rough terrain.

Although all bears can swim, polar bears are the most skilled. With their powerful, partially webbed forepaws, they can swim up to six miles per hour and have swum distances of 60 miles without rest. They usually swim with the head up while dog paddling with the forepaws, but in rough water they will actually swim underwater in spurts. They have been found swimming 200 miles from the nearest land; they had probably drifted far out on a chunk of ice that subsequently melted. Polar bears can leap up to eight feet in the air from a swimming start. When black bears swim, they seem to head in a direct line toward their goal. Should an obstacle such as a boat lie in their path, these bears have been known to clamber right over it rather than swim around it.

Bear vocalizations generally consist of various grunts and growls. **Grizzly bears** (*Ursus arctos horribilis*) are reported to produce a warning "cough" prior to attack, and pandas frequently produce chirping sounds.

Bears are powerful animals and should always be given a wide berth, no matter how gentle or human they appear. Too many park visitors have been injured simply because they did not heed the "Don't feed the bears" signs. Bears can easily kill with a casual swipe of the paw; more often they deliver a vicious bite while on all fours. There are reports of grizzlies breaking the necks of full-grown bison with a single blow from a forepaw (which carries claws up to five inches long). They have carried

Sloth Bear (*Ursus ursinus*)

off carcasses weighing up to 1,000 pounds. Large brown bears and polar bears can outrun horses and caribou for short distances, and both have been clocked at 35 miles per hour. A polar bear's forepaw can weigh up to 50 pounds, and with it the bear has been known to kill 700-pound bearded seals with one blow. This enormous strength actually let one polar bear snatch, "one-handed," a 200-pound seal up through a small breathing hole; the seal's pelvis was crushed as its body was yanked through the small hole. Most polar bears appear to be left-handed.

Most bears, except brown and polar bears, can climb trees even as adults. Young brown bears can climb easily, and a few adults have been known to do so as well. Some, such as the **sun bear** (*Ursus malayanus*), smallest and most arboreal of all bears, will even build nests as high as 60 feet in the trees where it sleeps, basks, and feeds. Both the **spectacled bear** (*Tremarctos ornatus*) and the **Himalayan black** or **moon bear** (*Ursus thibetanus*) build stick nests in trees as well. The **sloth bear** (*Ursus ursinus*), named for its slow gait and appearance, will climb to eat fermented sap being collected in pots high in date palms. They frequently become drunk on the sap and are known to fall 20–30 feet, producing loud thuds but apparently suffering no harm. They are, however, thought to suffer hangovers.

One often hears that bears hibernate in winter. Actually bears do *not* hibernate in the strict sense of the word. They show only "winter dormancy." In true hibernation, complex physiological changes take place. The blood levels of electrolytes (such as potassium and magnesium) and certain hormones (such as insulin) increase. The resulting changes help to protect the body from cold and lack of food. They include decreases in heart rate, temperature, and body metabolism in general. While in such a state, animals do not eat, drink, urinate, defecate, or even respond to severe physical stimuli. In many true hibernators, the process is brought on by cold weather. As the animal's body temperature falls, the blood's concentration of large molecules increases to provide an effective antifreeze and prevent the crystallization of body fluids. About 90 percent of the white blood cells migrate out of the bloodstream to other parts of the body, particularly the gastrointestinal tract, to protect against infection. Hibernating animals produce increased amounts of fibrinolytic enzymes to keep their sluggishly flowing blood from clotting. True hibernation can be found in bats, birds, monotremes, marsupials, rodents (at least four families, including Cricetidae Gliridae, Sciuridae, and Zapodidae), and insectivores (family Erinaceidae). Since bears do not exhibit the above conditions to the same degree, they do not truly hibernate.

In winter dormancy, bears do exhibit some of the patterns of true hibernation, and some authorities insist that bears are true hibernators. Heart rates rall from 90 to as low as eight per minute, body

218

temperature may fall 10–12°F, and many bears do not eat, drink, urinate, or defecate when in this state. Some bears don't sleep at all during winter, while others have been known to lie dormant for as long as six months. There is wide variation within a species, depending upon the area. They metabolize primarily fat during dormancy to avoid a build-up of toxic metabolites (such as urea from use of proteins). What little urea that is formed has been found to be metabolized into new protein. Black bears have up to five inches of fat just prior to their sleep, and Alaskan brown bears have been known to put on as much as 400 pounds of fat. The gastrointestinal tract is emptied by not eating for a few days prior to dormancy. The last meal consists of pine needles, wooden splinters, and even the animal's own hair to form an "anal plug" measuring up to a foot long. This prevents the animal from defecating in the den or on itself. The stomach and intestines then shrivel to only a fraction of their normal size. The kidneys do continue to produce urine, but it is reabsorbed through the bladder. A bear may lose a fourth of its weight during winter dormancy. After "hibernation," the bears fill up on ants, grasses, and fruits, which act as laxatives to rid them of their anal plugs. Since the large calluses on their feet have softened through disuse, they avoid walking on rough ground for a while after emerging from winter's sleep. Black bears are even known to eat callus pads that fall off at this time. In early spring, most bears construct crude beds to on above the frozen ground.

Bears lay dormant in various places including caves, dens, windfalls, and even tree holes as high as 60 feet. Entrances are often plugged with vegetation to help exclude the cold. In many dens, the ground slopes up from the entrance to hold in heat and prevent water from dripping in. Grizzlies seem to have an uncanny awareness of when the first lasting snowstorm is about to take place. Each grizzly moves toward its den to "hibernate" just prior to the storm's onset; the storm then covers their tracks. If the storm is not the one that brings the first lasting snowfall, the bears often do not enter their dens. When sleeping, sloth bears are notorious for their loud snoring. Bears can be aroused from winter sleep.

During dormancy, some bears have been observed sucking their paws, although the reason for this behavior is not clear. It may be related to the Himalayan black bear's habit of licking its paws until a foamy paste forms. It then sucks the paste off. The reasons hypothesized for this act include the following: they may somehow gain nourishment in this manner, the bears are afflicted with bear fever, glands in the paws provide substances that aid in digestion, it is a form of communication between bears, it is a bad habit, or it may be self-play developed out of boredom. None of the theories has been tested or proven.

Like many other large mammals, bears will frequently engage in mud baths. This helps to protect them from insects, aids in cooling them off, and assists the bear in shedding old fur to make room for new. Black bears, like dogs, will roll their bodies in foul-smelling substances.

As a rule, bears are omnivorous. Pandas feed primarily on five of the 20 species of bamboo. Since bamboo is low in nutritional value, particularly protein, they spend up to 12 hours a day consuming as much as 90 pounds. They digest less than 20 percent of the bamboo they eat. This results in the production of large amounts of fecal material. Pandas also feed on a wide variety of other substances, including plants, fruit, mice, fish, bird eggs, insects, and honey along with a few bees, but these items make up only a small percentage of their intake. Brown bears feed on an even wider variety of food items, including pine cones, nettles, wasp nests, and even porcupines (which have caused the demise of more than one bear). Brown bears have been known to eat soap, toothpaste, and feathers from bedding. They may dig out a ton of dirt to get at a ground squirrel. While so doing the bear may reflexively pounce and "capture" a rock it has thrown out, mistaking it for an escaping squirrel. Brown bears have even been known to kill and eat black bears. Cannibalism has been reported in brown, black, and polar bears.

Black bears also have catholic tastes. One was found with more than two quarts of yellow jackets in its stomach. The stings do not apparently penetrate the bear's oral mucosa. Interestingly, heavy honey feeders are known to get cavities. Black bears occasionally feast on apples, whose juice may ferment in their stomach and lead to drunkenness. The sloth bear is a termite feeder and, like the aardvark, can close off its nostrils to ward off attacking termites. The bear forms its mouth and lips into a tube and sucks up the termites through a gap in its front teeth. In so doing, it produces a suctioning sound, similar to a vacuum cleaner, audible up to 200 yards away. Sun bears dig out termite nests and then repeatedly place their paws in the rubble. After the insects have swarmed onto them, the bear licks its paws clean.

Black bears are notorious for raiding the food packs of campers. Even suspended packs have been obtained, either by chewing through the rope or by climbing the tree and leaping to the ground, grabbing the pack on the way down. Grizzlies are known to store excess food, sometimes by burying it in huge holes and at times by lying down on top of the food until they become hungry again. One was observed to dig a hole that could accomodate two deer merely to bury a single marmot. In the few weeks prior to "hibernation," bears may eat up to 2,000,000 kilo-calories per day.

The most carnivorous of the bears is the polar bear. Although these skillful hunters would seem to be well camouflaged, some of them are apparently aware that their black noses ruin the effect. They have been observed to cover their noses with a paw or even with a small pile of snow or chunk of ice that is inched forward while the bear stalks its prey on its belly. Recent researchers have not witnessed nose camouflaging behavior.

When on thin ice, polar bears spread out as much as possible to avoid breaking through. They have been reported to sneak up slowly on sleeping walruses and bash their heads with chunks of ice held between the forepaws. At times, they will heave large rocks down on prey as well. If in the water, however, the bear usually comes out on the short end if it attacks a walrus; more than one bear has been killed by the deadly tusks.

Polar bears have been observed to catch floating seabirds when swimming. They can smell seal dens through 5–6 feet of snow, and fresh meat several miles distant (possibly as far as ten miles). If, after breaking through a seal's den, the bear finds only a young pup, it is not unknown for it to take it to the edge of the floe or to a breathing hole and hold the pup down with one paw. The female will rush to the pup's screams only to be caught in the bear's trap.

Another technique some females teach their young is to block up all nearby seal breathing holes save one where the bear takes up watch. When the seal appears, the bear takes a swipe at it, usually with the left paw, digs its claws into the blubber, and pulls up its dinner. Polar bears also have been known to catch a seal by diving underneath the ice and coming up through the seal's breathing-escape hole, beside which the seal is often basking. If the bear is noisy in its approach, the seal usually panics and literally dives into the bear's waiting arms. When stalking, the bears frequently will synchronize their approach with the seal's sleep-wake cycle, which recurs every minute or so. Occasionally a polar bear has been known to get stuck head first in a seal's breathing hole. On rare occasions this has led to death. Polar bears have been known to fly into a "rage" when a seal escapes. One was observed to smack its paw on some rocks, causing numerous fractures to paw bones.

Polar bears can eat up to 100 pounds at a single meal. When food is scarce, they will bury leftover food. They crave blubber and also like to chew on plastic. In summer, they are known to eat grass, berries, and seaweed. Some researchers feel that grass serves not only as a laxative for bears, but also possibly as an antacid. Polar bears will dive up to six feet deep to fetch seaweed when hungry. Many of them carefully wash after meals. They are fastidious, and some even defecate into the water by hanging their rear end over the ice edge.

Most bears live solitary lives within small territories marked with urine, glandular secretions, and scratch marks. They are generally creatures of habit and do very little roaming, although there are exceptions such as the polar bear. Bears have been known to get into territorial disputes. One large grizzly male was once observed to throw another 20 feet in just such a fight. Bears show submission in a number of ways, including the lowering of their heads, turning their sides towards the aggressor, or by sitting down in the presence of a more dominant member. Himalayan black bears have been known to roam about in small family groups, often in single file with the male leading the way. In all other species, the female drives the male away from the cubs. This is true even for the sloth and sun bears, which may mate for life. Polar bears occasionally mingle in large groups. If well fed they will often "play." Interestingly, large bears will allow smaller bears to win wrestling contests to keep the smaller ones playing.

Most bear species are capable of breeding only once every two years. Bear courtship often consists of mutual rubbing, licking, and even wrestling. Grizzly females (sows) have discouraged mating attempts by males (boars) by digging a small burrow and backing into it. On the other hand, female grizzlies have also been known to "back in" to males in invitation to mating. During copulation, male polar bears often bite the female's neck. Ovulation is triggered by mating.

Bear gestations last 6–9 months, and most species show delayed implantation. Females generally give birth to two or three cubs, more rarely one, four, or as many as six. In brown bears, twins are almost always of opposite sex. The young are usually born during the female's "hibernation." The offspring of most are blind and sparsely haired and weigh less than a pound. Newborn black bears weigh about half a pound. In pandas, the young weigh only 3–5 ounces, or about 1/800 of their mother's weight. These panda infants exhibit a ten-fold weight gain in their first month. The female panda holds her young in her forearms when nursing.

In polar bears, the young are born in the five-by-four-foot snow den. The female controls the den temperature by widening or covering over the ventilation hole in the ceiling. In bad weather, the den can be 40°F warmer than outside. Newborn cubs do not sleep on the cold den floor but in their mother's fur until their naked skin becomes covered with a protective coat of fur. The female sleeps through their first few months of growth but does not sleep through the birth process as once believed.

In many species, the females will absorb the fertilized eggs if they have not put on enough fat by the time of denning. Females that give birth in winter lose up to 40 percent of their weight before spring. As mentioned, the males of most species are driven off since they might consume the young if given the chance. Occasionally, male sloth bears are allowed near older young, and they may even help raise them. Black bear mothers are known to eat the feces of their young at times.

Female bears transport their young by placing the entire head within their mouth. Sloth bear females carrying their young on their backs even when climbing trees. The young of polar bears are also known to ride on the mother's back while on land and to hitch rides by hanging onto her tail when in the water. The females will sometimes push the cubs off in order to teach them how to swim. Young polar bears may stay with their mother for up to two years, keeping them from giving birth more often than every third year. They may live to be 30 years old or more. Brown bears in captivity can reach 50 years.

Female brown bears will chase their progeny into a tree when danger threatens. They will then either face the problem or attempt to lure it away. The cubs are taught to stay up until she gives them an all-clear signal. Females have been known literally to spank disobedient cubs. Female brown bears will not only nurse the young of others but will adopt them outright. On occasion, sows have reclaimed a previously lost cub.

Most bears perform playful antics both in the wild and in captivity. Brown bear females and cubs have been observed shooting rapids over and over. Polar bears will sometimes amuse themselves with stones, occasionally even balancing them on top of their heads. Himalayan black bears have the curious habit of curling into a ball and rolling down hills over and over again, and brown bears are known to repeatedly slide down snow banks on their haunches. Black bear cubs often climb up saplings, using their weight to ride them to the ground time after time. Captive pandas are quite playful, performing handsprings and rolls and successfully breaking just about every toy given them. Sloth bears have traditionally been circus favorites because of their ability to juggle.

Although it is true that black bears, grizzlies, polar bears, and Himalayan black bears have killed humans, their reputed danger to people has been far overrated. This is not to say that they are not dangerous, but that when confronted they will in all likelihood head in another direction. In attacks, bears rarely rear up on their hind legs and take swipes with their paws; they never kill by giving "bear hugs." Usually, they approach on all fours and attack with their teeth. There are reports of polar bears attacking boats if harrassed.

The livers of polar bears, as well as those of sharks, bearded seals, ringed seals, walruses, and some sea lions, are poisonous to eat. This is due to their high concentration of stored vitamin A. When consumed, it causes nausea, headache, and sometimes a dermatitis. Almost 60 percent of the polar bears in one study were infected with the parasite *Trichinella spiralis* and therefore trichinosis.

13

ORDER PINNIPEDIA
(34 Species)

There are three families of marine carnivores: the true seals (Phocidae), sea lions and fur seals (Otariidae), and the walrus (Odobenidae). We will consider all three families together because of their many similarities. There are, however, major observable differences among the families. The sea lions and fur seals have visible external ears while the true seals and walrus don't. True seals derive most of their propulsive force from the work of their rear flippers, while sea lions and fur seals obtain theirs primarily from the front flipper. Walruses use both. The sea lions, fur seals, and walrus have naked flippers and can tuck their hind limbs underneath their bodies so they can "walk" on land. The true seals cannot do this and must move about like caterpillars or inch-worms. Both the **crabeater seal** (*Lobodon carcinophagus*) and the **ribbon seal** (*Phoca fasciata*) are said to travel occasionally by twisting the body from side to side while pulling the body forward with the front flippers. The awkward gaits of the true seals are not inefficient, however; the crabeater seal and ribbon seal can at times outrun a human on the ground.

Pinnipeds range in size from the 4.5 foot, 140 pound **Baikal seal** (*Phoca sibirica*) to the huge **southern elephant seal** (*Mirounga leonina*) which weighs, 9,000 pounds and

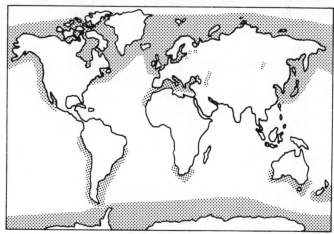

Order Pinnipedia

Northern Fur Seals (*Callorhinus ursinus*)

reaches lengths of 22 feet. Male pinnipeds are generally much larger or about the same size as corresponding females. Only in the **leopard seal** (*Hydrurga leptonyx*) is the female substantially larger than the male when full grown. The elephant seal exhibits the greatest difference in size between the sexes of all mammals. Generally, the males are 6,000 pounds heavier than the females. Elephant seals also pass their food through the longest—650 feet—intestinal tract of any animal (the human tract is only about 20 feet long), and their spines are the most flexible, being capable of bending a full 180 degrees back on itself.

Most pinnipeds inhabit only salt water, though there are, as always, exceptions. The Baikal seal lives in Siberia's freshwater Lake Baikal, over 1,000 miles from the nearest sea. One subspecies of the **harbor seal** (*Phoca vitulina*), the **Ungava seal** (*P.v. mellonae*), lives in the freshwater Upper and Lower Seal Lakes in Canada. Some ringed seals also inhabit freshwater lakes. A colony of harbor seals lives in freshwater Iliamna Lake in Alaska. Often individual **South American sea lions** (*Otaria flavescens*) can be found as much as 200 miles up the Rio de la Plata river. Harbor seals and others are also known to travel hundreds of miles up rivers. Pinnipeds do come ashore frequently, and the **Australian sea lion** (*Neophoca cinerea*), known for its cliff-climbing abilities, has been found as much as six miles inland. Some fur seals and monk seals have taken up residence in seaside caves living up to 100 feet inside.

All pinnipeds are well adapted to an aquatic life. All seals have nictitating membranes (as do most mammals except primates and cetaceans) to aid in underwater vision, although many lack lacrimal ducts. Thus, their tears must run down their faces, giving them a "saddened" appearance as though they have been or are crying (tears are *not* primarily a sign of pain, but a lubricant and cleanser for the eye). Many species also have a tapetum lucidum. In **harp seals** (*Phoca groenlandica*) and others, the corneas have a built-in astigmatism that increases visual acuity underwater when the pupils are dilated. On land, pupils constrict to a slit and eliminate the astigmatism to provide good visual acuity in air.

The valved nostrils of pinnipeds are closed as long as their muscles are relaxed; muscular contraction is necessary to open them. Pinnipeds do not breathe rhythmically but instead exhibit periods of 20–30 short breaths followed by apneic (no breathing) spells about ten minutes long. This pattern can be observed on land as well as in the water. The sense of smell is not well developed, but the rhinarium or snout pad may be able to pick up the scent of nearby fish schools and may be used by the females to locate their own pups. Eared seals (sea lions and fur seals) also must contract muscles to keep their ears open. Earless seals usually can contract their auditory canals to keep out water. Many pinnipeds have such supple backs that they can execute a 180-degree turn in less than one fourth of their body length.

Pinnipeds have many physiological similarities to the cetaceans. They have high concentrations of hemoglobin and myoglobin. The circulation to nonvital organs and tissues is decreased or shut down when diving. There is an increased tolerance to high levels of carbon dioxide and lactic acid in the bloodstream. They have rigid bronchi to hold residual air forced out of the lungs by water pressure; this prevents gaseous exchange at high pressures and thus averts the "bends." Also, blood sinuses bulge into the middle ear to compensate for air squeeze. When diving, their heart rate decreases to about a tenth its normal value through the mechanism known as the "seal reflex." The reflex, which exists in all mammals, including humans, is better developed in older, more experienced animals. The pulse rate of harp seals has been known to fall from 100 at the surface to as little as four during a dive.

In some diving pinnipeds, oxygen-depleted blood from the brain is stored in special liver sinuses until after the dive. It is thus taken out of circulation so that only oxygenated blood is pumped to the brain. The **Weddell seal** (*Leptonychotes weddelli*) is the champion diver of the order. It has been known to reach depths of almost 2,000 feet and can stay submerged, although not at that depth, for over an hour. It has been reported that elephant seals can dive to 2,000 feet, and they have been known to surface more than a mile from where they submerged. Probably the fastest of pinniped swimmers is the **California sea lion** (*Zalophus californianus*), which is capable of speeds reaching 25 miles per hour.

Pinnipeds use various means to regulate body temperature. The walrus and true seals have thick layers of blubber through which the blood flow and thus the amount of heat loss can be regulated. In times of extreme cold, their surface vessels are so constricted that these animals, and especially the walrus, appear very pale. Their outer layers of tissue can be maintained at a temperature so close to that of the environment that they can sleep on ice without melting it. If real cold, walruses will sometimes huddle together, and elephant seals are known to lie piled on top of each other; occasionally the seal on the bottom suffocates. California sea lions occasionally sleep on top of elephant seals for heat conservation.

Should the environment warm up, however, there can be trouble. The animals must then rid themselves of excess heat, and they can quickly melt away the ice. Elephant seals and others, particularly the young, have been known accidentally to commit suicide by sleeping too long in a snowy place; upon awakening, they find themselves at the bottom of a pit from which they cannot escape.

The animals can help rid themselves of heat by waving their well-vesseled flippers in the air. Fur seals sometimes urinate on their flippers to increase heat loss via evaporation. Some pinnipeds throw damp sand on their backs to enlist evaporation. Sea lions make their own damp sand by urinating on it. When dry, the lighter-colored sand also reflects the sunlight away and keeps the skin from drying to excess.

The fur seals do not have as thick a coat of blubber to protect them from the cold. Instead they rely upon two coats of fur. The outer coat is composed of long, coarse guard hairs that serve a

Northern Fur Seal (*Callorhinus ursinus*)

protective function. The inner coat is made up of short, thin fur that traps air and insulates the animals from the cold. They have up to 300,000 hairs per square inch, and as many as 50 hairs per follicle, a density that actually prevents water from coming into contact with their skin. This fine underfur also helps to produce a laminar flow of water so that a greater underwater speed can be obtained. Unfortunately, humans have treasured the underfur for clothing. Sea lions also have an underfur, but it is only about a fourth to a third as thick and is not of much commercial value.

The fur seals and sea lions regulate their temperature by panting and through their flippers, which do have sweat glands. It is interesting to note that, since fur acts as an insulator by trapping air, these animals do not dive as deeply or for as long as other pinnipeds. As one dives deeper, the increased pressure decreases the volume of air and thus causes a loss of insulation. For example, at 33 feet the air in the fur has only half its volume; at 200 feet it has only about one seventh of its volume.

Most pinnipeds are colored in shades of brown, gray, black or white, or some combination of these. There are often stripes, spots, or other markings, depending on the species. A few green **gray seals** (*Halichoerus grypus*) were once discovered in Canada, but they turned out to be covered with an alga (*Enteromorpha groenlandica*). This is also a common occurrence with the **monk seals** (*Monachus* spp.) which, along with some sea lions, are the only pinniped species to inhabit warm, tropical waters. Albinism has been reported in some eared seals.

True seals moult in patches. Sea lions and fur seals moult by losing and replacing individual hairs. When moulting in late summer, elephant seals and monk seals lose not only their fur but much of their skin as well. The irritating process seems to be somewhat relieved by wallowing in mud, and they frequently lie in large mud wallows for the whole moulting process. They even sleep stacked atop each other for long periods while submerged in foul-smelling mixtures of mud and their own wastes. They are very placid at this time and can be set on by humans without taking much, if any, notice.

One of the most interesting of pinnipeds is the **walrus** (*Odobenus rosmarus*). These pagophilic (ice-loving) animals can weigh up to 3,500 pounds. Their skin is usually about one inch thick, but it can reach three inches thick in the neck area. The skin overlies a layer of blubber up to six inches thick that can account for up to 900 pounds of the animal's weight. Their eyes are quite small and are usually bloodshot. To help keep itself afloat, this mighty beast can fill two large throat sacs with as much as eleven gallons of air, a process frequently engaged in while sleeping.

The best known part of the walrus' anatomy is its tusks. Since they are rootless, they never cease growing, but they are occasionally broken by accident. They are present in both sexes but are usually smaller and more divergent in the females. Tusks can grow to almost 40 inches in length and weigh up to 12 pounds each. The largest on record were only 32 inches long but 13 inches around at the base. The walrus can use these large tools as an aid in climbing from the water onto ice floes, to help keep breathing holes from freezing over, to defend itself from enemies, and possibly for sifting muddy seabeds in search of mollusks to eat. This last function is doubted by most zoologists, and there are some who believe the tusks serve primarily as a badge of dominance like antlers in deer. Walruses are capable of killing seals and even polar bears with their tusks, and some have been observed to sleep while dangling in the water, suspended from a floe by their tusks. When the animals are exhibited in zoos, the tusks are generally sawed off, since the bases often become infected under captive conditions and the animals' constant attempts at digging.

Seals sleep both on land and in the water. Many sleep submerged and periodically rise to the

228

surface for air. Others, such as the walrus and Weddell seal, "bottle" by inflating special throat pouches with air to keep the animal's nose above the water's surface and facilitate breathing. Only the adult walruses have throat pouches to hold air.

In the polar regions, pinnipeds spend much time during winter keeping their breathing holes open. As more and more ice forms, the holes turn into long tunnels just slightly bigger around than the seal's body. The walrus uses its thick skull to break through ice up to six inches thick. The tusks cannot be used to break open the ice but can aid in keeping it open. **Ringed seals** (*Phoca hispida*) also keep their breathing holes open by bumping cracks in the newly formed ice with their heads and scratching with their claws. Other Arctic species keep their holes open using their flippers and claws while the Weddell seal chews its holes open; it has strong incisors and a wide gape. Teeth can be broken or quickly worn by this activity. Many of these seals can survive long periods—possibly the whole winter—by breathing the air trapped between the ice and the water's surface.

Pinnipeds can be quite vocal, as illustrated by the bellowing of male elephant seals. Weddell seals produce whistlings that resemble the sounds of falling bombs. The **Arctic bearded seal** (*Erignathus barbatus*) produces melodious underwater songs detectable miles away, like those of some of the baleen whales. Many pinnipeds, including **southern fur seals** (*Arctocephalus* spp.), the harbor seal, the **Hawaiian monk seal** (*Monachus schauinslandi*), the **Ross seal** (*Ommatophoca rossii*), the Weddell seal, and others, use echolocation to catch prey, generally fish and squid. This enables them to feed at night and at the lower depths to which they descend and also explains the well-fed appearance of blind animals. Some sources report that many sea lion species, including the California sea lion, may catch prey by using both echolocation and sensitive whiskers which can detect underwater vibrations sent out by their prey.

All pinnipeds are carnivorous, and none eat any plant material intentionally. Most feed on fish, swallowing them whole and headfirst. Although several species are known to feed on warm-blooded prey, only one does so regularly: the leopard seal feeds primarily on krill, fish, and squid, but it will frequently feed on penguins and the young of other seal species. Leopard seals will leap up to ten feet onto a floe to get at penguins. After catching one, the seals are known literally to shake the bird out of its skin before eating it. This behavior is also seen in the **New Zealand fur seal** (*Arctocephalus forsteri*) when it infrequently feeds on a penguin. The trachea in the leopard seal is very flexible and can even collapse to accommodate the swallowing of such large prey. It sometimes seems that leopard seals enjoy killing Adelie penguins for the "fun of it." They will force the birds into the water, at times breaking apart the ice floe upon which they are standing by bumping it from below with their strong heads. Most pinnipeds get their water from prey, but a few otarids get theirs from the sea. The Weddell seal occasionally eats snow.

The walrus feeds regularly on shellfish, sucking their flesh from the shell. It does not swallow them whole and later regurgitate the shells. Walruses can dive up to 300 feet and stay submerged for as long as half an hour to obtain such fare. The tusks are probably not used to help dig out shellfish, and some experts claim the muzzle, equipped with 600–700 sensitive whiskers up to a foot long, play a major role in finding mollusks, which can be uncovered with jet streams of water emitted from the mouth. At any rate, they eat up to 170 pounds of food per day. Walruses are occasionally known to kill and eat small bearded and ringed seals and even small whales. They dispatch the seals by hugging them with their fore-flippers and then stabbing them with their tusks.

229

Crab-eating seals do not feed on crabs as their name implies. Instead, these most abundant of seals (about 40 million) have specialized teeth that form a plankton-filtering sieve when the mouth is closed. The seal feeds on krill by taking in a mouthful of water and krill and then forcing the water out much as do some baleen whales. The **common krill** (*Euphausia superba*) causes their feces to be somewhat pinkish in color. Falkland sea lions produce yellow feces after eating squid that contain ink of that color and sometimes red feces after eating certain crustaceans (*Mundia* spp.).

Many of the pinnipeds swallow stones. The exact reason for this is not known, although many theories have been proposed. They may serve as an aid in digestion, as ballast to decrease buoyancy, to kill parasitic worms, aid in regurgitating fish bones, or to ward off hunger during times of fasting such as migration and breeding season. It may even be done unintentionally when mouthing stones during inquisitive exploration as pups. Whatever the reason, one study found that the average fur seal contained about half a pound of stones; a similar study found a whopping sixteen pounds of pebbles and rocks, on the average, in each of several sea lions, with as much as 25 pounds in a single animal.

Even though some pinniped species spend most of their lives in the water—the elephant seal remains in the open sea for about seven months of the year—all species must return to land (or at least an ice floe) to moult and/or give birth. The **northern fur seal** (*Callorhinus ursinus*) is the greatest traveler, with migrations of up to 6,200 miles.

Breeding behavior ranges from the fur seals' polygamous harems to the small family units found in the walrus and most of the true seals. Some of the latter, such as the gray and elephant seals, are polygamous, with the males controlling harems of females. The **Caribbean monk seal** (*Monachus tropicalis*) is monogamous.

In the polygamous species, the males arrive at the mating areas first to stake out their territories. The stronger, more dominant males end up, often after many fights, with the large, prime beach areas favored by the cows. They thus obtain the mating rights to the greatest number of females. In the **northern elephant seal** (*Mirounga angustirostris*), the alpha (most dominant) male on a given beach usually mates with about half of all the females in that area. The more subordinate males, if they get any territory at all, end up inland or with beach areas covered over with water at high tide. At such times, they can be seen neck deep in the water guarding over their territory. Once the large bulls or beachmasters have obtained their territories and harems, they usually guard them quite jealously. In some, such as the fur and elephant seals, the large males go weeks and even months without food rather than leave their "property" unguarded. Male **Guadalupe fur seals** (*Arctocephalus townsendi*) will occasionally guard their harems from a position in the water, as will some others if high tide covers their territory. In one study done on the effects of stress caused by the quest for territory, it was found that male gray seals lived an average of ten years less than the females. Elephant seal beachmasters usually die 1–3 years after having obtained this status.

Territorial disputes are occasionally resolved through fighting. The most spectacular fights are seen in the elephant seals. During the breeding season, the male's snout enlarges through muscular contraction, inflation of air chambers, and blood engorgement. In the northern elephant seal, the nose enlarges to a length approaching 20 inches and does resemble a shortened elephant's trunk. The large beachmasters square off and engage in bloody battles that rarely, if ever, result in death to a participant. In fact, most confrontations end with postural threats wherein the animal forms a U-shape with its body, holding its flippers in the air. Submission is shown by contraction of the nose. When fights do break out, the resulting wounds can look appalling but will usually heal rapidly without complication. The seals are protected by increased blubber and a thick, nerve-poor hide in the neck area. Although a bull generally swings his trunk to the side before attempting to bite an opponent, he occasionally misjudges and bites his own nose. During fights, bulls bellow so loudly they can reportedly be heard two to three miles away.

Hooded or **bladdernose seals** (*Cystophora cristata*) possess a mechanism similar to the elephant seal's nose. Above their noses the males have a hood up to twelve inches long, nine inches high, and about six inches in diameter which they can inflate with two gallons, primarily air, not blood. There is also a red inflatable intranasal "balloon" that can be protruded through one of the nostrils, usually the left. The stimulus for inflation of both the hood and nasal air sac is still uncertain, but it is probably related to dominance hierarchies.

While harems generally consist of 3–50 females, some large male fur seals have a hundred or more females at one time. One northern fur seal male had 153 in his harem. One might think that this would present him with an insurmountable task at mating time. There is no problem, however, since the male northern fur seal can mate three times an hour and has been known to copulate an average of once per hour for three weeks straight.

Many of the males keep careful guard over their females. Some have been known to keep their harem herded together on shore during violent storms and bodily prevent the females from escaping to the sea. Male fur seals are known to grab and drag females to their territories forcefully and may even inflict extensive lacerations in the process. One northern fur seal grabbed one of his harem members by the neck and threw her 15 feet back into his territory when she attempted to escape. It is even known for two male northern fur seals to take hold of the same female and pull her in opposite directions toward their separate territories. The sea lions are generally not as possessive and often prefer to feed rather than fight over a few females that might take off. Bull fur seals do separate fighting females and will even chase away very irritable females that appear to be troublemakers.

Females generally return to the beaches where they were born. In northern fur seals, the females hold small territories within the confines of the males' larger territories. They will not allow even stray pups in their area and will bodily throw them out when they wander in. Females come into heat about one week after giving birth and remain in heat for a day and a half or until copulation, whichever comes first. At this time, fur seal females are very drowsy, almost comatose, and do not fight off the males' advances as they normally do. Female sea lions stimulate males by biting about their necks, often pulling out large tufts of fur in the process. Females can be belligerent and are capable of driving the male away by snapping at his sensitive whiskers. In the excitement of mating, many pups from the previous year's mating get crushed to death by the adults, particularly the excited males. Male elephant seals have accidentally killed females in their attempts to mate as well.

In the nonpolygamous species, such as the harp seal, the females come ashore first to give birth, followed later by the males. Each male generally mates with one female, with the process usually occurring in the water. Crabeater seals do it on land. Male ringed seals develop a noxious odor similar to gasoline at breeding time.

Most of the pinnipeds show delayed implantation (exceptions include the walrus). The delay can be as long as six months in some, such as the **South African fur sea** (*Arctocephalus pusillus*). Pinnipeds give birth to a single calf or pup. True seals give birth every year, though the bearded seal

231

does so only every other year. Walruses give birth every third year. Walruses and some seals, such as the harbor seal, will on rare occasions give birth while in the water, but the vast majority do so ashore. The Baikal and ringed seal females are the only ones that dig dens in the snow to give birth. Female pinnipeds rarely eat the placenta.

Many pinniped young are born with a white baby coat of fur known as lanugo. They may wear this for 2–6 weeks depending on species. In the walrus, the lanugo is shed and eaten by the fetus before birth. Newborn walruses are thus dressed in brown and their intestines are generally filled with white hairs. In harbor seals, the lanugo may or may not be shed prior to birth, depending upon where they live. They and the other true seals will also shed their milk teeth in utero or shortly after birth. Females recognize their young first through voice, confirming it through smell.

Of a female fur seal's 125 days of suckling only about 35 are spent ashore. Usually she will go to sea for 3–6 days and then return to suckle for two before leaving again. Baby walruses suckle for up to two years. The milk of pinnipeds is about 10 percent protein and up to 50 percent fat. In comparison, cow's milk is about four percent fat and human milk is about three percent fat. The nutritious quality of the milk is reflected in the tremendous growth of the young. Weddell seal pups can gain seven pounds a day, while southern elephant seal pups gain as much as 20 pounds per day. Since many female pinnipeds (but not the female fur seals) fast during the suckling period, the mothers may exhibit enormous weight losses. For example, the female gray seal loses 100 pounds, the female Weddell 300 pounds, and the southern elephant female as much as 700 pounds before weaning. Generally, the pups gain about a pound for every two the female loses. Many of the females can breast feed while underwater, but should a pup attempt to nurse from a female other than its mother it will be promptly turned away, sometimes forcefully.

The young of monk seals, harp seals, and others are grossly obese at the time of weaning. In fact female Galapagos fur seals have been observed suckling pups larger than themselves. This is necessary because they must live off their fat while they learn to hunt on their own; female pinnipeds never bring prey to their offspring. Mortality rates are high, running 30–50 percent. In one study of northern fur seals, there was a 72 percent mortality rate during the first year of life. Infant walruses have been found inside the stomachs of adult bulls, and up to 20 percent of elephant seal young are accidentally crushed to death by fighting males. Female **Hooker's sea lions** (*Phocarctos hookeri*) have been known to snatch pups out of the path of a charging bull. One female was even known to solicit sexually a bull which was lying on her pup. As soon as the pup was freed, she grabbed it and rebuffed the male. Australian sea lion females are known to babysit each other's pups.

Walruses, sea lions, harbor seals, and others are known to carry their young on their backs or necks occasionally when in the water. *Phoca* species and northern fur seal pups can swim at birth and receive no training, but many seal pups cannot swim instinctively; they must be carried to the water by their mothers for that first awkward dip in the sea. Female Weddell seals will cut out a "ramp" with their teeth for their young to enter the water. Sometimes female walruses will grasp their young in their flippers and take them down on a dive for food. Baby seals will sometimes suck on one of their front flippers when sleeping.

Pinnipeds have very few natural enemies. The two major ones are killer whales and sharks. Both of these can be and are escaped by going ashore—sometimes in great haste. Some pinnipeds have been known to jump to safety by leaping vertical heights of ten feet from the water's surface. When landing,

232

they are protected from injury by the flexibility of their cartilaginous ribs. Gray seals have jumped onto rocks with humans present, and some seals have even jumped into boats to escape the dreaded killer whale.

Elephant seals protect themselves from pestering insects by shoveling a blanket of sand over their bodies. They have also been known to use their front flippers to fling sand and small stones at intruders that venture too close. If irritated, captive walruses will often spit a forceful stream of water at the target of choice.

Humans are by far the worst enemy of the pinnipeds, for humans are very good at finding reasons to destroy wildlife. In the case of the northern fur seals, it is claimed that the animals compete for salmon with fishermen, many of whom kill the seals on sight. However, studies have shown that less than two percent of the animals killed had the remains of salmon in their stomachs. On the other hand, the seals *are* known to cause extensive damage to fishermen's nets. Many other species are killed for their coats. The clubbing to death of baby seals for their white lanugo is well known.

One last note is that the "seals" so often observed in seal shows are not seals at all, but sea lions. They are robust animals that are quite playful in the wild, making it easy to teach them entertaining tricks. In the wild, they can be observed playing catch with fish and rocks, surfing in waves, playing tag with each other, chasing their own bubbles, and so on.

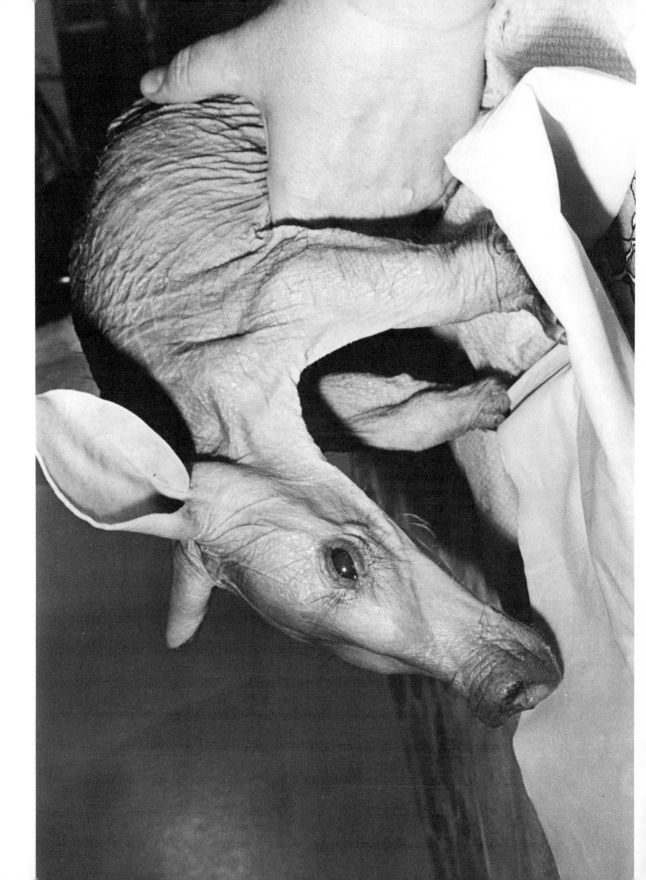

14

ORDER TUBULIDENTATA
(1 SPECIES)

Family Orycteropodidae (aardvark—1 species)

The tubulidentata is the only mammalian order that contains but a single species, the **aardvark** or **Earth pig** (*Orycteropus afer*). This species' isolation results from its unusual dental characteristics. Each tooth is composed of a number of fine tubules bunched together and surrounded by a thin layer of cement, hence the order's name of "tubular toothed." None of the teeth have roots or enamel. The functional teeth grow continuously throughout life, but the forward teeth never develop or may fall out early.

The aardvark's diet consists largely of termites, but it will feed on other soft-bodied insects and some fruits such as the "aardvark pumpkin," a wild gourd. The digestive system seems unable to deal with the hard exoskeleton of the true ants. Their hearing is so acute they can tell if an active colony is present in a termite mound. They obtain large numbers of termites by seeking out the nightly migrations of *Macrotermes* spp. and *Hodotermes* spp. and by digging into the large mounds of other species. In the latter case, the

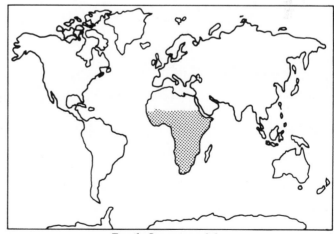

Family Orycteropodidae

insects swarm to the site to defend against and repair the damage. Many are then met by the aardvark's 18-inch sticky tongue. Sometimes the aardvark pokes its pig-like muzzle into the hole and sucks out large numbers of termites at a time. At rest the tongue occasionally hangs out of the mouth with the end wound up on itself. Up to 47 percent of their stomach contents can be composed of dirt and sand taken in as the termites are gathered. The aardvark is protected from insect bites by a thick, tough hide that is only sparsely haired. The nasal passages can be sealed off and are furthered protected by many stiff, bristly hairs at the entrance. The ears can be closed over by laying them back.

Aardvarks are incredible diggers. Some of the termite nests they tear open with their powerful front claws have quickly tired men digging with pick-axes. The aardvark is capable of burying itself in soft dirt within one minute, and it is reported that a single animal can outdig a team of six men equipped with shovels. Some excavate huge underground burrows reaching 20 feet below ground with tunnel systems spread out over 500 square yards and containing as many as ten entrances. For unknown reasons, they often turn a somersault as they enter their burrows.

Aardvarks are solitary except for brief pairing during the breeding season. They are also nocturnal and spend daylight hours holed up in their burrows. Occasionally, they block the entrance to keep out intruders. Some have been known to burrow into the side of a termite nest so that a supply of food is readily at hand. They are scrupulously clean and bury their wastes much as do cats. This often results in wild gourds growing near the burrows to supply them with food and moisture because the seeds pass through their intestines without harm. They are also surprisingly good swimmers. The young are born with very baggy, almost naked skin.

Aardvarks watch out for danger by standing on their hind legs and resting on their tails kangaroo-style. They defend themselves with their feet, sometimes rolling over onto their backs in order to kick with all fours. They are also known to whip attackers with their strong tail. Their head region is apparently very sensitive since they can be easily killed by a moderate blow to this area. They probably lack keen eyesight as they often crash into bushes and trees when fleeing. Their flesh tastes like formic acid as a result of their diet. They can also exude a foul, yellow fluid from anal scent glands.

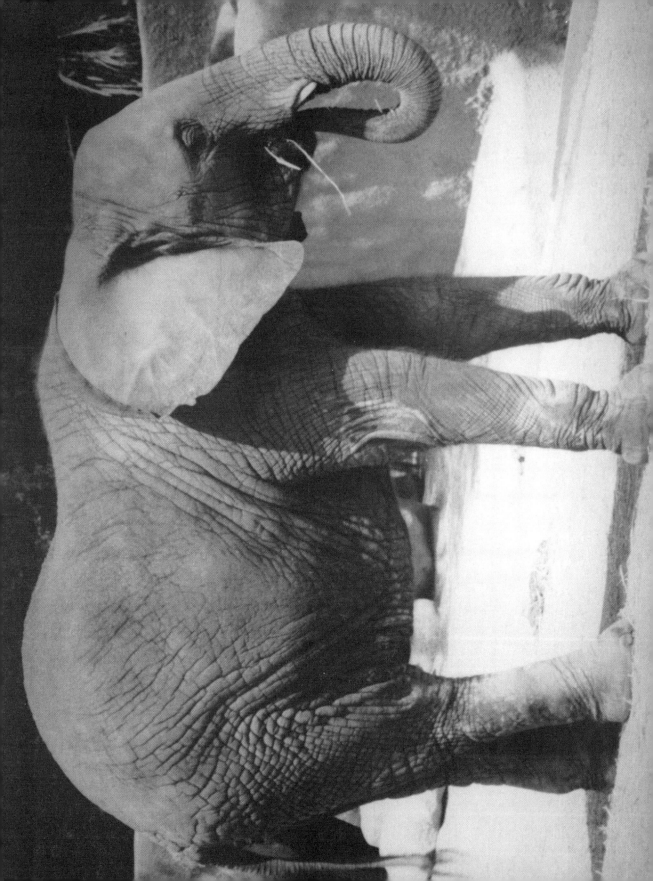

15
ORDER PROBOSCIDEA
(2 SPECIES)

Family Elephantidae (elephants—2 species)

The largest living terrestial animal is the **African elephant** (*Loxodonta africana*). Average adult males of the **bush** or **savannah elephant** (*L.a. africana*), the largest subspecies, stand 10.5 feet at the shoulder and weigh about 5.5 tons. The largest single specimen measured 13.67 feet at the shoulder after being felled by a hunter in 1974. It thus probably measured about 13 feet at the shoulder when standing since they decrease about five percent in height when on their feet. This huge specimen weighed an estimated 13.5 tons. The second largest specimen on record is mounted in the Smithsonian Institute. Its skin alone weighed over two tons. A large elephant's tail weighs up to 22 pounds, and its ear up to 110 pounds. There are **pygmy elephants**, but they are not a distinct species. They are actually just very small individual African elephants that only grow to a height of 4–6 feet.

The **Asiatic** or **Indian elephant** (*Elephas maximus*) is smaller and can be easily distinguished from its African counterpart by a number of characteristics. The African species has larger ears and tusks, a sloping as

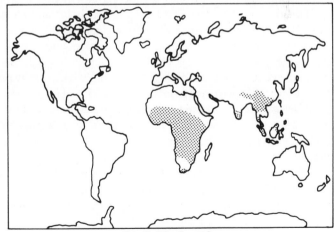

Family Elephantidae

African Elephant (*Loxodonta africana*)

239

opposed to a bulbous forehead, a concave rather than convex back, usually three toenails instead of four on the hind feet, transverse wrinkles located further up the trunk, and two finger-like lips rather than one at the tip of the trunk. The Asian elephant's ear tops turn forwards where those of the African turn backwards. In addition, the Asian elephant has fewer pairs of ribs and more tail vertebrae than the African elephant. Asiatic elephants also tend to develop unpigmented areas on their trunks and ears with increasing age.

Other than their size, the most notable aspects of elephants are their tusks and trunks. The tusks, which are actually their upper incisors, can grow to just under 11.5 feet and can weigh up to 258 pounds. Today, however, it is an exceptional tusk that reaches 75 pounds. There are recorded instances from the past of elephants with such large tusks that advancing age forced the animals to walk backwards because otherwise the weight of the tusks would bend the head down and cause them to plow into the ground. The skull is filled with many air pockets and sinuses to help relieve the neck muscles of excess weight. Tusks grow throughout the lifetime of the elephant and can be found in both sexes of the African species, although the males' tusks are usually larger. In Asiatic elephants, a tusker is almost always a male, but females may have small tusks. In two subspecies, the **Ceylonese** and **Sumatran** elephants, both sexes are usually tuskless. Quite frequently, a female African elephant can be found with only a single tusk, and some elephants have been found with as many as seven tusks. Just as people are right- or left-handed, elephants are right- or left-tusked. Most seem to be left-tusked since this is the one that is worn down most from use. Constant wear prevents most tusks from reaching more than about half of their potential length.

The elephant's trunk is very sensitive to smell and can detect odors up to two miles away upwind. The trunk contains 40,000 separate muscles and can be used for a wide variety of purposes. It is strong enough to uproot small trees and agile enough to pick up coins on the ground or even a single blade of grass. The trunk is used to transfer food to the mouth and to squirt up to five gallons of water into the mouth at a time (elephants do not drink through their trunks). It can work like a hose to coat the body with dust or mud to protect the inch-thick skin from insects and the sun, and it can even be used as a snorkel when the elephant is submerged or swimming, which they do surprisingly well. Elephants also trumpet through the trunk and produce hollow thumping sounds by banging it on the ground.

The trunk is very strong and can reportedly lift up to a ton in weight. One unfortunate man was thrown 40 yards by the trunk of an irate elephant. An interesting food-gathering technique is that of wrapping the trunk around a small shrub or plant and then giving the trunk a kick with the forefoot to uproot the morsel for eating. The trunk of an adult weighs about 300 pounds and if fully relaxed will hang down to the ground, although the elephant will usually not allow this to happen. It prefers to hold the trunk in the mouth, rest it on the tusks, curl it on itself, or lay it on the back of another elephant. The trunk's position is also used in intraspecific communication. For instance, a curled trunk indicates uncertainty or fear, while a pointed trunk reveals anger or frustration. An elephant will show submission by placing its trunk in the mouth of a dominant conspecific. Indian elephants are known to use their trunks to masturbate.

Elephants' ears comprise up to a sixth of their body surface area and are their primary means of heat regulation. Blood can be channeled through the many capillaries near the surface by opening and closing certain arterioles. The efficiency of this mechanism is increased by the fact that the ears can be flapped and also doused with water by the trunk to increase heat loss through evaporation. In heat distress, without readily available water, elephants have been known to stick their trunks down their throats and substitute digestive juices for water. It has been found that the blood leaving the ears can be 30°F cooler than the blood entering. The large ears can also be used to fan away bothersome insects, and dominance is exhibited through ear flapping as well.

Like most mammals, elephants have hair. It is coarse and wiry but only sparsely distributed over the animal's body. The eyelashes grow to about five inches in length. The toenails are generally kept shortened by wear, but occasionally an elephant will trim them by rubbing them against rocks. The skin is an inch or more thick and often forms folds that are known to house various species of insect. Due to their small surface to volume ratio, the animals do not require an insulating coat of fur or blubber. Mud baths help them to keep cool, and when the dried mud falls

Asian Elephant (*Elephas maximus*)

off it sometimes takes ticks and other parasites with it. Elephants are often coated with dirt that obscures their true skin color.

Running and jumping are both impossible for elephants. They can, however, walk—or pace (see horse)—at speeds up to 25 miles per hour, fast enough to outdistance a running human. They easily climb very steep inclines and do so with surprising quietness.

Elephants require about 30 gallons of water each day. Some large bulls take in up to 50 gallons. With such a large intake of fluid, it is not surprising that they urinate 10–15 times daily. Occasionally elephants drink from mud holes but some are particular about the water they drink and will refuse to take it if it is too muddy or polluted. At such times or during dry spells, the elephants will use their forefeet and tusks to dig "wells" up to ten feet deep in order to obtain fresh water. These new water holes are invaluable to many other species in the same area that might die of thirst without them.

The amount of vegetation required to keep a herd of elephants fed is phenomenal. Even though they only eat about four percent of their body weight daily, this means hundreds of pounds of food for each. A large male can pack away 600 pounds or more of food each day; the average elephant consumes only about 300 pounds. Their digestive system is quite inefficient and thoroughly processes only about 40 percent of the bulk. This large food intake requires them to spend as many as 18–20 hours each day feeding and necessitates the deposition of dung at approximately two-hour intervals. Elephants, particularly the young, will eat dung occasionally to obtain cellulose-digesting microbes. Interestingly, the seeds of some plants will not germinate unless they have passed through an elephant's gastrointestinal tract to remove a protective coating.

The continuous grinding of vegetation puts so much stress on the teeth that they are gradually worn away and eventually fall out in pieces. They are replaced in "conveyor belt" fashion, a process unique to elephants, kangaroos, manatees, some pigs, and the **little rock wallaby** (*Petrogale concinna*). The number of replacements is limited to six new teeth on each side of both upper and lower jaws, making a total of seven sets in all (including the milk teeth). Although there is much variation, the replacements generally appear at about the ages of three months, three years, five years, ten years, twenty years, and thirty years. After the final and largest set, measuring over a foot long, are worn away the elephant is doomed to slow death from starvation.

Elephants have been known to get drunk by feeding on the ripe fruit produced by the marula tree and borassus palm. What's worse, they also seem to suffer from hangovers and have been known to go on rampages during such times. Elephants have also dug out large caves in their quest to feed on salt-enriched earth.

There has long been a controversy over whether elephants stand or lie down to sleep. African elephants usually sleep mostly on their feet after reaching maturity, but they will still lie down for two hours or so for deep sleep, particularly after midnight. Asiatic elephants seem to spend more sleep time lying down at all ages. Both species are known to form themselves a pillow of vegetation before lying down, and each sleeps 2–5 hours in each 24 hour period. Both species are also known to snore.

As a rule, there are two types of elephant herd. The first is composed of nonmating bulls that have been ousted from the family herds after they reach puberty at about 12 years of age. The second is the matriarchal family herd, a family group led by a large, dominant female. The family herd may occasionally contain a large mating bull, although these are usually solitary in nature and only temporarily associate with the females.

Males rarely fight each other for dominance. When they do, the pair will square off, charge each other, butt heads, and wrap their trunks around each other. Occasionally a tusk will be broken in such duels; very rarely one bull may be killed when a tusk penetrates his brain.

Elephants sometimes exhibit a state known as musth. It has been observed in both species, both sexes, and occasionally juveniles, but it is best known and most prominent in Asiatic bull elephants. It was once thought to be a period of sexual excitement, but is now thought to be related to high temperatures and the dry season, without connection to sexual function. It is marked by the presence of waxy secretions running down the face. These emanate from special temporal glands near the eyes. The behavior of the elephant at this time has been frequently described as "bizarre" and "insane." They are extremely dangerous to approach in such a state and are very unpredictable.

Elephants, like humans, giraffes, and others, can breed all year long. In uncrowded areas, females become sexually mature at about ten years of age; in overpopulated regions, they don't become receptive to the males until almost 20 years old. The courtship is heralded by nudging, caressing, and trunk intertwining. The males' testes are located inside the abdominal cavity. The erect penis measures up to five feet long and weighs as much as 60 pounds. During intercourse, only the tip enters the vagina, and the penis thrusts independently as there is no pelvic thrusting. Copulation lasts 5–15 seconds. Mating is even known to occur with the pair positioned in deep water.

Elephants have the longest gestation period of all mammals—20–25 months, although the average is about 21 months. The females give birth every 4–9 years. During her labor and delivery, the mother is often aided by another female from the herd. These "midwives" have actually been observed to deliver the baby with their trunk and are known to clean off the afterbirth. At birth, the young are frequently dropped onto damp earth or a bed of leaves. The thickly haired newborn elephant stands about three feet high and weighs 200–250 pounds. Elephants grow continuously throughout life since their epiphyseal or growth plates remain open, a rare occurrence in mammals but also seen in some

242

cetaceans, kangaroos, opossums, porcupines, beavers, and some voles. Female elephants, along with female humans and rhesus monkeys, are among the very few mammals known to go through menopause.

Elephant young nurse for up to eight years. Female elephants are known to nurse the young of other females and will frequently "baby-sit" the calves of several others. The cows take turns with this chore, often with obvious reluctance, while the others spend the time eating. At times the baby-sitters are even known to prechew food for their younger charges. This practice of baby-sitting many young at once is also seen in other animals, including Cape hunting dogs, flamingoes, hippos, penguins, wapiti, wolves, and others. Female elephants call their young to them by slapping their ears against their head. Mothers and baby-sitters generally keep careful eyes on the young and are quick to discipline those that get out of line. Female African elephants have been observed to carry their young with their tusks and even their trunks when the offspring are quite young. It is interesting to note that the trunk does not seem to be very useful to baby elephants, which spend their first few years trying to keep it out of their way and inevitably end up tripping over it numerous times. This is understandable, for it must take a fair amount of time to develop skill in using so complicated an appendage. For instance, it takes months to years just to learn to drink with it properly. Some calves are known to suck on their trunks.

An interesting phenomenon in elephant life is their purring or rumbling. These noises can be heard up to a quarter mile away. They were first thought to be related to the digestive system and were labeled borborygmus, a name for the noises produced by gas moving through the intestinal tract. They are now known to be part of an alarm system used to warn each other of danger when the animals are not in visual contact. The noises are produced by vibrations of the larynx. As long as the elephants sense no danger and can hear their neighbors purring, they know that all is well. Should one of them sense trouble it will suddenly go silent. The others follow suit so that soon the whole herd becomes alerted. When the threat has passed, the elephants resume their rumblings once again. Wildebeests have a similar defense system in that they produce continuous low grunts when all is well and become silent if alarmed. Asiatic elephants will warn each other of danger by tapping the end of their trunks against a hard surface to produce a hollow resonance.

One often hears about "rogue" elephants, which are said to kill several hundred people each year. It is popularly believed that a rogue is an old, solitary bull, and it is true that the majority probably are. However, it has been found that many rogue elephants suffer from some infirmity such as bullet wounds, illness, or some sort of infection. The atrocities attributed to these individuals in their attacks on humans can be gruesome. The bodies are often trampled beyond recognition or even impaled upon the tusks of the "mad" beast. Some have been known to tear their victims apart limb by limb. However, these attacks are rare exceptions to normal behavior. Most elephants become alarmed at human scent and wisely move away. Even though elephants are the strongest of all terrestial mammals, they are pound for pound weaker than humans.

Like gorillas and rhinos, elephants often give bluff charges. These are usually sufficient to ward off an intruder. The display begins with raising of the trunk and spreading of the ears. This produces an impressive animal that can measure ten feet across, but its main purpose is to help locate the source of danger accurately. It may then display by shaking its head and slapping the ears against its side. In a charge, the trunk is held against the chest and the head is lowered. There is often a loud trumpeting and much kicking up of dust. As we have seen, elephants are quite capable of defending themselves, and some individuals are well known to follow through with most of their charges. Elephants have been known to drag crocodiles from the water, stamp them to death, and then toss the remaining carcass into a tree. As mentioned, some will mutilate their victims, even impaling them on their tusks or chewing on dismembered limbs. They are also known to throw objects such as stones and branches with their trunks or to squirt a jet stream of water through their trunk at any animal irritating them. If an elephant becomes panicky when threatened, the dominant female may calm it by forcing food into its mouth.

Another myth often repeated is that elephants are afraid of mice, supposedly fearing the small animals will run up their trunk. Should this happen in reality, the elephant could probably expel—and even kill—the mouse with a single blast of air. It has been shown experimentally that elephants are indifferent to mice. Surprisingly, they are fearful of mongooses and dachsunds, although why this is so is unknown. It may be related to their quick, darting motion.

Elephants generally live to an age of 45–55 years, although exceptional individuals have made it past 70. As they near death, they do *not* leave the herd to die in a secret elephant "graveyard." Any area where many carcasses have been found can be attributed to the result of a herd being trapped in a swamp, surrounded by a fire, a gathering at a final water hole during drought, or exterminated en masse by humans. In fact, it has been observed that when a member of the herd dies, the other elephants in the herd first attempt to arouse it through physical aid, feeding, threatening, and even sex. In attempts to lift a fallen conspecific, its tusks or the helper's tusks may get broken or pulled out in the process. When it becomes apparent that nothing is going to help, the group actually appears to mourn its death for hours and even days prior to leaving the body. Often a crude attempt at burial is made. This has been performed on other species including sleeping humans by well-meaning elephants. Occasionally, they will return at a later date to scatter the remains, break up the tusks and bones, carry them off, or even perform a burial.

Elephants seem to be universally liked. One reason is that they are quite intelligent and can appear to be very human. They offer each other aid during birth, sickness, and when wounded. Disabled elephants have been brought food and water by conspecifics for weeks on end. On one occasion, when a calf misbehaved, the mother promptly disciplined it by spanking it with a small tree held in her trunk. In India, some captive elephants once used their trunks to plug their warning bells to sneak into a plantation and feast on bananas. Some will use sticks as back-scratchers. Zoo specimens have feigned cramps to receive treatment consisting of a bucket full of gin, water, and ginger.

One bit of folklore that is apparently true is the elephant's memory. They do seem to possess an excellent memory, but like all animals they can and do forget. They have scored well in memory tests, and some individuals have been known to recognize certain humans after years of separation.

Although elephants are huge and incapable of jumping even an inch, they are extremely agile animals. Researchers are continually amazed at how silently these huge beasts can slip through forests and bush country without being detected. To get down steep hillsides, elephants will sometimes extend their legs and slide down on their bellies.

244

16
ORDER HYRACOIDEA
(7 SPECIES)

Family Procaviidae (hyraxes—7 species)

The **hyraxes** or **dassies** (3 genera) are, surprisingly, the closest living relatives of elephants, despite their overgrown guinea pig appearance. The various species usually inhabit trees, rocky areas, or caves. Some are solitary, but most live in colonies numbering from four to hundreds of animals. They are known to use communal latrines where large amounts of fecal wastes build up. In the **rock hyrax** (*Procavia capensis*), this material was once collected because one of the fecal ingredients, hyraceum, was used to make perfume.

Hyraxes are extremely agile climbers, being able to climb nearly perpendicular rock faces with the aid of fleshy foot pads that both are coated with sweaty secretions and can form a vacuum when muscles in the central area of the pad contract. At times, their suction-cup feet have been known to hold **gray hyraxes** (*Heterohyrax* spp.) in place despite death. Hyraxes occasionally are seen crouching on rocks and staring at the sun. Their eyes are protected from damage by a special membrane known as the umbraculum. They also

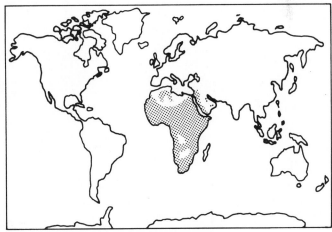

Family Procaviidae

Rock Hydrax (*Procavia capensis*)

247

Rock Hydrax (*Procavia capensis*)

like to roll in the dust often. Long tactile hairs cover their body and aid movement through dark tunnels and narrow crevices.

Both sexes of most species have a gland in the mid-back region. Its presence is marked by surrounding white or black hair, and it may have something to do with territory, mating, or defense. It is most developed in **tree hyraxes** (*Dendrohyrax* spp.), producing brownish or milky liquid secretions said to smell like leather or burnt sugar, depending on the species. Hyraxes show aggression by exposing their teeth, grinding the teeth together, and thumping the ground with their forefeet. When they are excited, the hairs spread away from the gland on their back as well. Hyraxes often huddle together to conserve heat.

Hyraxes feed on vegetation and actually chew a cud like the ungulates. The upper incisors grow continuously, and the animals are known to feed on plants that are poisonous to most other mammals. While feeding, they are said to post lookouts to bark out prompt warnings at the approach of enemies. When the going looks safe, an all-clear whistle is given and repeated throughout the colony. Male tree hyraxes are also known to produce howling calls at night. The eerie cry starts out low and builds to a piercing scream that can be heard, at its peak, up to a mile away. The cry may serve to announce territory ownership as well as to attract females, much as do the songs of birds. Often, a single adult rock hyrax will care for all the young while the other adults go out to feed.

PHOTO: WILLIAM VOELKER

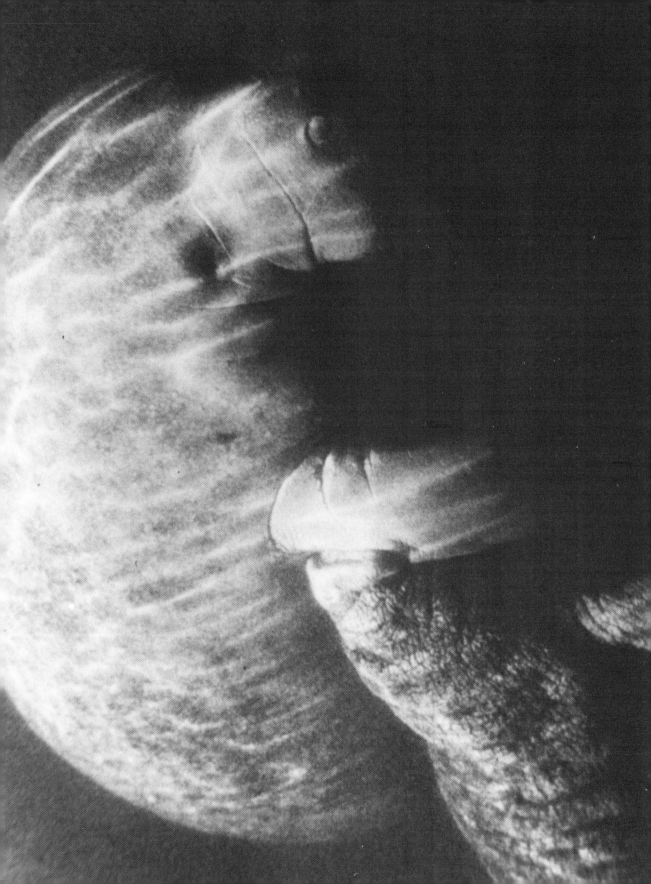

17

ORDER SIRENIA
(4 SPECIES)

Family Trichechidae (manatees—3 species) and
Family Dugongidae (dugong—1 species)

The **dugong** (*Dugong dugon*) and the **manatees** (*Trichechus* spp.) differ in a number of characteristics. The dugong is marine, has a notched tail fluke, a single upper lip, no toenails, and seven cervical vertebrae. It cannot completely close off its nostrils and does not stuff food into its mouth with its flippers, and the males have large incisor tusks that grow up to 12 inches in length, although they are mostly concealed by the upper lip. In contrast, the manatees take more to fresh water, have rounded tail flukes, and split upper lips. Most have flipper nails and only six cervical vertebrae. They can completely close off their nostrils, use their flippers to put food in their mouth, and do not have incisors. Manatee skin can measure up to two inches thick.

Sea cows are the only mammalian herbivores that spend their whole lives in the water. There are reports of dugongs leaving the water for short distances to graze on washed-up seaweed and to give birth. This is rare behavior since they are prone to sun-

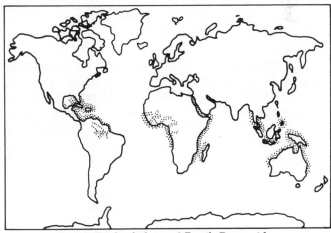

Family Trichechidae and Family Dugongidae

Manatee (*Trichechus manatus*)

251

burn, and when their weight is not supported by water, they find it hard to breathe. Manatees are not known to crawl completely out of the water, but they will put themselves out a foot or two to obtain food or to rest. They will often mate with their bodies half on dry ground.

Manatees are larger, reaching lengths of 15 feet and weights approaching a ton. In all three species, the teeth are replaced in an unusual way similar to that of elephants and some kangaroos. The teeth erupt in the rear and slowly make their way forward; upon reaching the front, they fall out if they have not already worn away. Manatees sometimes "brush" their teeth with their flippers. In the dugong, the forward molars wear out quickly. By the time a dugong reaches adulthood, it must thus make do with horny ridges overlying the gums. The two rear molars grow continuously. The dugong is also the only mammal that has a bony covering on the tip of its tongue.

Sirenians are not as well adapted to an aquatic life as the cetaceans. They do have valved nostrils to keep water out of their respiratory tracts, and their bones are mostly solid (most of their bones lack marrow) to increase density and thus decrease buoyancy. Unlike the cetaceans or pinnipeds, the sirens take a deep breath prior to submerging. They have only small amounts of myoglobin in their muscles, but they compensate for this lack with a much lower rate of metabolism. Although their normal respiratory rate is about one breath per minute, they can stay underwater for up to 20 minutes. Sirenians apparently are unable to breathe through their mouths, and their eyes are protected by thick, oily secretions.

The sea cows are generally given credit for stimulating the mermaid myth, but mermaid legends are said to predate European knowledge of sirenians. These large, placid creatures are known to "walk" on their flippers in shallow water, participate in frequent "hugging and kissing," and occasionally nurse or hold the young clasped to their chest in a human manner. Unlike their namesakes, the sirens of mythology who produced beautiful songs, modern day sirens usually limit vocalizations to occasional grunts and squeals. Some produce whistling sounds or screams when alarmed.

Sea cows are strict vegetarians and feed on seaweed, algae, and water plants. Crabs have been discovered in the stomachs of dugongs, but such invertebrates are probably eaten incidentally with the vegetation. Manatees may occasionally take fish from nets. Sea cows spend up to a quarter of the day consuming as much as 100 pounds of vegetation daily. They are said to find food by sight, smell, and touch. The dugong usually shakes its food vigorously to rid it of any sand or coral that could cut into its mouth when chewing. They sometimes stack food on shore for later feeding. Manatees are known to feed not only on their own feces but on those of other animals as well. A trait they share with elephants is the production of large amounts of intestinal gas.

Dugongs mate for life. In courtship, the males may occasionally leap from the water. They are one of the few animals to mate front to front, like humans. A single young is produced only once every two to three years. Sirenians, whales, and the hippopotamus are the only mammals to give birth underwater. Both headfirst and tail first births have been observed in sirenians. In all sea cows, the teats are located close to the armpits to facilitate nursing below as well as above the water's surface. Most lie in the water while nursing, but there are occasional reports that they can nurse while upright, clasping the baby to the breast, but this is rare behavior. The young are known to nurse until 18 months of age. In manatees, the males help in child care and will carry the young about for hours at a time while the females feed. Young sea cows use their flippers for locomotion; adults use only their flukes.

Female sirens frequently carry their young on their backs.

Sea cows are preyed upon by a few large predators. There are accounts of dugongs banding together to ward off sharks. Their greatest enemy by far is humans, who kill them for their flesh and accidentally cut them up with the propellors of motor boats. Their primary response to danger is to flee, which they can do at speeds up to 20 miles per hour. Lately, humans have attempted to employ manatees in the clearing of waterways clogged with growing vegetation such as the water hyacinth. The idea has met with at least partial success in Guyana. Manatees will sometimes "body surf" repeatedly just downstream of opened floodgates at powerplants. They frequent such areas because of the warm waters given off as a by-product of power production.

18

ORDER PERISSODACTYLA
(17 SPECIES)

Family Equidae (horses, asses, and zebras—8 species)

The Equidae have proved to be one of the most useful groups of mammals to humans down through the ages. With the exception of zebras, which generally are more stubborn than mules and absolutely refuse to do any work, the equids have been some of humanity's best workers. In return, people have wiped out all but one of the horse's ancestors, have put many of the ass and zebra species on the endangered list, and annihilated others. **Przewalski's wild horse** (*Equus przewalskii*) is the only true wild horse today; unfortunately, it is now thought to be confined primarily to zoological parks. Like all truly wild equids, it sports an erect mane. Some domestic breeds have erect manes, too, and some isolated zebra populations are maneless. (The "wild" mustangs are not truly wild; instead they are said to be feral—domestic animals that have gotten loose and taken to the wild.)

Humans have developed many breeds from the **horse** (*Equus caballus*) and classified them by appearance, color, and ability. This has resulted in a wide variety—over 60—of animals. They range in size from the **dwarf Falabella pony** that stands

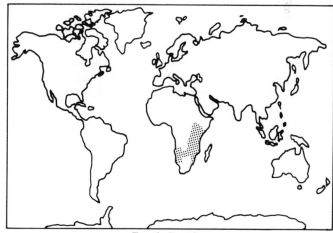

Family Equidae

Indian Rhino (*Rhinoceros unicornis*)

15–30 inches at the shoulder and weighs 40–80 pounds when full grown to the largest **Belgian** that stood over 6.5 feet at the shoulder and weighed 3,200 pounds. It is interesting to note that the Falabella pony is extremely fast for its size and can outrun a racehorse for the first 100 yards.

Various breeds have performed many remarkable feats. Racehorses can run up to 45 miles per hour, and surprisingly, it is said that a horse can be goaded into faster speeds with a rider on its back than without. Horses can make horizontal jumps of 30 feet and vertical ones of eight feet. Large draft horses have pulled 55 tons of railroad cars for 20 miles within six hours. Asses, too, have been bred for thousands of years to produce many breeds of domesticated donkey.

Horse terminology is quite extensive and involves the anatomy, age, sex, and work of the various breeds. A few common terms are:

"foal" = newborn

"colt" = male under the age of four

"filly" = female under the age of four

"horse" = male

"mare" = female

"stallion" = mature male

"gelding" = neutered male, and

"pony" = a horse that stands less than 58 inches or 14.5 hands (a hand = four inches) when full grown.

These terms do overlap and are used in different ways by various people. Two more common terms are "mule," for the result or hybrid of a male ass ("jackass") and a female horse, and "hinny," the hybrid of a male horse and a female ass ("jenny"). Both animals are sterile and unable to reproduce further (except for a few rare fertile mules found in the Orient). The hinny is smaller and inferior to the mule and is hence less common. Mules are known not only for their stubbornness, but also for their hard work. In addition, they once were frequently used to pull fire engines because they are much less fearful of fire than horses.

This family offers an appropriate place to discuss locomotion, for horses use more of the forms of locomotion (gaits) than other mammals and even exhibit special ones of their own that are too detailed to describe. Most four-footed animals, at least when walking, exhibit what is called "diagonal sequential movement" or the "cross walk." In this form of locomotion, the limbs are raised and moved in the following order: right front leg, left rear, left front, and right rear. In "ambling" or "pacing," both legs of the same side are moved simultaneously. It is best seen in camels, elephants, and giraffes. When diagonal legs are moved together, the gait is known as the "trot." In a "gallop," there is a four-beat gait with only one foot touching the ground at a time. The sequence of foot movement is variable, depending on species. It has been shown that during both the trot and the gallop there are moments when all four feet are in the air. The same number of steps will take a horse two and a half times as far when galloping as when walking.

The "stott," also known as the "pronk" or "spronk," is made by a simultaneous spring from all four legs. The animal generally lands with the legs held in extension and all four feet close together. The stott is usually seen in deer and antelope but occasionally in foxes and maras. The "ricochet" is bipedal hopping, as seen in kangaroos and many rodents. Of course, many other forms of locomotion, such as brachiation in apes and flying in bats, have been outlined earlier.

256

Equid tails are generally long and frequently serve the animals as fly-swatters. Przewalski's wild horses are even known to run in chain formation so that the beating tail of the horse in front keeps insects away from the follower's face. They reportedly will also form a huddle with heads toward the center while the tails beat furiously to keep insects from the area. Often the young will be in the middle, particularly in cold habitats, for protection from predators and the cold. Equids also protect themselves from insects and aid the moulting process by rolling in mud. The thick mud coating protects their hide, and when the dried chunks fall off they take loosened hair with them.

A horse's age can be estimated by looking at its incisors. Milk teeth are lost between ages 2.5 and 4.5. A horse older than 5 years has black depressions in the incisors. These disappear by 6–8 years, after which shape of the incisors is the key. Between 7 and 12 the upper jaw incisors are oval. They then become rounded, until after 18 or so years they are triangular. In extreme age, they again turn oval. Zebra teeth are open-rooted and thus grow continuously. The longest-lived horse was 62 at death; most live less than 30 years.

All equids have excellent hearing and a keen sense of smell. Horses are also one of the few mammals that possess color vision. Although many mammals rid themselves of heat through panting, a few, including horses, humans, zebras, gazelle, ibex, and many other hoofed animals, sweat and use evaporation to dissipate excess heat.

Much can be told of a horse's "feelings" through its facial expressions. Ears back with bared teeth and lowered head is a good indication of aggression or anger. Friendliness is exhibited by erect, forwardly placed ears. A lowered head or nibbling mouth, particularly directed at another horse's tail, is a dead giveaway of submission.

Zebras (*Equus* spp.) look like striped horses. Just like human fingerprints, no two zebras have identical patterns. The stripes, which seem to make the animal extremely conspicuous, actually interfere with the depth perception of predators. Lions and the like perceive the wider haunch stripes as being closer than they are; they thus occasionally misjudge distance and miss their strike. The stripes also make efficient camouflage, particularly at a distance in open grasslands near dawn or dusk. They may also make it difficult for a predator to single out one animal from a fleeing herd. Recently it has been found that the stripes actually

Damara Zebra (*Equus burchelli*)

protect the zebra from some biting insects that zero in on large patches of dark or light but not stripes. It is interesting that Africans consider zebras to be black animals with white stripes while Europeans almost always assume them to be white with black stripes.

Zebras are a good example of the biological rule which states that whenever two species inhabit the same area, behavior must be different enough to avoid competition that would lead to the demise of one of the species. Both gnus and zebras occupy the same grassland areas. To avoid threatening competition, they have taken to feeding on different parts of the vegetation. Zebras feed primarily on taller, older grasses, while the gnus feed mostly on shorter, young grasses. Unlike any zebra species, the **African wild ass** (*Equus asinus*) will raise up on its hind legs to reach food in tree branches. **Mountain zebra** (*Equus zebra*) are the only zebra to dig out water holes, which also benefit other wildlife in the area. African wild ass young require no water for their first six months and are even known occasionally to die if it is given to them during their first few months. The adults, too, can go long periods without drinking. The **Asiatic wild ass** (*Equus hemionus*) relies much more on a daily supply of water but can survive on highly saline water, as can other asses.

Like many other large mammals, the equids sleep while standing or lying down. They usually sleep standing, though, because when they lie down they get cramped easily and experience some difficulty in breathing. In any mammal, the amount of sleep and its depth are directly related to the animal's choice of sleeping site, the level at predation it faces, and its ability to defend itself if attacked. Some equids require that their sensitive ears be closed over during sleep, just as most humans must close their eyes to block out arousing stimuli.

The equids generally travel in herds composed of either a group of bachelor males or a stallion and his harem, including offspring. The stallion carefully watches over his charges, leads them from danger, and fights to protect them. Often two stallions will fight for dominance and thus harem possession. They raise up to kick, slam their bodies together, and bite. Rarely is one seriously hurt in such an encounter. Should one of the members get lost from a zebra stallion's harem, the other members will spend hours and sometimes even days searching for it. In **Exmoor ponies**, the herd is led by a dominant mare, and in wild asses the sexes generally form separate herds and come together only for mating.

The **Grevy zebra** (*Equus grevyi*) and the asses are the only equids to stake out and hold territories. In fact, they hold the largest territories of all ungulates, measuring up to four square miles or more in size. The areas are marked out with large heaps of dung. These are also used by the animals as navigational signposts. Interestingly, the territories are not guarded against other males, who are free to come and go unless a receptive mare is nearby. The owner will then chase other males away from the female, quite often by driving him deeper into the territory. Grevy zebras form many different types of grouping, including large herds of mixed animals during times of migration. In equid herds, the maturing males generally leave the herd voluntarily. Young mares, on the other hand, are frequently kidnapped by a stallion from another harem or bachelor herd. Often her original stallion will put up a fight to prevent the theft. All equids display the "flehman" gesture (see antelope) after smelling certain odors, such as that of urine or feces.

Mating attempts with females not quite in estrus lead to "courting fights." That is, the females respond with kicking to which the males may return bites. Some blood is often drawn in such encounters. Females in heat stand in a characteristic manner with their hind legs apart and their tails in

258

the air or to the side. They submit readily to the male's advances. During actual copulation, a stallion may bite the mare's mane. The females can postpone birth temporarily for several hours if disturbed, and they do not eat the afterbirth following parturition. The young are capable of standing and running within minutes of birth. Some zebras are born brown and white but turn black and white with age.

Humans are the major enemy of most equids, but the zebra's chief nemesis is the lion. Some, such as the Grevy zebra, have been known to drive lions away with their powerful hind legs. As a rule, however, zebras and most other prey offer very little resistance once they are brought down. They are known to succumb quickly to shock.

Family Tapiridae
(tapirs—4 species)

The tapirs resemble a cross between a pig and a hippopotamus. The females are usually larger than the males. All are somewhat brownish in color, except for the **Malayan** or **saddle-backed tapir** (*Tapirus indicus*), whose markings exhibit disruptive coloration. It is covered with a striking black and white coat capable of distorting its shape and size, particularly when out at night (all tapirs are nocturnal). The flashy coat

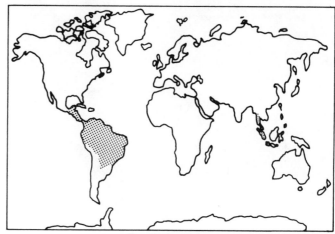

Family Tapiridae

blends in well with boulders and shadows, and the animal is said to resemble a pile of rocks when lying down.

The well-worn trails of the **Central American** or **Baird's tapir** (*Tapirus bairdii*) frequently become roofed over with vegetation to form long tunnelways stretching for miles. When land was first cleared in Central America, these trails were often used as the basis for roadways. All tapir species depend on water and remain quite close to it. Riverbanks are often marked with tapir "stairways" and "slides" that are used for quick entry. Most of the tapirs are said to be able to walk along the bottoms of rivers. They frequently take mud baths to protect their skin from insect pests.

Although they have keen senses of smell and hearing, their eyesight is very poor. In captivity, they feed on almost anything given to them, including old rags. In the wild, some dive to river bottoms to feed on aquatic vegetation.

Tapirs are loners and pair only for mating. In Malayan tapirs, the female will often initially chase the male. Mating is accompanied by mutual biting, kicking, snorting, frequent spraying of urine, and the production of bird-like trills and whistles. All of the tapir species produce young that are spotted and horizontally striped for their first 6–8 months. This is to help camouflage them from predators.

When threatened or attacked, tapirs will usually flee towards water, crashing through the brush as they go. Such crashing has actually dislodged attacking jaguars and tigers and is even known to rid the tapir of leeches, biting insects, and other parasites. If cornered, they will attack but will often just

Central American or Baird's Tapir (baby) (*Tapirus bairdii*)

trample the attacker under foot as they make good their escape.

Family Rhinocerotidae (rhinoceroses—5 species)

At up to four tons, the **white** or **square-lipped rhinoceros** (*Ceratotherium simum*) is the largest of the rhinoceroses and the second largest land mammal behind the elephant. Both the white and the **black** or **hook-lipped rhinoceros** (*Diceros bicornis*) are gray, although they may appear to be various colors of brown, black, or red depending upon the color of the mud caked to their gray skin. The name white actually developed as a misinterpretation of the Afrikaan word "wijt" and the Dutch word "weit" which both mean wide and refer to the animal's wide mouth. Asiatic rhinos differ from the two African species by having more hair, large skin folds, tusk-like lower canines known as "tushes," and usually only one horn. The **Sumatran rhinoceros** (*Dicerorhinus sumatrensis*) has two and is covered with long, shaggy hair.

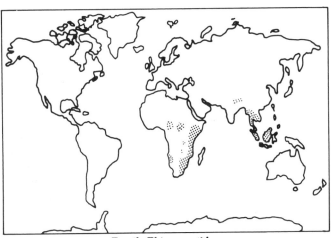

Family Rhinocerotidae

Rhinoceros horns are not made of bone like true horns. Instead, they are composed of keratin in the form of matted, compressed hair. In many instances, the horn layers unravel to form small tufts of hair; however, the horn fibers lack the cuticle of true hair. The **Indian rhinoceros** (*Rhinoceros unicornis*) and the **Java rhinoceros** (*Rhinoceros sondaicus*) each have one horn while the other three species have two. While female Javan rhinos are frequently hornless, the black rhinoceros may often have three or more—up to five—horns.

260

Some rhinos have been known to sprout horns on their rumps and even their flanks, although this is rare. The longest horn on record measured 62 inches; it was on a female white rhinoceros. In some instances, Indian rhinos have had their horns torn off in a fight. Rhino horns do regrow completely in the young and to a certain extent in adults. Horns can grow up to three inches per year.

Rhino horn is considered by many Africans and Orientals to have special powers. Some believe it to be a general antidote to all poisons, others use it medicinally, and a few believe it to have aphrodisiacal powers. The latter may be related to the fact that Indian rhinos often make love for an hour or more, during which time the male ejaculates every few minutes. Rhino horn has also been used as a cure for snakebites, aches, arthritis, cancer, fever, dysentery, kidney disease, leprosy, tuberculosis, and other diseases. Of course, none of these powers truly exist, but rhino horns bring as much as $500 per ounce on the black market, and their value is responsible for most of the illegal poaching of rhinos throughout the world. Most recently, horns are in demand for use as dagger handles in the Mideast.

The eyesight of all rhino species is very poor. They may not recognize a stationary object 75 feet away, and some experts state they can't see anything clearly farther away than 20 feet. During a charge, they are incapable of seeing directly in front of themselves because of the horns and the side placement of their eyes. Some species hear well while others don't. All rhinos have a good sense of smell and can detect upwind humans as far away as 2,500 feet. Their nostril cavities exceed the brain in size. Rhinos can run up to 35 miles per hour, and a group of them is known collectively as a "crash."

The Asian rhinos are much more aquatic than the African species. They are also covered with a thicker hide. The skin of the Indian rhinoceros resembles metal armor, rivets and all. The Indian species forms large, winding tunnels beneath the thick, up to 25-feet-high, elephant grass found in its habitat. Such pathways take on the form of gigantic rodent runs. The Sumatran rhinoceros is a good swimmer and his even been observed swimming at sea.

Rhinos, like many other large mammals, including buffalo, elephants, zebras, and others, appear to enjoy taking mud baths. These serve a number of purposes. For instance, mud bathing aids in temperature regulation by cooling them down. This is important because even though rhinos can tolerate a 7°F rise in body temperature without ill effect, they do not have any sweat glands to help them cool off. Mud bathing also protects their hides from biting insects and ticks. If no mud is available, they will take a dust bath, but this is much less efficient. African rhinos are host to many parasites including 26 species of ticks. They are aided by oxpecker birds on land and river turtles in the water. Both of these other animals feed on ticks and other parasites on the rhino's skin. Egrets, frequently found sitting on the backs of rhinos, do not help keep their skin clean but instead feed on insects disturbed on the ground by the rhinos' walking.

Rhinoceroses are vegetarians. Black and white rhinos can easily be told apart if one looks at their mouth, since each is adapted to its style of feeding. The black has a hooked upper lip for browsing on small trees and bushes. In comparison, the white rhino has a broad, squared-off mouth used for grazing on grasses. It is the only species of rhinoceros that is a grazer. Black rhinos occasionally eat their own dung or that of other animals such as wildebeests. The Indian rhinoceros eats about 30 pounds of vegetation and drinks 20–25 gallons of water daily. Rhinos occasionally use their horns to plow through the ground to feed on mineral salts thus exposed.

The rhino's vocal repertoire includes snorts and whistles. Females mew to their offspring. Infants

White Rhino (*Ceratotherium simum*)

scream. Bellows signal either attack or mating.

Rhinos are known to build dung piles up to four feet high and 20 feet across. These heaps mark the center of their territory, serve to aid navigation, and identify the owner's sex to other rhinos. The presence of dung is actually a stimulus for passing rhinos to defecate. Even rhinos in flight cannot resist the urge to stop and add their contribution to the pile. This same behavior of territory marking is seen in civet cats, four-horned antelope, guanacos, hippos, nilgai, steinbok, tapirs, vicunas, and others.

African rhinos frequently trample their feces and then deposit their scent on their trails. The pathways also reek of urine, which can be sprayed as high as seven feet into the vegetation. Black rhinos urinate with a powerful, backward-directed stream that has occasionally showered unwary zoo visitors; it can be sent in excess of 12 feet. Indian rhinos have special scent glands on their forefeet that are used to mark out their trails. Indian rhinos will urinate in their wallows and then cover their bodies with the scent of their urine. Much of this gets rubbed off inside their vegetation tunnels; when one walks inside these tunnels, the smell of urine can be overwhelming.

Excluding white rhinoceroses, all species are solitary except for mating purposes. Mating time brings both humor and occasional tragedy. Male rhinos are known to fight fiercely at mating time. This has been known to cause fatalities in both Indian and white rhinos. In black rhinos, the females are known to whistle to attract males. Although male black rhinos rarely fight for females, occasional sparring matches will be broken up by the females' choosing of one male over the other. The females make the advances and have even been observed to chase reluctant males for miles. This reluctance is understandable since once the pair comes together the female will usually charge the male and smack her body into his, often with enough force literally to knock a belch out of him.

Black rhinos also perform other courtship antics. They take turns running circles around each other. This is followed by a running dive in which the animal lands upon its throat and skids to a stop. Each may repeat this for up to half an hour before they become tired and then mate. It is said that these maneuvers synchronize the pair for mating.

Both Indian and white rhino pairs "fight" during courtship. In Indian rhinos this may rarely lead to the death of the female; in the white rhinoceros death, although still rare, most likely comes to the male. Actual mating lasts 20–80 minutes while the male ejaculates every three minutes or so; males have been observed to ejaculate as many as 56 times in a single mating. In copulation, the male rhino

262

PHOTO: LOS ANGELES ZOO

may actually ride totally on top of the female with all four legs straddling her body. Gestation lasts 18 months, and females give birth every three or more years.

Since they are such large, powerful animals, the rhinoceroses have few enemies with the exception of humans. Rhinos have been killed by elephants, hippos, and even crocodiles, and lions have been known to kill young ones. Although black rhinos are the most unpredictable, their reputation for ferociousness is often exaggerated. The white rhinoceros is generally more peaceable than its black cousin.

Both species are alerted by the egrets and parasite-picking oxpeckers which screech out warnings at the approach of danger. These birds are very helpful allies for an animal with such bad eyesight. When they do pick up a strange scent, rhinos will often make headlong dashes, occasionally towards the origin of the smell. The charge is usually preceded by much snorting and kicking up of dust and is almost always stopped short of the target, although some rhinos have been known to run for several miles before coming to a halt. Some charges have resulted in considerable damage to cars, and on at least one occasion a train was derailed by a charging rhinoceros. As mentioned, they can run up to 35 miles per hour and are said to be able to out-accelerate a truck to the 50 yard mark. A few people have been unlucky enough to end up impaled on a rhino's horns. When this happens, the rhino generally wanders about for awhile before attempting to remove the extra load. Asian rhinos will usually use their teeth rather than their horns as their major offensive weapons.

19

ORDER ARTIODACTYLA
(192 SPECIES)

The artiodactyls comprise the order of even-toed hoofed mammals. This order contains not only our most important domesticated mammals, but also the majority of the large mammals in the world. They have in common the feature of walking on an even number of toes, either two or four. They are further divided into suborders by gastrointestinal features. Those with a two- or three-chambered, nonruminating stomach include the families Suidae (two chambers), Tayassuidae (two chambers), and Hippopotamidae (three chambers). The Camelidae and Tragulidae have three-chambered ruminating stomachs, while Antilocapridae, Bovidae, Cervidae, and Giraffidae have four-chambered ruminating stomachs.

Family Suidae
(pigs—8 species)

The pigs come in many sizes. The smallest pig, the **pygmy hog** (*Sus salvanius*), stands only a foot high at the shoulder and weighs up to 20 pounds. The **giant forest hog** (*Hylochoerus meinertzhageni*) stands almost four feet high and can weigh in excess of 600 pounds. Some domestic breeds reach more than 1,000 pounds, with the record being 2,552 pounds for a **Poland-China hog** that

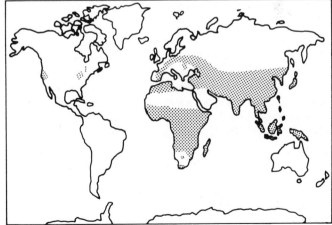

Family Suidae

American Bison (*Bison bison*)

measured nine feet in length. Male suids are boars, females are sows, and the young are piglets.

Warthogs (*Phacochoerus aethiopicus*) are named for four large warts prominently displayed on the male's face. The function of these protuberances is not known, but they can also be found in other species, including the **bearded pig** (*Sus barbatus*), giant forest hog, **Javan pig** (*Sus verrucosus*), and others. They may serve a protective role in fights with conspecifics to shield the animals' faces from each other's tusks. Pigs use their tails to shoo away bothersome insects. Both the warthog and the giant forest hog carry theirs erect when running.

Pigs keep their teeth honed razor sharp with a self-grinding mechanism. The upper and lower canines file each other's edges every time the mouth opens and closes. The warthog sports the largest canines, with the record length being 27 inches (including the root buried in the gum and jaw). The tusks of all pigs grow continuously and are usually hollow. The lower canines are the ones used primarily for attack or defense since the upper ones grow upwards and are often useless for such measures. It is felt by some that the large upper tusks of warthogs actually serve to protect the face and eyes from sharp branches as the pigs run through thick underbrush.

In the male **babirusa** (*Babyrousa babyrussa*), the upper canines actually grow up through the cheeks just below the eyes. These and the lower tusks both curve backwards over the face. In some individuals, the upper tusks swing all the way around to repenetrate the skull, thus forming two circles above the snout. They reportedly are easily broken. Natives claim that the babirusa hooks its tusks over brush or branches to support the weight of the head when sleeping. Some zoologists feel that the tusks serve as emblems of rank and sexual adornment. They also serve as a protective "cage" over the face. The upper tusks may be used to ward off the lower tusks of conspecifics when fighting, and the lower tusks are reportedly sharpened by rubbing them on trees.

Pigs are omnivorous and will eat a wide range of substances. Many spend much of their time rooting with their muzzles and digging with their tusks. The latter are sometimes used like hoes, a practice that aids reforestation by allowing seed burial and also aerates the soil. **Wild boars** (*Sus scrofa*) are also helpful in that they frequently feed on large numbers of injurious insects, including the larvae of sawflies, cockchafers, pine moths, and others, but they are also known to kill and eat fawns. Warthogs often graze while kneeling on their padded front "knees" (actually wrists), since their necks are too short to reach the ground effectively. They are also known to feed on dead wildebeests and other animals. **Bush pigs** (*Potamochoerus porcus*) will topple small trees to get at birds' eggs or fruit. They have also been known to work cooperatively to move fallen logs in order to feed on the abundant invertebrates and fungi underneath. Female pigs are greatly attracted to truffles because its scent is similar to a male pig's sexually alluring odor. Contrary to folklore, wild suids rarely overeat.

Most pigs are good swimmers. Wild boars can swim distances over a mile without problem. Most like to bathe frequently in water, but they will also wallow in mud baths by the hour. Mud bathing helps to keep the pigs cool and protects their skin from sunburn and insect bites. It has been shown that a mud bath is eight times as effective as a douse of water in keeping an animal cool.

Most suids are nocturnal. The warthog is often diurnal. Wild boars and others sleep huddled together to keep warm. Many make themselves a bed of vegetation before lying down. Although the suids do some digging of their own, most take shelter in burrows dug by other animals. For example, warthogs often take over aardvark dens. The last animal in usually goes down backwards so that its formidable tusks face any intruder. Bush pigs and wild boars also form vegetative roofs over their dens or beds and sometimes tunnel pathways to their dens by forcing up the thick vegetation from the forest floor. The giant forest hog digs holes at the base of trees for use as latrines. Bush pigs mark trees with their tusks.

Although often maligned, the **domestic pig** (derived from the wild boar species) is quite clean and has proved to be one of the more intelligent of mammals. It has been used to clear land, locate truffles, tread corn, round up cattle, pull carts, plant wheat (the hooves produce a hole just the right size), retrieve game, etc. More commonly, the pig is converted to meat and cooking fat.

Suids are gregarious and usually associate in family groups known as sounders. A sounder may have over 100 members on rare occasions. In many species, the males are solitary and only join females during the breeding season. Just prior to the rut, male wild boars develop a thickening of their skin over the shoulder area to help

Warthog (*Phacochoerus aethiopicus*)

shield themselves from the tusks of rival males. These thickenings are sometimes reinforced by a coating of resin obtained by rubbing up against trees. Male pigs produce a pheromone in their saliva at mating time. During this period, the males drool all over themselves and the ground. The scent that emanates from this attracts females for mating. A male pig's eighteen-inch penis has a corkscrew tip which twists into the female's uterus. The male ejaculates several times during each mating episode, placing up to a pint of semen into the female. The last ejaculation produces a jelly-like plug to prevent sperm leakage.

When pregnant, female suids are said to be in "farrow." Bush pig females build "bowers" or grass nests for their young. Although they usually give birth to half a dozen or fewer young, the champion producer was a domestic pig that gave birth to 34 piglets in one litter. As a rule, female suids do not lick the young or consume the afterbirth. With a few exceptions, including the babirusa, some domestic pigs, and the warthog, piglets bear stripes until about three months of age, when they begin to fade. These markings may serve less for camouflage than as stimuli to inhibit aggression in the adults. The young themselves are aggressive and in many species are known to cut each other with their canines as they fight over their mother's teats. When the female leaves her very young offspring, she will often cover them with vegetation. With large litters, particularly in hard times, it is not unknown for the female to consume some of her offspring. This form of cannibalism is also seen in other species of mammals. The infant mortality in suids runs about 50 percent in the first six months of life.

In most of the pig species, the general reaction to danger is to raise the mane or back hair and then usually flee. Warthogs are known to panic and exhibit what appears to be hysteria as they try frantically to escape. They can run up to 35 miles per hour. A cornered boar or a sow with young can become a formidable foe. The **soor** or **crested wild boar** (*Sus scrofa cristatus*) has even been known to kill healthy tigers.

Suids exhibit symbiotic relationships with other species. For example, bearded pigs allow **crowned wood partridges** (*Rollulus roulroul*) to pick ticks from their skin. They are also known to

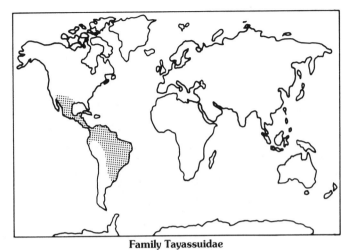

Family Tayassuidae

follow arboreal troops of primates to feed on fruit the monkeys drop.

Family Tayassuidae (peccaries—3 species)

Peccaries are the only artiodactyls with a vestigial fifth toe located on their hind feet. For many years, there were thought to be only two species, the **javelina** or **collared peccary** (*Tayassu tajacu*) and the more aggressive **white-lipped peccary** (*Tayassu pecari*). In the late 1970s, a third species, thought to be extinct, was discovered. This **Chacoan peccary** (*Catagonus wagneri*) is the most recently discovered large mammal.

Peccaries like pigs, are omnivorous. They feed on vegetation, insects, amphibians, reptiles, birds, and small mammals. Like warthogs, they have short necks and must often kneel to feed. In winter, they seem to savor prickly pears, spines and all, and their feces are frequently little more than a mass of prickly pear spines. Javelinas will even eat the feces of coyotes that had been feeding on prickly pear cactus. They are capable of

Collared Peccary (*Tayassu tajacu*)

smelling bulbs six inches below the surface and can easily dispatch and eat rattlesnakes since they are at least partially immune to the venom (perhaps because of delayed absorption through the thick layer of fat under their skin). Soil is occasionally eaten probably for its minerals.

Peccaries can raise the hair on their lower backs to expose scent glands capable of producing an obnoxious, skunk-like secretion. The substance is thought to aid peccaries not only in the marking of territory, but also in the marking of each other, presumably to decrease intragroup aggression and to

268

help keep tabs on each other. The latter should pose no problem since the odor can be detected by humans 100 yards or more away. Some experts feel that it may also be used to deter predators or even to warn each other silently of a predator's presence. Home areas are marked with dung heaps.

At one time, peccaries were known to wander in bands or "sounders" of 1,000 or more animals. Today, groups of 5–100 are more common. Before bedding down, often side by side, collared peccaries will take part in exuberant bouts of play for 15–20 minutes, flopping around, chasing each other, and mock fighting. Should one be attacked or wounded, the rest of the group reportedly comes to its defense.

Females usually give birth to twins. The young are not striped, and although the females have four nipples, only the rear two produce milk. The young are nursed from a standing position behind, rather than at the side of, the female. Sows are known to nurse any hungry youngster from the herd.

When alarmed or threatened, peccaries clap their jaws together, a warning that should be heeded promptly since even a single peccary is usually more than enough for a dog or bobcat.

Family Hippopotamidae (hippopotamuses—2 species)

The **hippopotamus** (*Hippopotamus amphibius*) is the third largest land mammal, although it is primarily aquatic. It exhibits several physical adaptations to such an existence. It has webbed toes and valved nostrils to keep out water. The eyes, ears, and nose are all located high on the head so that when the hippo submerges they may be the only parts exposed above the surface. Its specific gravity can be regulated to let it float at the surface or actually run along the bottom of the river. On land, the hippo can run up to 30 miles per hour. There are even recent reports that suggest hippos may use an echo-location system while in the water.

Hippos sleep underwater, rising periodically for fresh air. They usually breathe at least once every 2–5 minutes, but there are reports of their having stayed submerged for as long as 30 minutes. Upon surfacing, the ears are shaken independently to clear them of water.

Hippos "sweat" a red substance from their 1–2 inch thick skin. This oily material resembles blood from a distance. Actually, they have no sweat glands and the substance is produced by mucous glands located below the skin. This secretion screens out ultraviolet light to protect against sunburn; it also serves to lubricate the skin and may possess antiseptic properties. When it dries, it forms a protective lacquer on the skin and may then help retard water loss. Their skin is very permeable to water; thus, if they are away from it for too long, they can suffer from dehydration.

Hippos generally leave their home rivers at night to feed on vegeta-

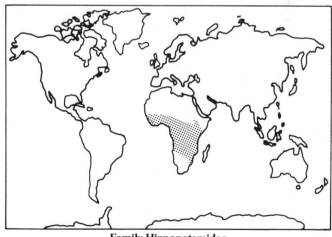

Family Hippopotamidae

269

Pygmy Hippo (*Choeropsis liberiensis*)

tion up to five miles away. In one night, an adult hippopotamus can eat 150 pounds or more of food. This is easily accommodated by a stomach that can be over ten feet long in a large animal. The trails leading to the feeding grounds are well worn after generations of use and are generously marked with large amounts of feces. The marking technique used by mature hippo males is unique. They vigorously wag their short, thick tails while defecating so that the wastes are flung for some distance, a behavior that has startled some zoo visitors into fleeing. It has been suggested that this is done not only to mark territory, but also to intimidate male challengers and impress females. The **pygmy hippopotamus** (*Choeropsis liberiensis*) is also known to defecate in this manner but will add a backwards spray of urine. Dung is also deposited by hippos in large marking mounds and in the water where it serves as the base for a huge pyramid of life. Often shoals of fish linger near the hippo's backside awaiting to feed on freshly discharged feces.

Hippos are gregarious and form groups known as "schools." The central area of the territory is usually held by the more dominant females. Males stake out their small areas on the periphery. When a male visits the central female area, he must abide by protocol or he will be attacked by a mass of angry females and driven out. For example, when a female rises up he must exhibit subordinate behavior by lying down.

Males sometimes engage in bloody battles to determine mating rights with the cows that wait complacently nearby. The frequent yawns so characteristic of hippos do not indicate sleepiness. Instead, they signal defiance and issue a challenge (as also seen in crocodiles, elephant seals, lions, monkeys, owls, tigers, walruses, and others). Throwing water with the mouth and the blowing of bubbles underwater are also threat gestures. The males fight each other primarily with their large canine tusks. The tusks and the incisors grow continuously and are kept razor sharp by wear against the upper teeth. Counting the roots, which may be about 50 percent of the length, these tusks can grow to over five feet long. Severe gashes are inflicted during confrontations, but they seem to heal rapidly; fatalities are not common. When death does occur, it may result from a tusk through the heart or from a broken foreleg, which prevents feeding. Many tusks are accidentally broken during these encounters.

Hippos are one of the few mammals to mate in the water. (Others include beaver, cetaceans, muskrats, otters, sirens, some tapirs, and some seals.) The female hippo gives birth to her 100-pound offspring either in the water or, more commonly, on the ground. In the latter case, the infant is usually dropped into a nest of reeds the mother has piled and trampled into a bed. Female hippos do not eat the placenta. Within five minutes, the newborn can walk, run, and swim. Other mammals that are known to give birth in the water include cetaceans, sirens, and the sea otter.

Baby hippos frequently swim before they walk and are often nursed while underwater, requiring them to take breathing breaks about every 30 seconds or so. The females are known to carry their young on their backs when in the water for their first few months of life. Mother hippos teach their young to fight and require strict obedience to all their "commands." Disobedient offspring are promptly disciplined with head swats. Hippos also use a system of baby-sitting to give mothers a chance to feed or mate. The communal area used for this purpose is sometimes referred to as a "refuge." There is a 20 percent mortality during the first year of life. Babies fall prey to lions, leopards, and crocodiles.

Although usually shy, hippos have attacked and sunk numerous small boats that have invaded their territory. Their powerful, sharp tusks can easily rip through the armor plating of crocodiles. More than one human has been bitten cleanly in two.

Hippos are quite beneficial to rivers because they both keep them cleared of vegetation and fertilize them. Hippos have developed a mutualistic relationship with various herons that both clean between the hippos' teeth and help to keep battle wounds free of debris and insects, thus enhancing the healing process and helping to prevent infection. The birds also screech out warnings at the approach of danger. The skin of hippos is additionally cleaned of parasitic algal growth by the fish *Labeo velifer.* One flatworm (*Oculotrema* spp.) resides in the eyes of hippos where it feeds on their tears. Young crocodiles occasionally bask while perched on a hippo's back.

Although less aquatic than the hippopotamus, the 500-pound pygmy hippopotamus must constantly remain near water since its skin can easily dry out and develop fatal sores. Its oily secretions are clear rather than red, and it tends to be much more solitary, being seen either alone or in pairs. It can close off its ear canals to keep out water. When the females come into heat, the males are known to smack their lips. In courtship, the pair spread the mucus from their skin glands on each other and cause it to lather, making them look as if someone had sprayed them with foam. They mate with the female on her stomach. The young are born on land since birth in water would drown them. The young are suckled on land.

Family Camelidae (camels and camelids— 6 species)

Camels

The two species of camel can be easily told apart by the number of their humps present, even though both are humpless at birth. Occasionally a freak individual with as many as four humps can be found. The **two-humped** or **Bactrian camel** (*Camelus bactrianus*) is the shorter

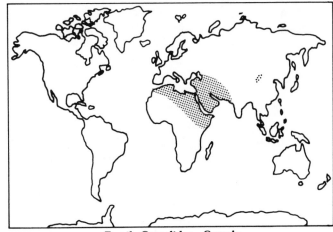

Family Camelidae—Camels

271

Bactrian Camel (*Camelus bactrianus*)

of the two and is often covered with a shaggy coat that enables it to withstand temperatures as low as -20°F. The coat is shed in the warm season when temperatures hit 100°F. The **one-humped** or **Arabian camel** (*Camelus drome-darius*) is often referred to as the **dromedary camel**, as implied by its species name, but most experts insist that the name dromedary be reserved for a special breed of Arabian camel used for riding.

The camel is well adapted for desert life. Each eye is protected by three eyelids; extra-long, double lashes; and thick eyebrows. Since camels can partially see through their third eyelid or nictitating membrane they may travel about in sand storms with "their eyes closed." The ears and nostrils are valved to keep out sand. Camels have large protective callouses on their chest, elbows, knees, and wrists. Although these appear to be the result of wear and tear, they can be found to some extent in one-month-olds and rarely in embryos. Camels have splayed feet for increased traction in sand.

Camels are best known as beasts of burden and have often been called "ships of the desert." One can carry 600 pounds of goods up to 30 miles a day. Some have carried 1,000 pounds for shorter distances. They are also well known for their ability to go long periods without a drink. The apparent record for a working camel fully packed is 537 miles in 34 days without water.

This ability to go extended periods without water has long intrigued humans. It was first thought that the camel stored water in its stomach. Although one of the stomach pouches usually contains a greenish mixture of digestive juices and food, the fluid smells foul and amounts to less in volume than can be found in other species such as cattle.

The humps do not store water either, for they are fat reservoirs. Since one of the by-products of fat metabolism is water, it was thought that the fat might provide enough water without drinking. The breakdown of 90 pounds of fat yields about 12 gallons of water. However, the process requires oxygen, and breathing itself loses water more readily than catabolism produces it. It takes a 13 gallon loss through respiration to metabolize that 90 pounds of fat, yielding a one gallon net loss.

The ability to go long periods, up to ten months if there is plenty of green vegetation and dew to feed on, has been found to result from a number of physiological adaptations. One of the major factors is that camels can lose up to 40 percent of their body weight (or about ⅔ of their total body water, since total body water equals 50–60 percent of body weight depending on sex, age, and amount of fat) without showing any ill effects. This can be as much as 600 pounds of water in an adult camel. In comparison, a human who has lost 12 percent of his body weight in water (20 percent of his total body

272

water) is in severe distress and may even die. A camel can smell water several miles away, possibly as far as six miles with the right wind. They have been known to drink 30 gallons (240 pounds) of water within ten minutes and as much as 50 gallons over several hours, turning a skinny, emaciated beast back into a plumply normal camel almost instantaneously. Camels can without problem drink brackish water and seawater and are capable swimmers. They often refuse to drink four-smelling water.

Other factors enabling the camel to go so long without water supplement its ability to withstand dehydration. The camel can vary its body temperature by as much as 14°F (92°F to 106°F) without suffering any harm. This conserves water since it cuts down on the amount of sweating (which is delayed until the body temperature reaches 104°F). Since the camel has few sweat glands, it sweats very little anyway. It also excretes only a small amount of urine since some of its waste products, such as urea, can be recycled to produce new protein. This also makes the animal more resistant to starvation. In dry times, a camel can lose as little as a pint of urine per day. This is equivalent to the urine production of a dehydrated human, who weighs about an eighth as much. Camels are capable of absorbing some water in their nasal passages to reduce respiratory water loss, and a groove leads to the mouth so nasal moisture can be swallowed to further conserve water.

The camel's fat is mostly concentrated in its hump rather than as a subcutaneous insulating layer that would tend to hold heat inside. The rest of the body can thus act like a radiator to shed excess heat. In addition, camels align their bodies with the sun to expose the least amount of surface area to its heat. The four-inch-thick wool on the camel's back serves as a barrier to heat, helping to keep the skin cool. On hot days, this fur has been measured at temperatures of 175°F on the surface. Another adaptation allowing the camel to withstand large fluid changes is that their red blood cells are oval rather than round (unique to the Camelidae) and can swell to 2.5 times their normal size without rupturing.

Camels have a reputation for being stubborn and bad-tempered. When bothered, they often bite, and they are known to spit saliva, stones, and even regurgitated stomach contents with pretty fair accuracy. Camels will eat and apparently digest almost anything, including sharp thorns that can puncture car tires, alkaline plants that are poisonous to other animals, blankets, bones, fish, rope, and even leather. They may gnaw on bones for their mineral content.

During the rut, males often fight. At this time, Arabian camels will inflate their soft palate, causing it to protrude from the side of the mouth. The protrusion can be as large as a human head and is referred to as a "dulaa." It is sometimes displayed in courtship and mating as well. They simultaneously produce a tremendous gurgling roar. Male Bactrian camels grind their teeth and froth at the mouth. They also secrete a sweet-smelling, black substance from scent glands on the backs of their heads. The secretions are spread onto the front hump; they are also used to mark out territory, as is urine, which is flicked about with the tail. The tail is also used to spread urine onto the back hump during dominance displays and occasionally in courtship.

When a camel pair mates, it has been observed, other camels in the herd will sometimes form a circle around them. All members of this family mate in a lying position and are the only ungulates to do so. Mating is thought to induce ovulation. Gestation is 13 months, and females give birth only every second year or more. The calf is dropped from either a standing or lying position. The newborn is not licked, and the afterbirth is not eaten. The young can run awkwardly 2–3 hours after being born. Camels live up to 40 years.

Family Camelidae—Camelids

Camelids

Although quite similiar in appearance, the **alpaca** (*Lama pacos*), **guanaco** (*L. guanicoe*), **llama** (*L. glama*), and **vicuna** (*Vicugna vicugna*) are usually considered to be separate species. Some experts feel they are merely breeds derived from a single species, most likely the guanaco. All are known to interbreed and produce fertile offspring. The alpaca and llama are domesticated.

The vicuna is the only ungulate whose lower incisors grow continually throughout its lifetime, as do the tusks of boars, elephants, hippos, and walruses and the incisors of rodents. Vicunas and guanacos are said to grow the finest quality fur of all mammals. The vicuna's is said to be a little better and individual hairs are .0005 inch thick. This is about half as thick as the finest sheep wool. The llama's red blood cells can absorb more oxygen than those of any other mammal, and their red blood cells' lifespan of 240 days is longer than in any other mammal. It is 2.5 times as long as that of human red blood cells.

Castrated male llamas are usually the ones used as pack animals. They can be packed with up to 100 pounds, and up to 200 in some, but if given too much or if not part of a group of other llamas they may lie down or refuse to budge. Some feel that the camelids are even worse tempered than the camels. They are said to be more prone to biting and spitting. Like the camel, they may spit rocks or regurgitated food. Males are even said to spit at each other and into the faces of reluctant females at mating time. The key to knowing when they will spit is to watch their ears. If the ears are laid back, look out.

Guanaco (*Lama guanicoe*)

Vicunas were much sought after in the past because of their high incidence of bezoar stones (see goat). They prefer moist ground and can develop a fatal foot disease if they lack it. Guanacos seem to

274

enjoy standing and lying in cold mountain streams, but can go indefinitely without drinking.

Camelids are found in female herds headed by a large stallion. The other males either are solitary or form their own male herds. When two male guanacos fight for dominance, they run at each other, draw up their front legs, thump their chests together, and then bite at each other's necks. Such antics may be accompanied by high-pitched screams. The harem consists of up to 20 mares but usually less. In fact, the male is known to drive off new females attempting to join the harem if the territory is low on feed. He also chases away offspring when they are about a year old. Should the stallion be killed, the mares will often remain with the body for some time. Vicuna stallions are aggressive and in captivity are known to attack just about anything. Camelid herds usually defecate in the same spots to mark their territory. This leads to dung heaps up to seven feet in diameter and a foot high. Baby vicunas or "crias" can outrun humans a short time after birth.

Family Tragulidae (chevrotains—4 species)

The **chevrotains** or **mouse deer** (*Tragulus* spp.) have no horns or antlers in either sex. The males do grow long canine tusks which protrude somewhat sideways. These are used in both intra- and inter-specific fighting. The smallest ruminant in the world is the **lesser Malay chevrotain** (*Tragulus javanicus*), which measures eight inches at the withers and weighs up to six or seven pounds. Chevrotains are primarily vegetarians but are known to eat

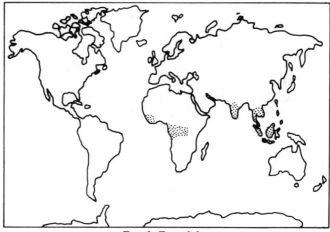

Family Tragulidae

insects, fish, and meat on occasion. They occasionally climb trees, perhaps to find a place to sleep, sunbathe, or escape predators. These small ungulates construct tunneled runways under the thick vegetation in the areas they inhabit. They are solitary except during the mating season. The females give birth to a single young but must lift one of their hind legs to nurse it. The afterbirth is eaten. When threatened, some species take to the water while others are said to play dead.

Family Cervidae (deer—38 species)

The smallest deer in the world is the **pudu** (*Pudu mephistophiles*). It stands a little over a foot high at the withers and weighs up to 18 pounds. It reaches full growth in as little as three months. In contrast, the largest deer in the world is the **moose** (*Alces alces*), standing up to 7.67 feet at the shoulder and weighing up to 1,800 pounds. When young it can gain 4.5 pounds per day, the greatest growth of any artiodactyl.

A unique feature of deer is the presence of antlers, although not all species possess them (the

275

Chinese water deer [*Hydropotes inermis*] and **musk deer** [*Moschus* spp.] lack them). In the **reindeer** or **caribou** (*Rangifer tarandus*), both sexes grow antlers, although the females' are smaller. In all other deer, only the males grow antlers, although rarely female **roe deer** (*Capreolus capreolus*), moose, and others are known to sprout small antlers, probably due to hormonal disorders. The rare male unable to grow antlers is called a "hummel" or "nott."

Antlers and horns differ in construction and characteristics. Antlers are composed of solid bone; horns have a bony core overlain by keratinous growth. Antlers are shed and regrown every year while horns are sometimes shed once in adolescence; thereafter they increase in size each succeeding year. (Pronghorns are an exception in that only the outer horny covering is shed every year.) Horns are much more commonly found in both sexes.

The fact that most deer shed and regrow their antlers annually is truly amazing in larger species such as the moose. The moose grows both the largest and the strongest of all antlers. A large rack can measure over six feet in width and weigh in excess of 80 pounds. In some species, such as the **Pere David's deer** (*Elaphurus davidianus*), the males are known occasionally to grow and shed two separate pairs of antlers each year, although the second growth is much smaller. In other species, such as the **red brocket** (*Mazama americana*), the males shed their antlers less frequently than yearly and possibly not at all. In castrated deer, the antlers also become permanent and are never shed. In temperate species, the antlers are usually shed in winter, while tropical species, such as the **swamp** or **marsh deer** (*Blastocerus dichotomus*), the **chital** or **axis deer** (*Axis axis*), the **pampas deer** (*Ozotoceros bezoarticus*), and others, shed their antlers year round (mating also takes place year round in most of these species). The two antlers of a pair are usually shed within hours or at most 2–3 days of each other.

Antlers grow in only 3–4 months. **Wapiti** (*Cervus elaphus*) and caribou antlers grow approximately half an inch per day. During growth, antlers are covered in velvet, which contains a rich supply of fine blood vessels. These can radiate away excess heat. Some experts believe that this function is the antlers' true purpose. Parasite infestation or injury early in growth can produce deformed antlers. They will in all likelihood, however, grow out normally the next year. At the end of growth, the velvet apparently causes severe itching and is persistently rubbed off on trees and brush. Many species consume the shed velvet. During velvet shedding, the antlers often appear bloody. It is interesting to note that velvet contains a substance called panocrin which speeds healing in wounds and ulcers. There is no certain correlation between age and antler size. In fact, after the prime of life (4–5 years in many species), the antlers decrease in size and quality. Poor nutrition may also lead to small antlers.

In antlers, the lowest branches

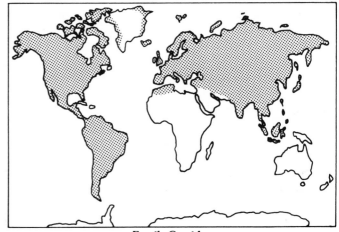

Family Cervidae

276

projecting over the head are called the brow tines. The second branches are called bay or bez tines. The brow and bay tines together are referred to as the war tines or dog killers. The third group are tray or trez tines, and the fourth are royal tines or dagger points. The antlers terminal branches are known as sur-royals. In Wapiti, a male with 12 tines is a royal stag and with 14 tines becomes an imperial stag.

One theory for why antlers are shed is that after the rut or breeding season the males are in a weakened state and would stand out as easy prey to predators. With shedding, the sexes both look the same. This may be supported by the fact that males often take on female behaviors such as crouching to urinate, although these may result from a decrease in male hormone levels at this time.

Just prior to shedding the antlers, deer reabsorb some of the valuable minerals in the bases. Some deer species such as the caribou and the **red deer** (*Cervus elaphus*—the European equivalent of the wapiti) are known occasionally to consume a portion of their shed antlers, particularly in areas where such minerals are hard to come by. Deer must consume up to 100 or more pounds of calcium in several months to produce their antlers. In addition, many of the shed antlers are consumed by various species of rodent.

Antlers vary in shape depending on the deer species. Those of the **tufted deer** (*Elaphodus cephalophus*) can be hidden by the long tufts of hair on its head. Antlers may be mere unbranched spikes as in the brockets, or huge palmate racks as sported by moose. Some species have fairly uniform antler growths while others are quite variable. Those of the Pere David's deer appear to have been put on backwards. Hunters like to take specimens with large well-branched antlers. The record number of points in a **white-tailed deer** (*Odocoileus virginianus*) is 28, while that for the **mule deer** (*Odocoileus hemionus*) is more than 50. Sometimes deer will grow long, straight antlers with few branches. These individuals are much more liable to kill opponents because they lack blocking tines and more readily pierce a rival's skull. **Thamin** or **brow-antlered deer** (*Cervus eldi*) and caribou males frequently lose eyes due to the shape of their antlers and their pugnacious nature. (With their good sense of hearing, blind individuals can usually stay with the herd and survive.) In some species, a tine comes out over the eye and is thought to protect the eyes when pushing through thick brush. When sporting antlers, wapiti and others are known to run with their heads up and antlers back, presumably to prevent entanglement with vegetation. It is known for deer to entangle their antlers not only with vegetation but with a rival's antlers, leading to death from starvation. The fights with conspecifics can become violent enough to rip off antlers in wapiti and others. At times a piece of the skull is lost with the antler.

Many functions have been attributed to antlers. They do have a role in dominance as "badges" of rank, and they are used in jousts. As mentioned, they can radiate away excess heat, and some feel that in some species they may serve as an erotic stimulus to females. Rarely, it has been reported that antlers have been manipulated as tools. Some deer use them to scratch their backs, red deer have extended their reach with the antlers to get at apples out of browsing range, and, possibly, caribou may on occasion use theirs to scrape away snow in order to feed, although their hooves usually perform this function. Antlers are ground up and taken as medicine in many areas of the Orient. Surprisingly, they do contain sex hormones and may be beneficial to some old age symptoms.

Deer have no upper incisors, but instead a hard pad against which the lower incisors bite. In the antlerless Chinese water and musk deer, the males are armed with large canine tusks that protrude from the upper jaw and can be observed even when the mouth is closed. In the musk deer, the four-inch

White-Tailed Deer (*Odocoileus virginianus*)

tusks can be pulled further forward by special muscles that contract when the mouth is opened. Long, tusk-like teeth are also found in **guemals** (*Hippocamelus* spp.), **muntjacs** or **barking deer** (*Muntiacus* spp.), and tufted deer. Wapiti have two small tusks that once were coveted by Indians and are the symbols of the Order of the Elk today. Muntjacs have the fewest chromosomes of all mammals; the females have six and the males seven.

Nomenclature is frequently confusing and frustrating. In fact, inappropriate use often becomes so much the rule that it is finally accepted. For example, the wapiti is often referred to as the elk in America and the red deer in Europe. What makes things worse is that the moose is known as the elk to all but Americans. Technically, male moose, wapiti, and caribou are referred to as bulls and the females as cows. In other deer, the males are bucks and the females are does. The red deer is an exception in which the males are stags and the females are hinds. Today many people use the terms interchangeably.

Caribou hooves can splay apart to act as natural snowshoes. In addition, pseudo-claws are present to help distribute the weight further and prevent sinking. These same adaptations are present in marsh deer, moose, musk deer, and others that often inhabit swampy areas. The caribou and Pere David's deer can be heard up to 100 yards away when they walk because of a clicking tendon in the caribou's ankle and cracking toes in the feet of Pere David's deer.

The **fallow deer** (*Dama dama*) is said to vary more in color than any other wild mammal. It is found in white, cream, sandy, silver, gray, spotted, and black. Although most deer species are spotted when young, a few species, including the axis deer, fallow deer, **hog deer** (*Axis porcinus*), some **sambar** (*Cervus unicolor alfredi*), and some **sika deer** (*Cervus nippon*), retain their spots into adulthood. The fallow and sika deer have spots during summer, when sunlight speckles through the leaf-covered trees, but lose them during winter, when the trees lose their leaves and more uniform light reaches the ground. The coats become

278

uniformly brown or only faintly spotted then. The spots return with the leaves in springtime. Some sika in evergreen forests keep their spots year round.

Various deer species occasionally bear albinos. In the case of white-tailed deer, these individuals are usually deaf to some extent. The caribou's outer coat hairs are hollow, as are the winter hairs of many deer such as the white-tail and moose. This generally lightens the coat color, even turning it white. The air-filled hollows improve insulation against the cold and increase buoyancy when swimming. In frontier times, the skin of deer, referred to as buckskin, was valuable and even used as money. From this use originated one of the nicknames for the dollar, the buck.

Surprisingly, most deer are excellent swimmers. Caribou are probably the best swimmers and can be found swimming at sea in herds. Wapiti can easily swim lakes over a mile wide. The moose, too, is an avid swimmer and is also occasionally found swimming at sea. They can reportedly swim at six miles per hour for two hours or more. White-tails have been found five miles offshore and have been reported to swim up to 13 miles per hour.

The white-tailed deer is a speedster on land. It can run up to 35 miles per hour and hurdle obstacles eight feet high, and it has made horizontal leaps of 30 feet.

Deer are vegetarians. The white-tailed deer feeds on more than 600 species of plants, including poison ivy and mushrooms. In hard winters, they are known to feed on tree branches that have been cut, nibbled, and dropped by porcupines. Musk deer will climb angled trees to feed on the tops, and sika deer sometimes eat algae along the seashore. Although they are primarily browsers, moose will sometimes graze. At such times they may kneel on their front legs since their necks are so short. They may push over small trees up to 3 inches in diameter to feed on the upper leaves or break into muskrat lodges to obtain a winter meal. They also will submerge themselves in water as much as 18 feet deep to feed on water vegetation. This ploy is also used to get away from insects. Moose consume up to 60 pounds of food daily. In winter, moose will cooperate to trample down large areas of snow and form "feeding yards." When these become depleted, they move on to form new ones. Caribou and elk usually use their hooves to dig down as much as two feet in snow to get at food. Caribou may on rare occasions use their antlers to aid in digging. They are also known to trample and eat lemmings during these rodents' irruptions.

Most deer lie down to sleep. The muntjacs probably go with as little sleep as any mammal. It has been found that they sleep only occasionally each 24 hour period and then only for single "winks" of 25 seconds or less.

Most cervids form herds or at least family groups, although there are some solitary species such as the moose, muntjacs, musk deer, and water deer. While grouped, there is usually a dominance hierarchy. Although antlers are known to play a role in some species, antlerless males are sometimes dominant in other species. In reindeer, the size of the antlers determines rank in the herd. The males are dominant to the females at first, but since they drop their antlers earlier, a phase of female dominance follows. Both red deer and reindeer are known to fight at times to determine dominance hierarchies. In **barasingha** (*Cervus duvauceli*), the older, more dominant bucks become subordinant to younger males when they drop their antlers at an earlier time. Herds are usually led by the dominant doe. Artiodactyl submission is almost universally displayed by assuming an exaggerated feeding position or behavior.

Most male deer hold territories. The exceptions include white-tails, mule deer, reindeer, and a few others. Often the size of the territory is related to the size of the deer. Territories are marked with glandular secretions, urine, brush and tree marks, and ground holes dug up with the hooves. Depending on the species, special marking glands are located on the face, feet, perianal area, or some combination of these. The foot glands of male pampas deer give out a strong garlicky odor that can be detected a mile or more downwind. In some, such as the white-tail, the interdigital glands on the hooves mark their tracks with their scent. Male white-tails and mule deer supplement this by urinating on their tarsal glands frequently. White-tails, mule deer, moose, and others will slash bushes and trees with their antlers in a behavior known as "horning." In musk deer, (also roe deer, and others) both sexes set up territories. Each marks out its boundaries with secretions from tail glands. The male has an extra gland on the belly that secretes the musk that has served as an important ingredient in expensive perfumes. Some captive deer are even known to mark their keepers with head gland secretions to "claim" them as their property. Like cats, musk deer bury their feces.

Males become quite aggressive during the rut and are responsible for more attacks on humans than any wild species of predator in the United States today. The males increase their marking behavior and are known to dig holes in which they urinate and then roll to mark themselves. Male chital, barasingha, red deer, wapiti, and others will drape vegetation over their antlers to enhance their display value. In most species, the males engage in combat with rivals to gain control of female harems. In some instances in numerous species, including moose, red deer, wapiti, white-tailed deer, and others, this has led to locked antlers and starvation for both. Various species, including moose, musk deer, and wapiti, have been known to fight to the death on rare occasions. In wapiti, the thunderous crashes have been known to break the neck of one of the contestants. In fighting, most bucks use primarily their antlers, but some, like Pere David's deer and red deer, will box with their forelegs while standing on their hind legs and will clash with their teeth as well. In red deer, one in twenty males at some time in his life will receive a serious injury while sparring with a rival. In some species, such as the wapiti, the females are also known to fight each other for territory. When agitated, female moose are known to urinate on their hind legs.

In most cervids, the males attempt to add as many females as possible to their harems. Although some, such as the roe deer and white-tail, rarely produce sounds, most males will roar, bugle, bellow, bark, or somehow call out to females. The wapiti and red deer can be heard for distances up to a mile. One overzealous male red deer was known to take over a herd of dairy cows as his harem and even chased the perplexed owner away from his cows. Male wapiti rarely hold their harem for the full breeding season. As they fight off other males, they exhaust their fat stores and become somewhat emaciated. At this point, they begin to metabolize muscle tissue, and protein metabolites show up in their urine. When other bulls detect this, they more readily challenge him and eventually overthrow him to take over his harem.

Mating in deer is usually preceeded by some form of ritual. One of the most elaborate of such rituals is seen when roe deer produce their well known roe or "witch's" rings in small areas usually centered on a tree or bush. The buck chases the doe around the center, often in circles or even figure-eights. Eventually, they wear a path into the ground to form the rings in which the two often mate when the chase finally ends. If a bull moose is too aggressive towards a cow's calf from her previous pregnancy, the cow may leave him without mating. This is one of the very rare cases in which offspring influence a parent's courtship.

280

Gestation usually takes 6–9 months. Roe deer are the only cervids to use delayed implantation. Most species, including axis deer, barasinghas, brockets, caribou, fallow deer, muntjacs, musk deer, pampas deer, pudus, and most **true deer** (*Cervus* spp.), bear only a single young. In most of the remaining species, the female bears a single young her first year but mainly twins thereafter. The Chinese water deer may have as many as eight young in one litter, the most of any deer, but two seems to be more usual. Many females consume the afterbirth to build up their strength.

The fawns of most deer are camouflaged with white spots. Exceptions to this include moose, caribou, and sambar. The fawns of many stay put in a concealed area their first week of life. They are protected both by their camouflage and by the fact that they have little or no scent because their scent glands do not function for several weeks. They may not urinate or defecate for their first week of life, and in some species the female consumes her offspring's wastes. In many, the female comes around only to feed the fawns, otherwise keeping her distance.

Baby ruminants do not require four stomach chambers to digest milk. It has been found that the mother's milk has salts that stimulate a special nerve in the babies' mouths. This causes the esophagus to form a special tube leading directly to the fourth chamber, by-passing the first three.

The pampas deer, musk deer, and other females are known to protect their young by feigning injury or sickness. They thus lead predators away from their offspring's hideout before taking flight to save themselves. Some young are quite precocious. A day-old baby caribou or fallow deer, a four-day-old moose, and a one-month white-tail can all outrun a human. In roe deer and others, the white mark under the tail lets fawns recognize their mothers and follow her easily. Baby chitals feed by standing between the doe's forelegs facing in opposite directions as they nurse. Male pampas deer are the only species known in which the males stay and help care for the young. In most cervids, the young are usually on their own right about a year's time.

Deer generally respond to danger by warning others of their kind. This is usually accomplished by flipping up their tails to expose a patch of white fur (fallow deer, red deer, roe deer, white-tails, etc.) or by barking out a warning (roe deer, muntjacs, tufted deer, wapiti, moose, etc.). Muntjacs are also known to rattle their teeth when excited. Mule deer will stamp their feet to give warning, and some, such as caribou, may secrete odors to give warning. When alarmed, caribou and others perform an "excitation leap." The animal places its weight on the hind legs and leaps into the air. Such maneuvers visually alert others in the area. With the leap, the hind hooves are splayed out to release scent which lasts for hours and serves as a chemical alert. Almost all cervids will attempt to flee danger. Some, such as marsh deer, may "freeze" until the last second. If cornered, though, most deer can put up a surprising defense. At this time, many will turn on their attackers with tusk, hoof, and antler. A wapiti has been known to break the back of a wolf with a single kick, and moose have even killed grizzly bears. There are records of moose having attacked cars, trains, and at least one Boeing 720 jet. More than one moose hunter out for a rack has literally become racked himself. Many deer species are known to kill snakes, including venomous species, with their sharp hooves.

Probably the best cervid defense mechanism is their extreme wariness. During rainstorms, many deer will head for meadows or clearings where visibility is better, the raindrops make less noise and they can better detect an approaching predator. Some also take to water and thus destroy their trail of scent. Moose, axis deer, sambar, and others will escape pursuers by diving into rivers or lakes and then swimming away. Some deer run to areas usually populated by others of their kind. This may transfer the chase to one of its conspecifics or at least confuse the predator with multiple scent trails.

Deer are frequently bothered by insects. Many species habitually take mud baths to rid themselves of parasitic ticks and lice as well as to prevent gad flies from laying eggs between their toes or other flies from biting into their hide. Some submerge in lakes or lie in streams to discourage parasitic flies. Others take refuge on high snow patches. In one study, it was found that caribou could lose up to a quart of blood per week to mosquitoes and other blood-sucking insects. Some caribou hold their noses to the ground to prevent the entry of parasitic nostril flies, and there are reports of their running off cliffs in suicidal leaps in their panic to escape insect tormentors. Mule deer chew the ticks from each other's backs when the parasites are plentiful. Magpies, too, will pick ticks from their backs. Barasingha also get their coats cleaned of parasites by birds such as mynahs, but they may find that the birds leave droppings on their backs and occasionally pluck out hair for nesting purposes.

Reindeer are the only members of this family that have been widely domesticated. A large one can carry up to 90 pounds on its back or pull 450 pounds on a sleigh. In areas where reindeer have been introduced for herding, the local wild caribou are persecuted since they sometimes make raids to lead the "domesticated" reindeer off into the wild. Reindeer have been the mainstay of some peoples for many years. All parts of the animal are used, including the stomach contents. Humans cannot directly consume the mosses and lichens upon which reindeer feed, but after they have been partially broken down in the reindeer's stomach they may be eaten as a salad known as "nerrock."

Many deer migrate during hard times. Some, like the caribou, travel hundreds of miles in numbers up to 200,000, primarily in single file. A migrating herd of caribou may string out to a length of 185 miles and may take weeks to pass a particular site. They have been known to suffer heavy casualties, greater than 500, when attempting to cross flooded rivers. Other deer simply migrate up and down the mountains of their home range. In very cold weather, moose may bury themselves under two feet of snow and remain there where the temperature may be 80°F warmer than the ambient air.

There are stories of caribou having "sanitariums" located on secluded, predator-free islands in the far north. It is true that specimens with broken legs and the like have been found on such islands, and some naturalists feel that they swam there in such a condition. Infection is prevented by fly maggots in the wounds since they feed on devitalized tissue and reportedly deposit an "antibiotic-like" substance.

Family Bovidae (antelope, cattle, goats, and sheep—128 species)

Antelope

The term antelope is applied to a large group of loosely related ungulates. They range in size from the five-pound, fully grown **Salt's dik-dik** (*Madoqua saltiana*) that stands 14 inches tall to the one-ton **giant eland** (*Taurotragus derbianus*) that stands six feet at the shoulder. The **royal** or **pygmy antelope** (*Neotragus pygmaeus*) is the shortest antelope since it only grows to 12 inches, but its stockier build gives it a slightly larger average weight than the Salt's dik-dik.

The bovids possess the most beautiful horns of all mammals. The smallest are the 1.25-inch spikes of the pygmy antelope, while the largest are the 6.5-foot curled horns of the **Marco Polo sheep** (*Ovis ammon*). The longest horns are found on the **Indian buffalo** (*Bubalus bubalis*), one of which had horns measuring almost 14 feet from tip to tip along the curve. In all species of wild bovid, at least the

282

male normally has two horns. Many breeds of domestic cattle, goat, and sheep, however, are hornless or "polled." The naturally occurring **four-horned antelope** or **chousingha** (*Tetracerus quadricornis*) has one small pair of horns just above the eyes and a larger pair on top of the head. Occasionally the front pair will drop off spontaneously. Four horns are not unique to this antelope, although certain domestic sheep, such as the **piebald** or **Jacob's sheep** (*Ovis aries*), have been bred to have four horns, and other domesticated sheep have been known to sprout as

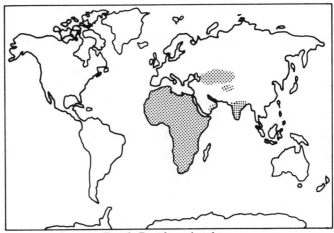

Family Bovidae—Antelope

many as eight horns. There is also a breed that grows only a single horn, fittingly known as the **unicorn sheep** (again *Ovis aries*, as are all domestic sheep).

In antelope, both sexes have horns except in dwarf antelope (including beira, dik-diks, grysbok, klipspringer, oribi, pygmy antelope, royal antelope, steinbok, and suni), common duiker, four-horned antelope, nilgai, reedbuck, rhebok, saiga, spiral-horned antelope (including bushbuck, kudus, nyalas, and sitatunga), Tibetan antelope, and waterbuck (including kob, lechwe, and puku). In most African gazelles (except the dibatag, gerenuk, and impala), both sexes have horns, while in Asian gazelles (including blackbuck, goa, goitered gazelle, Mongolian gazelle, Tibetan gazelle, and zeren), usually only the males have horns. Some species, such as the **klipspringer** (*Oreotragus oreotragus*), have races whose females have horns and other races whose females are hornless. As a general rule, species in which the females grow horns are often found in large herds or groups. Often in these instances, the horns of the males are larger or longer, although in the **common eland** (*Taurotragus oryx*), **Ankole cattle** (*Bos taurus*), and the **gemsbok** (*Oryx gazella*) the females are known often to have larger or longer horns than their male counterparts.

Many bovids shed their horns once in adolescence. They may fall off in sleeve-form as in the **kouprey** (*Bos sauveli*), or be shed as the "tip" of the new permanent horn as in the **blackbuck** (*Antilope cervicapra*). The shredded areas around the tips of the horns in the **European bison** (*Bison bonasus*), kouprey, and others are actually a remnant of early horn growth but may be exaggerated as a result of digging in the ground with the horns. With the exception of early sheddings and the pronghorn's unique growths, horns are permanent structures and are not dropped or lost except through trauma. At death, horns easily separate into a hard bony covering (used at times for bugles, powder horns, etc.) and an inner core of spongy bone. Parasitic black flies (Family Simuliidae) are partial to the blood-filled early growth of horns. Infestation can lead to stunted and deformed growth.

The function of horns has been debated for years. Although they undoubtedly serve many purposes, they are thought to serve primarily as a badge of rank and dominance. Only a few species use them to fight off enemies, as we will discuss later. The horns are used by many in sparring matches

283

with rival males. As in deer, some antelope, such as the **greater kudu** (*Tragelaphus strepsiceros*), have been known to lock their head gear together, leading to death from starvation or predators. Elands are known to use their horns to break off tree branches for feeding, and **bongos** (*Tragelaphus euryceros*) sometimes use their horns to dig up roots.

In antelope, the sexes are usually colored the same, although sexual dimorphism does occur in a few, including the blackbuck, four-horned antelope, kudus, nilgai, nyala, and waterbuck. In blackbuck only the most dominant male takes on a darkened appearance, while other males retain female coloration. Usually males are larger than females, although the reverse is true in the **common duiker** (*Sylvicapra grimmia*) and many of the dwarf antelope such as dik-diks. Bald areas are often present in the dorsal neck area of some antelope, such as the bongo, **bushbuck** (*Tragelaphus scriptus*), and kudus. These result from rubbing of the horn tips on the skin when the animal runs through the brush with its neck fully extended to keep its horns from entanglement. It is stated that the spiral horns found in these and other species are less apt to get caught in brush than are straight horns. Spiralled horns are also less likely to slip and cause injury during sparring matches.

There are a few anatomical peculiarities among the antelope. The eland's knees click as the animal walks and can be heard up to 100 yards away. The **saiga** (*Saiga tatarica*) can be readily recognized by its large, humped nose. It contains special sacs to warm, filter, and moisten the cold, dusty air found at the high elevations it inhabits. A similar but less conspicuous nasal apparatus is found in the **Tibetan antelope** or **chiru** (*Pantholops hodgsoni*). Several antelope species have hooves that are easily splayed to increase their surface area and therefore traction. The **sitatunga** (*Tragelaphus spekei*) needs them for walking in its swampy habitat, while the **addax** (*Addax nasomaculatus*), **Arabian oryx** (*Oryx leucoryx*), **goitered gazelle** (*Gazella subgutturosa*), and others need the traction in sand. The addax and oryx are also reported to use their hooves to dig small holes for shelter from sandstorms.

The antelopes are generally fast, agile, and excellent jumpers. One of the fastest is the blackbuck, credited with speeds of 50 miles per hour, a stride of 22 feet, and leaps up to six feet high, all phenomenal accomplishments for an animal that weighs an average of 80 pounds. **Impala** (*Aepyceros melampus*) can jump over ten feet high and have made leaps 40 feet long. Even the tiny, ten-inch royal antelope has been known to jump 9.5 feet. There are reports of saiga and springbok running as fast as 60 miles per hour. The klipspringer is said to be more agile and sure-footed than any goat. It is probably the world's leading high jumper since it has been reported to leap vertical heights of 20 or more feet. They are capable of jumping down precipices, landing with

Klipspringer (*Oreotragus oreotragus*)

all four feet on a spot no larger than a silver dollar. Their thick, brittle fur, which is said to rattle when shaken, is easily pulled away but grows back quickly. This is a major asset in that it enables the animal to escape predators and prevents it from getting snagged on bushes, thorns, or twigs which could cause a small but fatal loss of balance while bounding across steep cliffs.

The antelope are relatively quiet animals. Some species bark, some mew, and quite a few produce whistles, particularly when alarmed. Bushbuck are known occasionally to create bird-like twitters.

The addax, **beira** (*Dorcatragus megalotis*), **dibatag** (*Ammodorcas clarkei*), dik-diks, gazelle, **gerenuk** (*Litocranius walleri*), oryx, saiga, springbok, and others can go months and in some cases possibly indefinitely without drinking water. They do get substantial water in the vegetation they eat, and greater kudu are known to do so by eating melons. Some conserve water by allowing their body temperature to rise—11°F in the oryx—without setting off heat-loss mechanisms such as panting or sweating. In **Thomson's gazelle** (*Gazella thomsoni*) and the oryx, the blood vessels serving the brain are arranged to cool incoming blood. This is important because the body temperature can exceed the level harmful to brain cells. While the smaller ungulates pant to rid themselves of excess heat, the larger species must pant and sweat since enough heat cannot be dissipated through panting alone. It is also interesting to note that gerenuks are known to drink each other's urine when water is not available. This is particularly true of pregnant or nursing females.

As a rule, the antelopes are vegetarians. Bongos, bushbucks, and **duikers** (*Cephalophus* spp.) have occasionally been known to kill and eat small mammals and birds or will feed on carrion. Duikers are also known to feed on insects, with some individuals appearing to relish termites. **Red duikers** (*C. natalensis*) often climb slanted trees to feed on the foliage. Some duikers crack open monkey oranges (a hard-shelled fruit) by repeatedly butting them against trees or rocks until they split. Bongos occasionally eat dirt and charcoal from burnt trees, apparently for the salt content.

Some antelope are aided or hindered anatomically in their feeding. The gerenuk and dibatag are noted for their long, giraffe-like necks which, along with their habit of standing on their hind legs, enables them to feed on foliage 6–7 feet out of the reach of other small antelopes. On the other hand, the **white-tailed gnu** or **black wildebeest** (*Connochaetes gnou*) and **nilgai** (*Boselaphus tragocamelus*) both have such short necks that they often graze while kneeling on their forelegs. These species, along with the **hartebeests** (*Alcelaphus buselaphus*), **roan antelope** (*Hippotragus equinus*), and **sable antelope** (*Hippotragus niger*), often use the position when skirmishing with rivals.

Animals that graze in the same areas avoid competition by feeding on different parts of the same plants. For example, gnus feed on new shoots, zebras on more mature parts of the grass, and **topis** (*Damaliscus lunatus*) on the coarse, old stems. Unusual items can be found in the diets of some antelope. The Arabian oryx occasionally eats colocynth or "bitter apples," known to be poisonous to humans, and the saiga feeds on plants that are poisonous or too salty for other animals. During the mating season or rut, male saigas eat only snow. The sitatunga and the lechwe are often found spending the better part of the day up to their necks in water as they feed on aquatic vegetation, which they must completely submerge to reach.

Dik-diks and **steinbok** (*Raphicerus campestris*) will at times bury their feces. Some dik-diks, four-horned antelope, gazelle, hartebeest, nilgai, wildebeest, and others will defecate in the same areas over and over to produce large dung heaps that may be important in marking their territory. **Bontebok** (*Damaliscus dorcas*) also produce large dung heaps, on top of which they are known to lie.

The vast majority of antelope travel about in groups, although a few, such as the bushbuck, four-horned antelope, and steinbok, are for the most part solitary; the dik-diks and duikers are often solitary or found at most in pairs. Antelope groups may be small families or large herds composed of both sexes. Some group together in herds of a single sex, some are composed of a male with his harem, and some form female groups while the males are solitary except during the rut, when they join with the females for mating. Many species may show more than one type of group. Tibetan antelope males are known physically to prevent females from leaving their harems.

Most male antelopes hold territories, with the bushbuck and elands among the exceptions. The size of the territories is variable, being as small as 50 feet in diameter in the case of the **kob** (*Kobus kob*) to as much as 1.5 miles in diameter in **Grant's gazelle** (*Gazella granti*). Some such territories are held year round, while others last only for hours. The latter is seen in gnus, whose males stake out territories during migration, as do Grant's gazelle, kob, springbok, Thomson's gazelle, and topi. At rest stops, the males become territorial and hold their small areas for a few hours to several days. Usually the most dominant males hold land in the central part of the herd.

Territories are usually marked with urine, feces, and glandular secretions. Most antelope except addax, dibatag, eland, impala, lechwe, and oryx possess small facial glands to serve this purpose. In addition, duikers, hartebeest, kudu, **nyala** (*Tragelaphus angasi*), **oribi** (*Ourebia ourebi*), roan antelope, sable antelope, and many others have glands on their feet to mark out their areas. Many dwarf antelope secrete tears that give a characteristic musky smell to their territories. It should be noted that most antelope have sensitive olfactory receptors able to assess hormonal concentrations in urine. They readily detect occupied territory and estrus in females since hormonal output is increased in both land-owning males and females in heat. Male hartebeests and kobs mark their territories primarily with their presence, a behavior referred to as static-optical demarcation. Each male positions himself on the highest pinnacle in his territory, such as a large termite mound, to advertise his ownership of the surrounding area. This also has the added advantage of giving the animal a good panoramic view so that it can quickly detect approaching danger. Kobs supplement this with frequent whistles to announce ownership to rivals. Rheboks supplement their markings with tongue clicks.

Dominance is the factor that usually determines which males will mate with the females. Except during the rut, most ungulates or hooved mammals are matriarchal in nature. Most herds exhibit a definite hierarchy of dominance. The dominant animals usually take the best eating and sleeping spots often by first ejecting a more subordinate animal. In some, such as the kudu, a dominant animal loses some of its superiority by lying down and only then is likely to be disturbed by a more subordinate animal. Dominance is usually determined through some sort of sparring. In males this is most commonly done with the horns. Sparring is also used to settle territorial disputes, to obtain females, or even for play, particularly in younger animals. In **Soemmerring's gazelle** (*Gazella soemmerringi*), the males spar by locking their hook-tipped horns and engaging in a tug-of-war. Most others butt each other with their horns and heads and engage in pushing matches. Some of the more aggressive species, known to have fought to the death on rare occasion, include **lechwe** (*Kobus leche*), **rhebok** (*Pelea capreolus*), oryx, saiga, Tibetan antelope, and others.

Reports of sparring deaths seem to come mostly from zoos, where a losing animal is denied a means of escape that would be open to it in the wild. The victor continues to press the attack until death results. Such fatalities are rarer under natural conditions.

286

Courtship rituals and displays vary greatly with species. In the majority of antelope, with a few exceptions including bushbuck, eland, gnus, kudus, hartebeest, and impala, the courtship includes a ritual referred to as the Laufschlag gesture or "mating kick." The male slowly kicks or extends his foreleg to the undersurface of the female between her legs or else upon her hindlegs. Many ungulates, including deer, duikers, gazelle, and others, including cats and bats, also use the "Flehmen" ritual. The male places his muzzle between the legs of a urinating female. He then stands back, curls the upper lip, and sniffs. In this manner, odor molecules are brought into closer contact with the Jacobson's organ in the upper jaw so he can carefully analyze the urine to determine if the female is in estrus. If the female is reluctant to urinate, the males will sometimes force them to do so, scaring them by leaping, roaring, or pawing at the ground. If the females are in estrus, the courtship can begin.

The many courtship and mating rituals involve various jumps, posturings, buttings, neck fighting, biting, and so on. Many of the rituals are thought to make the females more receptive to or synchronize them with the males. In duikers and gazelles, the males butt the females in the flanks and belly. Female oryx butt back. Each sex is somewhat protected by extra thick skin on the shoulders. During successful kudu courtships, the females are also known to butt the males back in the flanks. Should the female spurn the male's courtship advances, however, he becomes frustrated and extremely irritable, to the point of attacking anything close by, including other animals, trees, shrubs, or even the ground.

Many courtships include circle running. Hartebeest pairs do it around trees; male eland, roan, and sable antelope circle the females; and impala males are known to run around the remaining herd of females and young. The male oribi stimulates the female by lifting her hind end with his head and pushing her along as if she were a wheelbarrow. Courting duikers and steinbok exchange secretions by pressing their preorbital glands together and smearing the fluids over each other's faces, a ritual referred to as the "accolade ceremony." Many other antelope males, including gerenuk and dibatag, are known to rub preorbital gland secretions on the head and neck of the females. In the **Mrs. Gray's waterbuck** (*Kobus megaceros*), the male bends his head to the ground and then urinates onto his neck and throat. The dripping wet throat is then rubbed onto the female's forehead and back. Male eland rub their foreheads in urine and mud and will throw earth on their bodies with their horns during the rut. Male goitered gazelles' throats swell during this time.

Female antelopes are receptive to males for only short periods. The female hartebeest is in estrus for only a single day. As in almost all artiodactyls, the actual act of mating lasts for only 2–3 seconds. Gazelles usually mate while walking, without missing a step. Some duikers, dik diks, royal antelope, and klipspringers may mate for life.

Most antelope females give birth to a single young from a standing position. There are exceptions: the steinbok female lies down to give birth; and the goitered gazelle, Mongolian gazelle, nilgai, and saiga usually give birth to twins. Most lick the newborn clean and continue to do so throughout infancy. Licking of the anogenital area stimulates the discharge of feces and urine, while body licking helps to cement the pair-bond between young and mother and helps to keep the young odor-free. In gnus and others, the afterbirth is not discharged for several hours. This is about how long it takes for the newborn to learn to run efficiently, although they can walk and follow the female within five minutes of birth. If the afterbirth were to be discharged immediately, it might attract hyenas which could pose a threat to the newborn. For this same reason, female gazelles and many other antelope females eat the placenta. Female duikers, kudus, gazelles, and others also consume their youngs' waste products to

287

keep the nest area clean and to dispose of all clues that might tip lurking predators to the babies' whereabouts. Female duikers mark their young with preorbital gland secretions, apparently to aid in identification.

Most bovid (and other artiodactyl) young can be classified as "followers" or "hiders," depending on the species. Followers will within hours trail after their mothers; hiders remain hidden when very young, exposing themselves only at nursing time. Since they have little scent, they approach the female rather than vice versa so she won't leave a scent trail leading to the young's hiding place.

Most antelope young are coated in a subcutaneous layer of special fat known as "brown fat." This type of adipose tissue can convert fat into heat energy 20 times faster than can ordinary fat tissue; it thus serves as an emergency supply of heat. Brown fat is also found in bats, bears, hedgehogs, and others, and it is particularly plentiful in those that hibernate, since it can be used as an emergency heat supply to prevent freezing. When first discovered in such animals, it was labeled the "hibernating gland."

Females often attempt to protect their young, but males rarely do. In fact, male Thomson gazelles have been observed to attempt to prevent females from leaving their territories when the females were actually trying to prevent jackals from getting her young. In gnus there is an 80 percent mortality of calves in the first year of life.

Antelopes broadcast danger in a number of ways. The **springbok** (*Antidorcas marsupialis*) everts a large pocket on its back to expose a wide area of white hair. This is often accompanied by vertical bounds 10–12 feet into the air, a behavior known as "pronking" or "stotting." It seems to be contagious and often leads to the whole herd pronking away. Pronking seems to be used as an alarm, to confuse or distract predators, and even to lure predators away from young. More recent investigators feel that antelope (gazelle, oribi, and springbok) pronk to get a better view of predators.

When excited or threatened, **Speke's gazelle** (*Gazella spekei*) inflates a loose flap of skin on its nose. The resulting spherical "balloon," about the size of a tennis ball, can be quickly deflated to produce a loud popping noise similar to a gunshot. Some antelopes, such as bushbuck and dik-diks, bark when alarmed, while many others, including duikers, klipspringers, reedbuck, springbok, steinbok, and others, emit whistling alarm calls. Alarmed sitatunga and hartebeests give a warning "sneeze." Thomson gazelle can transmit warnings by ceasing all movement, including the almost continually active tail. When the wagging stops, others become alarmed. When fleeing, oribis and others lift their tail to flash a visual warning of danger by exposing a white patch of hair underneath.

The wariness of dik-diks has been their downfall. They are often the first to detect danger and will give out an alarm bark that puts all game within earshot on the alert or on the run. Since this has frustrated many hunters, the latter occasionally have taken to mass slaughter of these small antelope. It was also once thought that female gazelle were more wary than males, since predators took more males than females. Actually, subordinate males holding poor territory are the easiest to take and thus account for the discrepancy.

When alarmed or confronted by an attacker, antelope react in various ways. Many, including **bohor** (*Redunca redunca*) and steinbok, will initially "freeze" flat against the ground. This effectively does away with any tell-tale shadows and can be seen in deer, squirrels, and other mammals. If danger gets too close, the majority will take flight. Grysboks and some young antelope, such as the saiga, will initially run away but then drop to the ground, giving the appearance of suddenly vanishing. Since

288

predators such as cheetahs often single out one specific animal, antelope in herds have developed the defensive mechanism of bounding in and out of the fleeing herd. Some, such as the impala, even jump over conspecifics. These maneuvers often confuse and frustrate the predator.

Others have postulated that the zig-zag pattern of flight shown by many antelope may be a way to keep both eyes alternately on the predator to keep from running in circles. Others have reported that antelope will sometimes make a quick dash in front of their pursuer to see if it is the object of the chase. If so, the animal then goes all out, but if not, it often slows down. This is also given as an explanation for the quick dashes of such animals in front of speeding autos. When chased, steinbok have been known actually to take shelter in aardvark burrows, which are also used as "nurseries" for the young. The bushbuck, greater kudu, lechwe, sitatunga, and waterbuck have all been reported to escape enemies by diving into the water and submerging all but their nostrils. The sitatunga is even capable of swimming underwater. Antelope such as kudu have been known to take refuge next to humans when chased by wild dogs. Surprisingly, this has taken place with success on numerous occasions. Some gazelle have escaped the final rush of a predator such as a lion by actually jumping over the attacker.

Although most animals fight back if cornered, there is a small handful of antelope, including the bushbuck, oryx, roan antelope, and sable antelope, which frequently stand their ground and fight initially. Each is equipped with a formidable pair of horns. In several instances, confrontations between oryx and lions have ended in a draw with the lion hopelessly impaled on the long horns. The oryx, too, were doomed since they were unable to rid themselves of the lion's body. Eland and bongo have been reported to stand their own against leopards and cheetahs. It has been found that when resting, **lesser kudus** (*Tragelaphus imberbis*) keep trees and bushes at their backs to thwart sneak attacks from predators.

Many species exhibit mutualism with oxpeckers and other birds. The antelopes rid themselves of parasites while the birds obtain easy pickings and will even occasionally pull out some hair for lining their nests. In addition, the antelope may get early warning of the approach of danger through flight and alarm calls from the birds.

Such huge herds of springbok once existed that it was actually known for lions, leopards, humans, and smaller animals, and even some springbok to be trampled to death in the passing of "animal waves."

Occasionally, human observers find it difficult to ascribe functions to animal behavior. These observers then often theorize that the behaviors are forms of play. This applies well to kudu, which do appear to play by throwing branches into the air and then catching them as they fall. The **reedbuck's** (*Redunca arundinum*) unusual dancing rituals may be related more to territorial herding behavior. The old males lie around the edges of an "arena" and spectate. Inside the arena, the females wander about while the yearling males do the actual dancing, consisting of various types of leaps and jumps.

Cattle

Cattle range in size from the 300-pound **anoa** (*Bubalus depressicornis*) to the greater than one ton **gaur** (*Bos gaurus*). The largest specimen of **domestic cattle** (*Bos taurus*) was a large **Hereford-shorthorn** bull. It was over 16 feet long, almost 14 feet around, and stood 6.33 feet at the forequarters. It weighed 4,750 pounds.

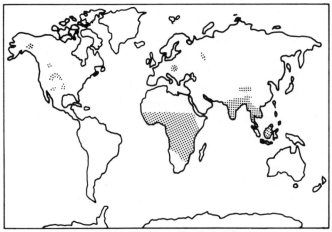

Family Bovidae—Cattle (wild)

Humans have kept domestic cattle for centuries, exploiting them for milk, meat, and hides primarily, but also as a source of power, fat, fertilizer, glue, soap, and utensils. The record milk producer is the **Friesian** breed, exceptional cows of which can give up to 4,300 gallons of milk per year. Mozzarella cheese is made from the milk of the **water buffalo** (*Bubalus bubalis*).

Some of the terms used for cattle include bull for a male; ox and steer for a castrated bull; cow for a female; heifer for a virgin female; calf for the young; and polled for hornless breeds.

The **yak** (*Bos grunniens*) lives higher than any other living mammal, inhabiting altitudes of 20,000 feet where it withstands temperatures of -45°F. It gets its water by eating snow. Most cattle are heavily dependent on a readily available supply of water. One reason is that most tend to drool heavily from their mouths. The **American bison** (*Bison bison*), more commonly but incorrectly referred to as the **buffalo**, is said to be able to smell water up to three miles away.

Cattle have long been used in entertainment. They play important roles in rodeos and in the despicable sport of bullfighting. **Iberian**

Gaur (*Bos gaurus*)

bulls are frequently used in the latter because of their aggressive nature. A common myth is that bulls are aggravated by the color red. In fact, domestic cattle are color blind and see only in shades of gray.

(Some wild bovids are reported to have color vision.) What actually causes the bull to charge is the manner in which the matador shakes the cape.

The **zebu** (*Bos taurus*) is a hardy breed valued for its many fine characteristics. Like the **banteng** (*Bos javanicus*), gaur, and **kouprey** (*Bos sauveli*), it has a large hump on its back in the shoulder area. This is composed of muscle rather than fat. The zebu's good qualities include heat tolerance and disease resistance. They can withstand high temperatures because of their many sweat glands, although this holds down milk production. They avoid disease because their sweat repels insects and their dense, short hair resists ticks. Most other cattle species will wallow to some degree in mud to rid themselves of heat and to protect their hides from biting insects and possibly solar radiation. Bison wallows at one time measured up to two feet deep and 50 feet in diameter. In addition, the skin of some cattle is sometimes cleared of parasites by oxpeckers and other birds, just as with antelopes.

The hooves of some, such as the water buffalo, easily splay out to increase traction in wet, marshy areas. Smell is the best developed sense in cattle, with hearing a close second. All cattle use long tails, tufted at the end, to keep away pesky insects. Cattle experts can differentiate more than ten separate meaningful grunts, moos, and roars. Bison can produce a rumbling bellow audible for distances of three miles. Most sleep briefly for 2–10 minutes at a time totalling about an hour per day.

All cattle are vegetarians. They and other ruminants often feed by grasping grass between their lower incisors and tongue since they lack upper incisors. Large domestic cattle eat about 150 pounds of grass each day. Bison will plow through as much as four feet of snow to get at the grass below. Recently it has been found that substances in the saliva of grazing animals such as cattle actually stimulate the growth of grass.

Cattle are the best known of the ruminants or cud-chewers. Cud chewing allows the animals to spend less time grazing in the open, where they are more vulnerable to attack from predators. It also helps special bacteria and protozoa in the stomach break down and digest cellulose. Hundreds of species of bacteria and protozoa have been found in the many ungulates. The microbes multiply quickly, and many are themselves digested by the host animal, thus increasing its intake of protein, fatty acids, vitamins and minerals. In addition, the exothermic (heat-producing) reactions of the microbes as they break down the cellulose give these animals built-in heaters. Rumination destroys many plant toxins, permitting the ingestion of many poisonous species.

When cattle eat, they first quickly crop and swallow the vegetation to the first stomach chamber (the rumen or paunch), where it is worked on by the symbiotic microbes. It then goes to the second chamber (the reticulum or honeycomb bag), which sends it back to the mouth to be rechewed as the cud. After being mixed with saliva, it is again swallowed to the third chamber (the psalterium, omasum, or manyplies) without passing through either of the first two chambers. The food then makes its way through the fourth stomach chamber (the reed or abomasum) for primary digestion before being passed into the intestine. There are other variations depending on the species of ruminant. Also, when drinking, the esophagus channels fluid in a groove past the first two "stomachs." In calves, the first three chambers are not well developed, so that milk travels straight through them.

Most cattle group together to form herds. They are often segregated by sex except during the rut. A few, such as the anoa, are predominantly solitary. Old bulls are usually loners. Males often spar to determine dominance. Sparring **Cape buffalo** (*Syncerus caffer*) occasionally knock off the tips of their horns and on rare occasions whole horns. Fighting bison and others have been known to be killed.

The large herds of bison at one time were composed of up to 4,000,000 individual animals covering 1,000 square miles of prairie. There were an estimated 60,000,000 in total population at their peak, making them the most numerous wild cattle ever present at one time in the world. Their systematic slaughter by invading Europeans is well known and almost led to their extinction.

During the rut, many cattle, such as the anoa and banteng, become very aggressive and will even attack trees and bushes. At this time bison take part in what is referred to as a "fighting storm," as the bulls usually refuse to yield to each other. Each challenge is generally countered with a charge as the dominance hierarchy is rearranged. It is at these times that well-matched opponents may fight to the death. The fighting takes a lot out of the bulls, which may lose up to 300 pounds in weight. Indecision during conflicts may lead to displacement behavior consisting of pawing the ground with hoof or horn or even wallowing, sometimes in their own urine. The thick fur on top of their heads and their "bells" (goatees) are thought to serve as badges of rank, like antlers in deer. The head fur also protects them in fighting and jousting, and in one bison it was so thick and strong that it stopped a .30/06 slug short of the bone. The animals show submission by turning their heads to the side or by running away.

Some cattle, such as the **wisent** or **European bison** (*Bison bonasus*), hold territory. The wisent marks its territory by urinating on the ground to form a small puddle in which it wallows with its shoulders. The animal then rubs the mud and urine mixture onto trees which it has previously stripped of bark with its horns.

Breeding seasons vary with the species and can depend on many factors. In domestic cattle, it has been found that the mere presence of a bull is enough to accelerate a cow's ovarian cycle. Most females give birth to a single young. Female bison usually leave the herd to calve. After giving birth, she sniffs, nuzzles, and licks the calf. This imprints them to each other to prevent loss or mix-up, and hence death from neglect. In gaur herds (as in some antelopes, such as the topi), the young are kept in nursing groups and carefully watched over by the females. Many young herbivores begin to chew vegetation prior to weaning. It is thought that this is done to learn the art of chewing, since the substances are not at first swallowed. In domestic cattle, the cows can be stimulated to give milk by simulating birth. This is done by blowing air into the vagina. Few male mammals take part in rearing their offspring. A few ungulates are among the exceptions; they include banteng, bison, and yak, whose males are known to defend young from predators.

Wild animals, including cattle, are careful to hide sickness and injury, since their survival often depends on their ability to fool any observing predators. It is said that gaur herds always post a sentry that whistles out an alarm at the approach of danger. They are also known to take advantage of the better hearing and sight of jungle fowl, with which they frequently associate. When alarmed, many wild cattle, including the water buffalo, stamp their feet. Water buffalo have been known to kill tigers single-handedly and with the help of conspecifics. Gaurs have been known to backtrack and attack hunters and have also been known to keep tigers at bay. It is interesting to note that when it comes to defense, many big game hunters consider the most formidable adversary to be the Cape buffalo. This is substantiated by the fact that they have killed more big game hunters than any other animal, sometimes mangling their victims gruesomely. Buffalo have used their victims as "balls," throwing them back and forth with their horns, even while the victims were still struggling for life. They are known to trample their enemies, have killed lions, and their tongues are very rasp-like, reportedly being able to lick the skin right off a person's body.

292

Goats and Sheep

There are a number of general differences between sheep and goats. Sheep have narrow noses with concave foreheads; goats have convex foreheads. Goats usually have beards and scent glands at the base of the tail. Sheep usually have face and foot scent glands. The horns of goats usually curve up and back. Sheep horns usually spiral around at the sides of their heads. Goats are also said to emit a much stronger stench than sheep. This is partially due to the fact that male goats, particularly during the breeding season, frequently urinate on themselves.

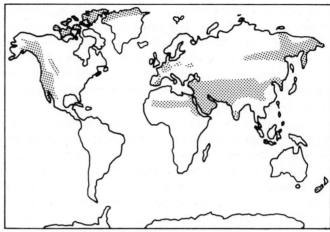

Family Bovidae—Goats and Sheep

In most species, both sexes bear horns, although a few, such as the **mouflon sheep** (*Ovis musimon*), the females may be hornless. In the largest sheep, the **argalis** (*Ovis ammon*), of which the **Marco Polo sheep** is a subspecies, the horns measure to six feet on the curl and may weigh up to 60 pounds. This is about 13 percent of their live weight and equal to almost twice their remaining skeletal weight. The age of **bighorn sheep** (*Ovis canadensis*), **chamois** (*Rupicapra rupicapra*), **Dall sheep** (*Ovis dalli*), **ibex** (*Capra ibex*), **mountain goats** (*Oreamnos americanus*), and others can be determined by counting the number of growth rings present on the horns. The bighorn's horns can grow to 30 pounds, about the same weight as the rest of its skeleton, and circle 1.5 times to occlude some of the animal's visual field. To help keep this to a minimum, the sheep will often rub their horns on rocks to keep them worn down. The horns of some Dall sheep occasionally become deformed and have been known to grow in through the skull causing death. Like antlers, sheep and goat horns serve primarily as badges of rank rather than tools or weapons.

Sheep and goats have been domesticated for centuries and have been transplanted all over the world. They are kept mainly for their milk, wool, and meat. **Domestic sheep** (*Ovis aries*), with over 800 breeds, have been bred to produce unusual characteristics. Many of them are hornless. The **Ausimi** and **Dumba** breeds store fat in their tails. At times the tails become so large, up to 40 pounds, that the sheep require the aid of a tail-carrying trolley in order to walk. In **Karakul sheep**, the female's rump can become so fat and hang so low that it prevents copulation. Roquefort cheese is made from the milk of sheep, and it has been shown that the best tasting mutton in some areas comes from sheep that graze in pastures containing many snails; the snails are inadvertently eaten along with the vegetation.

The finest sheep wool is derived from **Merino sheep**, and the **Corriedale** can grow 20 pounds of wool each year. **Domestic goats** (*Capra hircus*) are also important for their coats. The **takin** (*Budorcas taxicolor*) was supposedly the original owner of Jason's "golden fleece" and has an oily secretion on its fur thought to protect it from moisture in the foggy areas it inhabits. Nonmoulting **Angora goats** have been developed that grow hair up to two feet long. It is used to produce "mohair" for clothing. "Cashmere" is also derived from the hair of goats. The longest hairs of any mammal, however, are the 36 inch long coat hairs on the **musk ox** (*Ovibos moschatus*). These long guard hairs overlie the soft, valuable underfur or "qiviut" below. Qiviut is on a par with cashmere and vicuna wool. It is 8 times warmer than sheep wool. One pound of it can produce ten miles of yarn, and only about four ounces is needed to make a dress. Mature musk oxen can produce about six pounds of qiviut per year and can also get along with a sixth the feed needed by domestic cattle. Their thick hair protects them from cold and insects.

Special terms used for goats include the following: the young are kids; a female up to two years of age is a goatling and after that a nanny goat; a male up to two years of age is a buckling and after that a billy goat. In sheep, the male is a ram, the female an ewe, and the young are lambs.

Goats and sheep are well adapted for living in high, rocky areas. The chamois' lower legs have a mechanical spring-like action to serve as "shock-absorbers" in its spectacular leaps. They have been observed making vertical jumps of 13 feet and long jumps of 23 feet. Bighorns have dived down heights of 50 feet without problem and, like cats, can turn in mid-air to land on their feet. The hooves of bighorns, mountain goats, and others are double-shelled and are able to spread to a width that exceeds their length. The hoof can then be pulled together to grip onto ridges. There is a soft cushioning pad in the center to increase traction, and mountain goats can use their dew claws for added purchase. All these features enable these animals to make 20-foot jumps down onto ridges only inches wide. When heading along a narrow ledge that suddenly ends, mountain goats have been seen to stand on their hind legs to hug the wall as they about-face. At other times, they have been known to leap down narrow cliffs by alternately bouncing from side to side. Ibex have descended rock faces at rates of 500 feet per minute. Chamois have made quick descents on steep slopes by sliding down on their bellies.

Barbary Sheep or Aoudad (*Ammotragus lervia*)

As a rule, sheep and goats have excellent senses of sight and smell. They can smell humans a quarter mile upwind. Bighorn sheep and Dall sheep have sight equivalent to humans with 8X binoculars. They can see humans approaching up to five miles away.

Domestic goats have the highest average normal body temperature of all mammals, 103.8°F, with a normal range of 101.7–105.3°F. Desert bighorns survive the heat in much the same way as camels. During the day, their body temperature can rise from 99 to 105°F without problem. When the cooler night comes, the excess heat is radiated away. Mountain goats decrease their temperature by throwing cool soil over their bodies. This also provides some protection from biting insects. Dall sheep and bighorn sheep inhabiting colder areas are protected by hollow, air-filled guard hairs overlying a three-inch-thick under-wool that gives excellent insulation even at -50°F. The bighorn is also one of the few hooved mammals that will reuse the same bedding areas over and over. They form sleeping depressions over a foot deep. These often reek terribly since the animals often urinate there before leaving each morning. It is said that **true goats** (*Capra* spp.) require no deep sleep, but merely go through periods of drowsiness.

Sheep are known as the proverbial followers for a logical reason. When sheep travel in the wild, they often move in single file, and each sheep, since its field of vision is therefore limited, must do exactly what the one in front of it does or it may not survive. When the group encounters a gap, the lead sheep jumps and so do all the rest. If the sheep do not follow instinctively, they will not realize the need to jump until too late. This behavior has carried over into some domestic sheep. Whole flocks of sheep can be stimulated to jump by the action of one. In some, it has been found, a jump is preceded or signaled by the forward extension of its ears; if the ears are tied back, the sheep will not jump. For this reason, some shepherds have been known to tie back the leading ram's ears to keep the flock from jumping over fences.

Both goats and sheep are well known for their habit of cropping vegetation too close to the ground, leading to the formation of desert areas when overgrazing is allowed, despite the terrific refertilization power of sheep manure. Like all ruminants, goats and sheep have instead of upper incisors a firm pad. It is their cleft upper lip that allows them to crop vegetation so close to the ground. **Markhors** (*Capra falconeri*) and other goats will climb into trees to get at the edible leaves. Some will even climb onto the backs of other animals such as cattle and donkeys to reach higher tree leaves. One goat actually butted a donkey into position so it could climb up to eat. Domestic goats can eat the foliage of yews, known to be fatal to cattle and horses, and sheep can eat arsenic without harm. Mouflon can feed on spurge and deadly nightshade, both known to be poisonous to other animals. Both sheep and goats are quite capable of digesting paper and cloth with the help of cellulose-digesting protistans in their intestinal tracts. Goats cannot digest tin cans but are fond of the paper labels and adherent glue. They also eat straw and occasionally will use long stems of it to scratch their backs. Mountain goats are known to trample, dig, and lick areas where humans have urinated to obtain sodium and other minerals.

When undigestible substances such as hair and rocks are swallowed by the **bezoar goat** (*Capra hircus*), they become coated with lime and magnesium phosphate to form hard, smooth "bezoar stones." In some cultures, the stones are considered antidotes to poison and cure-alls for numerous diseases. Bezoar stones can also be found in the stomachs of cattle and gazelle and in the gallbladders and intestines of the **capped langur monkey** (*Presbytis* spp.), vicuna, and others.

In many species, it is dominance that determines which rams obtain breeding rights with the females. In some, such as the bighorn, there are no dominance hierarchies and the ewes are promiscuous breeders. The males of most goats and sheep joust. It is thought that jousting is related to dominance and breeding, but that it also serves to reduce tension and stress that builds during the rut.

When jousting, rams butt foreheads or the base of the horns. At times damage, sometimes severe, results to the horns as a result of poor aim or glancing blows. Serious injury is usually prevented by the thick hide and the double-layered skull (there is one inch or more of air space between the layers). Despite such protection, rams have been known to break horns, skulls, jaws, and noses. In addition, they may lose eyes or even die from brain damage. After the thunderous head-on blows, the two contestants will often appear dazed for a short time. This is quite understandable since the participants may weigh up to 350 pounds and may each be running at 25–30 miles per hour prior to the collision. A force in excess of a ton can result in a crash audible as much as two miles away. After a blow, the two rams often display their horns to each other. With experience, a ram can learn to judge the strength of another by its horn size. In this way, some confrontations are avoided, since the smaller one will generally back down. One evenly matched pair of bighorn rams jousted 25.5 hours straight with an average of five collisions per hour before one finally gave in. Defeated males behave like females and are even mounted at times by the winner. Dominant rams are even known to ejaculate during such encounters.

In bighorn sheep, it has been found that both rams and ewes joust. At times, a dozen or so rams will engage in a "battle royal." Sometimes young rams become overanxious to test their strength and enter the battles of older rams via hit-and-run techniques. The youngster sneaks up on a battling duo and quickly delivers a blow to one before it takes off. This has led to fractured ribs and can bring a battle quickly to an end. Older rivals are also known to sneak attack each other at times.

Jousting **tahrs** (*Hemitragus jemlahicus*) seem to lose some of their caution and will occasionally slip and fall to their death. When large musk oxen butt their heads, the resulting impact can be heard over a mile away. It is fortunate for them that they are protected by a nine-inch bony shield that is capable of stopping a large-caliber bullet. Even so, death has resulted from their jousting. The males are extremely irritable during the rut and charge anything in sight, including small birds that alight too close to a female musk ox. Musk oxen emit a powerful musky odor at this time; they can be smelled 300 feet away even though they have no musk glands. The odor emanates from their urine and facial glands. Most sheep and goats drench their bellies, rumps, and tails with urine. During the rut, male mouflon sheep bang their horns against rocks and trees, producing a racket detectable a mile away. This is thought possibly to represent an accoustical marking of their presence. **Barbary sheep** or **aoudads** (*Ammotragus lervia*), in addition to forehead butting, sometimes spar by locking horns while standing side by side and then attempting to throw each other. Mouflon sheep have been known accidentally to lock horns while jousting and have died from starvation when they were unable to get them apart. Mountain goats joust by standing side by side to attack each other's hind area. The skin on their rump, called dermal shields, is an inch thick to protect them in such battles, as well as from predators and falls. Mountain goats and Barbary sheep have also been known to die as a result of rival fights. Chamois may resort to goring each other if one is not intimidated by display.

Ibex males court the females by sticking out their tongues and waving their tails. When real excited they flap their tongues. Male mountain goats approach the females by crawling on their bellies and bleating in the manner of baby goats. The females often leave the group singly or in small groups to give birth. Dall sheep and others often give birth on small, inaccessible cliff ledges. Most species bear one or two young, and the females usually eat the placenta. Musk oxen are born during the winter which, if harsh, may cause the newborn to freeze to death before it is even dry.

Most young goats and sheep can leap and climb within one to two hours of birth. Baby Dall sheep obtain the bacteria they need for proper digestion by licking their mother's mouth. Young chamois and others develop coordination and practice by leaping over and onto the backs of their mothers. If a baby chamois loses its mother, another female may adopt it as her own. Young sheep and goats are known to frolic and play tag, follow-the-leader, and leapfrog by the hour. Mountain goats frequently turn somersaults and may even make games of this behavior. Bighorns jump and twist in mid-air in a behavior called "gamboling." They seem to do it for the fun or exercise. They are also known to climb hills repeatedly and sled down on their rumps.

Most goats and sheep are protected from enemies by inhabiting rocky areas where most predators do not venture. The number one killers of mountain goats are rock slides and avalanches. An avalanche can produce forces as high as 11 tons per square foot and reach speeds of 300 miles per hour on 40 degree slopes. Mountain goats have been lost to wolves, bears, and big cats. Sometimes large eagles will succeed in snatching a young goat or even knock an adult off a cliff. Some goats and sheep die yearly from accidental falls and from harsh winters. Male mountain goats have on rare occasion killed wolves and grizzly bears.

Goats and sheep respond to danger in a number of ways. Adult chamois and mouflons stamp their feet and give out a warning whistle. Gorals produce a loud hiss, and takins give out a warning cough, after which they sometimes charge en masse. Dall sheep have butted attacking wolves from cliffs, but tahrs have been known to faint when startled. Mountain goats appear to post sentinels to keep watch for danger. Female bighorns will straddle their young should they spot a golden eagle flying about. When a herd of musk oxen is threatened by an enemy such as wolves, they form what is known as a "phalanx." The large rams form a circle and face the outside while the females and calves remain protected in the middle. This defense is worse than useless against human hunters who merely sit and pick them off one by one. Baby musk oxen often take shelter under their mother's long hair, where they are barely visible. In bad weather such as blizzards, musk oxen huddle together to keep warm. They sometimes do this by forming a wedge or diamond formation facing into the storm. This particular grouping is sometimes referred to as a "yard." Again the young are kept in the more protected central area. In severe winters, their long hairs, which are even present on the females' udders, sometimes become so encrusted with ice that the animals become immobilized and may even starve to death as a result.

Family Antilocapridae (pronghorn antelope— 1 species)

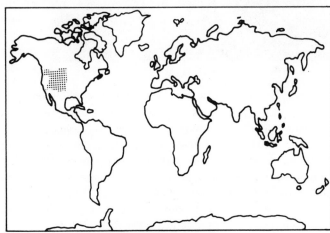

Family Antilocapridae

The **pronghorn antelope** (*Antilocapra americana*) is not a true antelope. It is the only mammal whose horns are naturally branched and are shed annually like antlers. Actually, the bony core remains permanent; only the sheath of fused hairs or keratin is dropped. This sheath is then usually eaten by gophers, mice, rabbits, or squirrels. The horns are most prominent in the males. Females' horns are usually smaller and unbranched, and some are hornless altogether.

Pronghorns are among the fastest of all mammals and deserve the title of fastest running animal

Pronghorn (*Antilocapra americana*)

over sustained distances. They can run up to 60 miles per hour, and some reports say they can do 70. They have been accurately clocked at 50 for distances nearing a mile, and they have run nonstop at 30 miles per hour or more for distances of 15 miles. While running, they have made leaps of 25 feet. They even seem to enjoy running games, playing follow-the-leader and often racing each other, horses, and cars for the fun of it. Such races frequently end when the antelope passes the competition and cuts across in front of its opponent.

Pronghorns are physiologically adapted for their running feats. They have relatively large hearts and lungs to keep their muscles well oxygenated. The trachea or windpipe is five inches in diameter to move large amounts of air in and out of the lungs. The feet are padded with cartilage, which lets them move more quickly and quietly.

Pronghorns have very keen eyesight, said by some to be the best of all mammals. The eyes are certainly large, being about the same size as an elephant's eyes. They are located far out on the head enabling them to see forwards and backwards simultaneously. They reportedly see as well as or better than a human using 8X binoculars and can detect small moving objects as far away as four miles.

Pronghorns are browsers and feed preferentially on cactus, sagebrush, thistles, weeds, and even plants that are poisonous to cattle. Their teeth grow continuously to counteract wear. They often scrape a hole in which they deposit their droppings.

At mating time, the males collect harems of up to fifteen females. In courtship, a pair will run in wide circles. Males occasionally fight for harem possession, with death resulting in rare instances. The females use delayed implantation. As in many deer, the females give birth to a single young the first time around and thereafter twins or, rarely, triplets.

The young are at first "hiders" and will flatten out against the ground at the approach of danger. The twins even hide in separate areas 250–300 feet apart so that predators do not usually take both together. The young, like other hiders, possess little scent, and females consume their young's wastes. At three days of age, they can outrun a human and at three months a horse. Mothers have been known to attack coyotes, eagles, and foxes in defense of their young.

Both males and females snort when angry or surprised. Alarm is signaled by erecting long, white rump hairs in a flash visible up to 2.5 miles away. In addition, scent is released from glands near the base of the hairs. Its odor can be detected by humans up to 100 yards away.

Pronghorn antelope often look rather stupid because of their habit of allowing a portion of their lower lip to droop. They are known to be extremely curious animals. This was exploited by early settlers, who would attract them easily to gun range by putting up a stick with a rag on it or even by lying on their backs and waving their up-stretched legs. Such tactics almost doomed the animals to extinction.

Family Giraffidae (giraffe and okapi— 2 species)

The **giraffe** (*Giraffa camelopardalis*) is the tallest animal in the world, reaching a height of 19 feet with weights of up to 3,600 pounds. They have one of the largest surface areas proportional to their weight of all large mammals. This is the reason they, unlike elephants, hippos, and rhinos, do not require mud baths to keep cool. Giraffes have on their heads two to five (depending on the

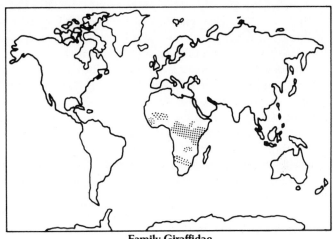

Family Giraffidae

subspecies) small horns or ossicones. Along with the male **okapi's** (*Okapia johnstoni*) horns, these are the only animal horns covered with permanent skin and hair. The horns are still strong and can easily put dents in timber. The giraffe's neck, despite being up to eight feet long, has only seven cervical vertebrae, but each one may be almost a foot long. Neck mobility is enhanced by the vertebrae being connected by ball-and-socket joints. (Most other mammals also have seven cervical vertebrae; the two-toed sloth and African manatee have six, the lesser anteater and occasional pangolins eight, and the three-toed sloth nine; some cetaceans, the marsupial mole, and various rodents have cervical vertebrae fused to variable degrees.) Giraffe coloration is variable with some individuals being nearly white, others almost black, and a few are occasionally unspotted.

Giraffes are able to change their head altitude quickly without developing ill effects due to some unique physiological adaptations. These include a large, 25-pound heart, two feet long with three-inch walls; valves in both arteries and veins; alternate routes for blood flow; and special blood-containing reservoirs in the head. These adaptations work so efficiently that cerebral blood pressure actually drops from 200 millimeters of mercury systolic to 175 instead of bursting the vessels and giving the animal a stroke when it bends down for a drink. They also keep the animal from passing out when it raises its head again.

The combined weight of the head and neck can exceed 500 pounds. The giraffe has the highest normal blood pressure of any animal, 260 over 160. The heart pumps 15 gallons each minute at rest,

299

the walls of the carotid artery are half an inch thick, and the five-foot trachea is the longest of any terrestial animal.

Giraffes are one of the few species of mammal to perceive color. Their visual field approaches 360 degrees from their high vantage point. They are not mute, as was once popularly believed. In fact, they have quite large voice boxes, even though they rarely vocalize. The young bleat while adults grunt or cough. Adults occasionally snore when asleep. Some experts feel that they may even communicate via ultrasonic utterances.

Giraffes are actually more agile than one would think to look at them. They can run up to 35 miles per hour and have been known to jump five-foot barbed wire fences. It is often stated that they are unable to swim, but there are reports from natives of their swimming across deep rivers with only their head and about a third of their neck above water.

Another debated subject is their sleep. They are known to exhibit periods of deep sleep lasting from one to five minutes, occasionally when lying down with the neck stretched out over the back. Deep sleep is said to take place for 20–60 minutes each 24 hour period. Some have claimed that giraffes are capable of sleeping with their eyes open. Most sleep takes place while standing.

Giraffes can go weeks without drinking. When they do drink, they leave themselves wide open to attack by predators. This is because they must spread their forelegs wide apart in order to get their heads down to ground level. They can drink up to ten gallons at one time.

Giraffes spend up to 16 hours consuming up to 100 pounds of food daily. They are browsers and feed primarily on trees, including the heavily thorned acacia tree. Although they are fairly well protected by their thick, 18–22 inch long tongue, leathery lips, and stiff mouth hairs, many giraffes wear thorns visible in the skin of their neck and legs. Sometimes an individual will have thousands in its skin. The tongue is black to protect it from the sun's ultraviolet light, and the nostrils can be closed off.

The trees giraffes feed on can be easily recognized. The very tall ones become hourglass shaped, being thinnest at the upper limits of the giraffes' reach; smaller trees often become flat-topped. It is interesting to note that the seeds of some acacia trees will not germinate unless they pass through the giraffe's digestive system. Giraffe saliva is very thick. Carrion is occasionally eaten for its salt and moisture.

A giraffe shows dominance by keeping its neck straight and tilting its chin up. To show submission, it lowers its chin. In dominance fighting, giraffes duel with their long necks. Each flings his neck across and down towards the other. The resulting blows can be heard almost a mile away. Usually no serious harm is done, but occasionally one comes across a giraffe with a dislocated neck. Rivals have been known to be knocked unconscious for as long as 20 minutes. The necks are also used for mutual caressing in premating rituals. Giraffes mate year round. The males gather a harem of females and their young. Males sometimes collect some of the female's urine in their mouths and then spit it out in a thin stream. They can thus detect hormone levels and determine her receptiveness to mating.

The females give birth to a single young from a standing position. The newborn falls 5–6 feet to the ground but often twists to land belly or feet first. At times, a female giving birth is surrounded by other adult giraffes from the herd. The newborn is usually about six feet tall and weighs about 150 pounds. They are the only mammals born with horns. These measure about one inch at birth. During their early growth spurt, giraffes average about 1.25 inches gain in height daily. Like dolphins, elephants,

Giraffe (*Giraffa camelopardalis*)

elk, lions, some monkeys, whales, and a few other species, giraffes engage conspecific baby-sitters for their young. Baby giraffes are known to take cover beneath their mother to allow her to kick out at attackers. The young suffer a 50 percent mortality during their first year of life.

Giraffes' hooves can be formidable weapons. On several occasions, they have literally beheaded attacking lions. They are also known to attack with their heads. The head blow from a giraffe has been known to break the shoulder bone of an eland. However, lions are capable of downing full grown giraffes. In one instance, an attacking lion successfully downed one but was killed when its prey landed on top of it.

The giraffe's closest relative is the okapi. Unlike its relative, only male okapis bear horns, though females often do have small knobs. The females are slightly larger than the males. Okapis are also browsers, and they are known to feed on charcoal from lightning-burnt trees. They are thought to be solitary. Okapis bathe themselves with their long tongue, which can even reach up to wash off the eyes. The tongue and neck in combination are long and mobile enough to allow the animals to reach almost their entire body surface. It is reported that the females still come into heat even though they are pregnant. Unlike the vast majority of other ungulates (as well as many rodents and primates), female okapis do not lick their offspring's anal area to stimulate defecation. The females are known to adopt the young of other okapis, and only the young possess conspicuous manes.

APPENDIX I

ANIMAL YOUNG

Bedlamer—Harp or Hooded Seal
Blinker—Mackerel
Bullock—Bull
Bunny—Rabbit
Calf—Antelope, Cattle, Elephant, Hippo, Rhino, Whale, Walrus
Catling—Cat
Cheeper—Grouse, Partridge, Quail, Goose
Chick—Chicken
Cockerel—Rooster
Codling—Codfish
Colt—Male Horse
Cosset—Sheep
Coy—Polar Bear
Cria—Vicuna
Cub—Bear, Shark, Lion, Tiger
Cygnet—Swan
Duckling—Duck
Eaglet—Eagle
Elver—Eel
Eyas—Hawk
Farrow—Pig
Fawn—Deer
Filly—Female Horse
Fingerling—Fish
Flapper—Wild Fowl
Fledgeling—Birds
Foal—Horse, Zebra
Fry—Fish
Gosling—Goose
Grilse—Salmon

Heifer—Cow
Hog—Unshorn Sheep
Joey—Kangaroo
Kid—Goat
Kit—Beaver, Rabbit, Cat
Kitten—Cat
Lamb—Sheep
Lambkin—Sheep
Leveret—Hare
Nestling—Birds
Owlet—Owl
Parr—Salmon
Piglet—Pig
Polliwog—Frog
Porcupette—Porcupine
Poult—Turkey
Pullet—Hen
Pup or Puppy—Dog, Seal, Rat
Puss or Pussy—Cat
Shoat—Pig
Smolt—Salmon
Sprag or Sprat—Codfish
Spike—Mackerel
Squab—Pigeon
Squeaker—Pigeon
Suckling—Pig
Tadpole—Frog
Tinker—Mackerel
Whelp—Dog, Tiger
Woolly—Sea Otter

302

APPENDIX II
ANIMAL GROUPS

Arribada—Sea Turtles nesting on shore
Army—Ants, Caterpillars
Ascension—Larks
Bale—Turtles
Band—Gorillas
Barren—Mules
Bed—Clams, Oysters
Bevy—Quail, Swans, Roebucks
Bloom—Jellyfish
Boil—Hawks riding a thermal
Bouquet—Pheasants
Brace—Ducks
Brood—Chicks, Hens
Building—Rooks
Bundling — Walruses on shore
Business — Ferrets
Camp — Flying Foxes
Cast — Hawks
Catch — Fish
Cete — Badgers
Charm — Goldfinches, Finches
Chattering — Choughs
Clamor — Rooks
Clapmatch — Breeding Female Seals
Clash — Bucks
Cloud — Gnats
Clowder — Domestic Cats
Clutch — Chicks, Eggs
Clutter — Cats
Colony — Rabbits, Ants, Bats, Bees

Company — Widgeon
Congregation — Plovers
Convert — Coots
Cover — Coots
Covey — Quail, Partridge
Cowardice — Curs (Mongrel Dogs)
Crash — Rhinos
Cry — Hounds
Deceit — Lapwings
Descent — Woodpeckers
Dissimulation — Birds
Dopping or Dropping — Diving Ducks
Down — Hares
Draft — Fish
Dray — Squirrels
Drift — Swine
Drought — Fish
Drove — Cattle, Sheep, Pig
Dule — Turkeys, Doves
Exaltation — Larks
Fall — Woodcocks
Farrow — Domestic Pigs
Fesynes — Ferrets
Flight — Birds, Swallows
Flock — Sheep, Birds, Swifts, Geese, Ducks
Gaggle — Geese on the ground
Gam — Whales
Gang — Elk
Gownder — Domestic Pigs
Grist — Bees

Harem—Female Seals, Female Deer
Harrass — Horses in a corral
Haul — Fish
Head — Curlews, Horses, Elephants
Hedge — Herons
Herd — Antelope, Asses, Cattle, Cranes, Curlews,
  Deer, Goats, Horses, Elephants, Seals, Swans,
  Whales, Wolves, etc.
Hive — Bees
Host — Sparrows
Hover — Trout
Horde — Gnats
Husk — Hares
Hutch — Rabbits
Kettle — Hawks, Swifts
Kindergarten—Young Elephants, Young Giraffes
Kindle or Kendle — Kittens
Kine — Cows
Knot — Toads
Labor — Moles
Leap or Leep — Leopards
Leash — Foxes, Greyhounds, Ducks on shore
Litter — Pigs, Dogs, Cats
Mob — Kangaroos, Camels
Murder — Crows
Murmuration — Starlings
Muster — Peacocks, Storks
Mute — Hounds
Nest — Vipers, Rabbits, Snakes
Nest, Nide, or Nye — Pheasants
Node—Mountain Goats
Ostentation — Peacocks
Pace — Asses
Pack — Hounds, Wolves, Mules, Rats, Grouse,
  Killer Whales
Paddling — Ducks in the water
Pair — Horses
Parcel or Passel — Hogs
Parliament — Rooks, Owls
Peep — Chickens
Pitying — Turtledoves
Plague — Locusts
Plump — Wildfowl
Pod — Whales, Seals, Pelicans
Pride — Lions

Quarrel — Sparrows
Rag — Colts
Raft — Sea Otters
Rafter — Turkeys
Richness — Martens
Rookery — Breeding Colony of Seals or Birds
Route — Wolves
Run — Fish
School — Fish, Hippos, Whales
Sedge or Siege — Cranes, Herons, Bitterns
Shoal — Fish
Shrewdness — Apes
Singular — Wild Boars
Skein — Geese on the wing
Skulk — Foxes
Sloth or Sleuth — Bears
Smack — Jellyfish
Sord — Mallards
Sounder — Pigs, Peccaries
Span — Mules
Spring — Teal
String — Horses, Camels
Stud — Mares
Swarm — Bees, Ants, Flies, Gnats, Rats
Team — Ducks on shore, Horses, Oxen
Tiding — Magpies
Town—Prairie Dogs
Tribe or Trip — Goats
Troop or Troup — Monkeys, Horses, Kangaroos
Unkindness — Ravens
Volary or Volery — Birds
Walk — Snipe
Watch — Nightingales
Wedge — Swans
Whiteness — Swans
Wig — Breeding Male Seals
Wing — Plover
Wisp — Snipe
Yoke — Oxen

SELECTED BIBLIOGRAPHY

Allaby, Michael, *Animal Artisans* (New York: Alfred A. Knopf, 1982).

Allen, Thomas B. (editor), *Wild Animals of North America* (Washington, DC: National Geographic Society, 1979).

Amos, William H., *Wildlife of the Islands* (New York: Harry N. Abrams, 1980).

Amos, William H., *Wildlife of the Rivers* (New York: Harry N. Abrams, 1981).

Anderson, Sydney (editor), *Simon & Schuster's Guide to Mammals* (New York: Simon & Schuster, 1983).

Attenborough, David, *Life on Earth* (London: Reader's Digest Association, Ltd., 1980).

Ayensu, Edward S., *Jungles* (New York: Crown, 1980).

Badino, Guido, *Big Cats of the World* (New York: Crown, 1975).

Baker, Robin (editor), *The Mystery of Migration* (New York: Viking Press, 1981).

Bertin, Leon, et al., *The New Larousse Encyclopedia of Animal Life* (New York: Larousse and Co., 1980).

Black, David, *Animal Wonders of the World* (London: Orbis Publishing, 1981).

Boorer, Michael, *Wild Cats* (New York: Grosset and Dunlap, 1970).

Breeden, Stanley and Kay, *The Life of the Kangaroo* (New York: Tapliner Publishing Co., 1967).

Breland, Osmond P., *Animal Life and Lore* (San Francisco: Harper and Row, 1972).

Bridges, William, *The Bronx Zoo Book of Wild Animals* (New York: Golden Press, 1968).

Buckles, Mary Parker, *Mammals of the World* (New York: Ridge Press, 1976).

Burn, David M. (editor), *The Complete Encyclopedia of the Animal World* (London: Octopus Books Ltd., 1980).

Burrud, Bill, *Bill Burrud's Animal Quiz* (New York: Grosset and Dunlap, 1979).

Burt, William H., and Richard P. Grossenheider, *A Field Guide to the Mammals* (Boston: Houghton Mifflin, 1976).

Burton, Maurice and Jane, *The Colorful World of Animals* (Norwalk, CT: Longmeadow Press, 1975).

Burton, Maurice, *Curiosities of Animal Life* (New York: Sterling Publishing Co., 1959).

Burton, Maurice, *The Family of Animals* (New York: Arco Publishing Co., 1978).

Burton, Maurice, *How Mammals Live* (London: Elsevier, 1975).

Burton, Maurice and Robert, *Inside the Animal World: An Encyclopedia of Animal Behavior* (New York: Quadrangle, 1977).

Burton, Maurice, *Just Like an Animal* (New York: Scribner's, 1978).

Burton, Robert, *The Life and Death of Whales* (New York: Universe Books, 1980).

Burton, Robert, *The Mating Game* (New York: Crown, 1976).

Caras, Roger, *Dangerous to Man* (Philadelphia: Chilton Books, 1975).

Caras, Roger, *North American Mammals* (New York: Meredith Press, 1967).

Caras, Roger, *The Private Lives of Animals* (New York: Grosset and Dunlap, 1974).

Caras, Roger, *Venemous Animals of the World* (Englewood Cliffs, NJ: Prentice-Hall, 1974).

Carrighar, Sally, *Wild Heritage* (New York: Ballantine, 1965).

Carrington, Richard, *The Mammals* (New York: Time-Life Books, 1963).

Chinery, Michael, *Killers of the Wild* (New York: Chartwell Books, 1979).

Clarke, James, *Man is the Prey* (New York: Stein and Day, 1969).

Coffey, D.J., *Dolphins, Whales, and Porpoises* (New York: Macmillan, 1977).

Costello, David F., *The World of the Porcupine* (Philadelphia: Lippincott, 1966).

Cousteau, Jacques-Yves, and Philippe Diole, *Dolphins* (Garden City, NY: Doubleday, 1975).

Cousteau, Jacques-Yves, and Philippe Diole, *The Whale: Mighty Monarch of the Sea* (Garden City, NY: Doubleday, 1972).

Craighead, Frank C., Jr., *Track of the Grizzly* (San Francisco: Sierra Club Books, 1979).

Crandall, Lee S., *A Zoo Man's Notebook* (Chicago: University of Chicago Press, 1966).

Crump, Donald J. (editor), *Book of Mammals* (2 vol.) (Washington, DC: National Geographic Society, 1981).

Curry-Lindahl, Kai, *Wildlife of the Prairies and Plains* (New York: Harry N. Abrams, 1981).

Davids, Richard C., *Lords of the Arctic* (New York: Macmillan, 1982).

Davis, Flora, *Eloquent Animals: A Study in Animal Communication* (New York: Coward, McCann, and Geoghegan, 1978).

Devoe, Alan, *This Fascinating Animal World* (New York: McGraw-Hill, 1951).

Dorst, Jean, and Pierre Dandelot, *The Larger Mammals of Africa* (London: Collins, 1972).

Dozier, Thomas A., *Whales and Other Sea Mammals* (Chicago: Time-Life Films, 1977).

Earnest, Don, *Life in Zoos and Preserves* (Chicago: Time-Life Films, 1978).

Earnest, Don, and Richard Oulahan, *Rabbits and Other Small Mammals* (Chicago: Time-Life Films, 1978).

East, Ben, *Bears* (New York: Crown, 1977).

Edey, Maitland, et al., *The Cats* (Chicago: Time-Life Films, 1976).

Ellis, Richard, *The Book of Whales* (New York: Alfred A. Knopf, 1980).

Ellis, Richard, *Dolphins and Porpoises* (New York: Alfred A. Knopf, 1982).

Elman, Robert, *The Living World of Audubon Mammals* (New York: Grosset and Dunlap, 1976).

Eltringham, S.K., *Elephants* (Dorset, England: Blandford Press, 1982).

Encyclopedia of Animal Life (6 vol.) (Hicksville, NY: Marshall Cavendish Ltd., 1973).

Encyclopedia of the Animal World (21 vol.) (London: Elsevier, 1972).

The Encyclopedia of Wildlife (Secaucus, NJ: Castle Books, 1974).

Fascinating World of Animals (Pleasantville, NY: Reader's Digest Association, 1971).

Fenton, M. Brock, *Just Bats* (Toronto: University of Toronto Press, 1983).

Fogden, Michael and Patricia, *Animals and their Colors* (New York: Crown, 1974).

Freedman, Hy, *Sex Link* (New York: M. Evans and Co., 1977).

Freeman, Dan, *Elephants: The Vanishing Giants* (New York: Putnam's, 1981).

Freeman, Dan, *The Love of Monkeys and Apes* (London: Octopus Books Ltd., 1977).

Frings, Hubert and Mable, *Animal Communication* (2nd edition) (Norman, OK: University of Oklahoma Press, 1977).

Gauthier-Pilters, Hilde, and Anne Innis Dagg, *The Camel* (Chicago: University of Chicago Press, 1981).

George, Jean Craighead, *Beastley Inventions* (New York: David McKay, 1970).

Grzimek, Bernhard, et al., *Grzimek's Animal Life Encyclopedia* (vol. 10–13) (New York: Van Nostrand Reinhold, 1972).

Grzimek, Bernhard, et al., *Grzimek's Encyclopedia of Ethology* (New York: Van Nostrand Reinhold, 1977).

Guinness, Alma E. (editor), *Joy of Nature* (Pleasantville, NY: Reader's Digest Association, 1977).

Haley, Delphine, *Sleek and Savage* (Seattle, WA: Pacific Search Books, 1975).

Halliday, Tim, *Sexual Strategy* (Chicago: University of Chicago Press, 1982).

Henning, Robert A., et al., *Alaska Whales and Whaling* (Anchorage, AK: Alaska Northwest Publishing Co., 1978).

Hirschland, Roger B., et al., *Secrets of Animal Survival* (Washington, DC: National Geographic Society, 1983).

Hopf, Alice L., *Wild Cousins of the Cat* (New York: Putnam's, 1975).

Hoyt, Erich, *Orca the Whale Called Killer* (Ontario: Camden House Publishing Ltd., 1984).

The Illustrated Encyclopedia of the Animal Kingdom (vol. 1–6) (Danbury, CT: Danbury Press, 1972).

The Illustrated Encyclopedia of Animal Life (vol. 1–7) (New York: Greystone Press, 1961).

Ingles, Lloyd G., *Mammals of the Pacific States* (Stanford, CA: Stanford University Press, 1965).

Jackson, Donald Dale, *Kangaroos and Other Creatures from Down Under* (Chicago: Time-Life Films, 1977).

Jensen, Albert C., *Wildlife of the Oceans* (New York: Harry N. Abrams, 1979).

Johnson, Cheryl (coordinator), *Maneaters and Marmosets: Strange and Fascinating Tales from the Animal Kingdom* (New York: Hearst, 1976).

Kavanagh, Michael, *A Complete Guide to Monkeys, Apes and Other Primates* (New York: Viking Press, 1983).

Keefe, James F., *The World of the Opossum* (Philadelphia: Lippincott, 1967).

King, Judith, *Seals of the World* (Ithaca, NY: Cornell University Press, 1983).

Lane, Frank W., *Animal Wonder World* (New York: Sheridan House, 1951).

Larousse Encyclopedia of the Animal World (New York: Larousse and Co., 1976).

Lawlor, Timothy E., *Handbook to the Orders and Families of Living Mammals* (Eureka, CA: Mad River Press, 1979).

Laycock, George, *North American Wildlife* (New York: Bison Books, 1983).

Leatherwood, Stephen, and Randall R. Reeves, *The Sierra Club Handbook of Whales and Dolphins* (San Francisco: Sierra Club Books, 1983).

Leen, Nina, *The Bat* (New York: Holt, Rinehart, and Winston, 1976).

Lockley, Ronald, *Whales, Dolphins, and Porpoises* (New York: W.W. Norton, 1979).

Lorenz, Konrad Z., *King Solomon's Ring* (New York: Thomas Y. Crowell, 1952).

Lorenz, Konrad Z., *On Aggression* (New York: Harcourt Brace Jovanovich, 1963).

MacClintock, Dorcas, *A Natural History of Zebras* (New York: Scribner's, 1976).

Macdonald, David (editor), *The Encyclopedia of Mammals* (New York: Facts On File Publications, 1984).

Marten, Michael, John May, and Rosemary Taylor, *Weird and Wonderful Wildlife* (San Francisco: Chronicle Books, 1982).

Marvels of Animal Behavior (Washington, DC: National Geographic Society, 1972).

Matthews, L. Harrison, *The Natural History of the Whale* (New York: Columbia University Press, 1978).

Matthews, L. Harrison, *The Whale* (New York: Simon and Schuster, 1968).

May, John, and Michael Marten, *The Book of Beasts* (New York: Viking Press, 1983).

McIntyre, Joan, et al., *Mind in the Waters* (New York: Scribner's, 1974).

McWhirter, Norris, *Guinness Book of World Records* (21st edition) (New York: Bantam Books, 1983).

Miller, Tom, *The World of the California Gray Whale* (Santa Ana, CA: Baja Trail Publications, 1973).

Mills, Enos A., *The Grizzly* (Sausalito, CA: Comstock Editions, 1981).

Milne, Lorus, and Margery and Franklin Russell, *The Secret Life of Animals* (New York: E.P. Dutton, 1975).

Minasian, Stanley, Kenneth Balcomb, III, and Larry Foster, *The World's Whales* (Washington DC: Smithsonian Books, 1984).

Mohr, Charles E., *The World of the Bat* (Philadelphia: Lippincott, 1976).

Morcombe, Michael, *Australian Marsupials and Other Native Mammals* (New York: Scribner's, 1972).

Morris, Ramona and Desmond, *The Giant Panda* (New York: Penguin, 1982).

Moscow, Henry, *Domestic Descendants* (Chicago: Time-Life Films, 1979).

Napier, John and Prue, et al., *Elephants and Other Land Giants* (Chicago: Time-Life Films, 1976).

Napier, Prue and John, *Monkeys and Apes* (Chicago: Time-Life Films, 1976).

Nayman, Jacqueline, *Whales, Dolphins, and Man* (New York: Hamlyn Publishing Group Ltd., 1973).

Neary, John, *Wild Herds* (Chicago: Time-Life Films, 1977).

Neff, Nancy A., *The Big Cats* (New York: Harry N. Abrams, 1982).

Nickerson, Roy, *Sea Otters* (San Francisco:

Chronicle Books, 1984).

Novick, Alvin, *The World of Bats* (New York: Holt, Rinehart, and Winston, 1969).

Nowak, Ronald M., and John L. Paradiso, *Walker's Mammals of the World* (4th edition) (Baltimore: Johns Hopkins University Press, 1983).

Our Amazing World of Nature Its Marvels and Mysteries (Pleasantville, NY: Reader's Digest Association, 1969).

Our Magnificent Wildlife: How to Enjoy and Preserve It (Pleasantville, NY: Reader's Digest Association, 1975).

Owen, Denis, *Camouflage and Mimicry* (Chicago: University of Chicago Press, 1982).

Owen, Jennifer, *Feeding Strategy* (Chicago: University of Chicago Press, 1982).

Park, Ed, *The World of the Otter* (Philadelphia: Lippincott, 1971).

Parker, Cecilia I. (editor), *Earth's Amazing Animals* (Washington, DC: National Wildlife Federation, 1983).

Perry, Richard, *The World of the Polar Bear* (Seattle, WA: University of Washington Press, 1966).

Peters, Barbara (editor), *America's Wildlife Sampler* (Washington, DC: National Wildlife Federation, 1982).

Ray, G. Carleton, and M.G. McCormick-Ray, *Wildlife of the Polar Regions* (New York: Harry N. Abrams, 1981).

Reardon, Jim (chief editor), *Alaska Mammals* (Anchorage, AK: Alaska Geographic Society, 1981).

Rensberger, Boyce, *The Cult of the Wild* (Garden City, NY: Anchor Press/Doubleday, 1977).

Ricciuti, Edward R., *Killer Animals* (New York: Walker, 1976).

Ricciuti, Edward R., *The Wild Cats:* (New York: Ridge Press, 1979).

Ricciuti, Edward R., *Wildlife of the Mountains* (New York: Harry N. Abrams, 1979).

Robinson, David, *Living Wild: The Secrets of Animal Survival* (Washington, DC: National Wildlife Federation, 1980).

Rue, Leonard Lee, III, *Complete Guide to Game Animals* (2nd edition) (New York: Van Nostrand Reinhold, 1981).

Rue, Leonard Lee, III, *Furbearing Animals of North America* (New York: Crown, 1981).

Rue, Leonard Lee, III, *The World of the Beaver* (Philadelphia: Lippincott, 1964).

Rue, Leonard Lee, III, *The World of the Raccoon* (Philadelphia: Lippincott, 1964).

Rue, Leonard Lee, III, *The World of the Red Fox* (Philadelphia: Lippincott, 1969).

Rutter, Russell J., and Douglas H. Pimlott, *The World of the Wolf* (Philadelphia: Lippincott, 1968).

Ryden, Hope, *Bobcat Year* (New York: Viking Press, 1981).

Sanderson, Ivan T., *Living Mammals of the World* (Garden City, NY: Doubleday, 1955).

Savage, Arthur and Candace, *Wild Mammals of Northwest America* (Baltimore: Johns Hopkins University Press, 1981).

Schafer, Jack, *An American Bestiary* (Boston: Houghton Mifflin, 1975).

Scheffer, Victor B., *A Natural History of Marine Mammals* (New York: Scribner's, 1976).

Scheffer, Victor B., *The Year of the Whale* (New York: Scribner's, 1969).

Secrets of the Seas (Pleasantville, NY: Reader's Digest Association, 1972).

Sielmann, Heinz, *Wilderness Expeditions* (New York: Franklin Watts, 1981).

Slijper, Everhard J., *Whales and Dolphins* (Ann Arbor, MI: University of Michigan Press, 1976).

Smith, Larry L., and Robin W. Doughty, *The Amazing Armadillo* (Austin, TX: University of Texas Press, 1984).

Smythe, R.H., *Animal Habits* (Springfield, IL: Charles C. Thomas, 1962).

Sparks, John, *The Discovery of Animal Behavior* (Boston: Little, Brown, 1982).

Spinage, C.A., *The Book of the Giraffe* (Boston: Houghton Mifflin, 1968).

Sutton, Ann and Myron, *Wildlife of the Forests* (New York: Harry N. Abrams, 1979).

Taglianti, Augusto Vigna, *The World of Mammals* (New York: Abbeville Press, 1979).

Tanner, Ogden, *Animal Defenses* (Chicago: Time-Life Films, 1978).

Tanner, Ogden, *Bears and Other Carnivores* (Chicago: Time-Life Films, 1976).

Tanner, Ogden, *Beavers and Other Pond Dwellers* (Chicago: Time-Life Films, 1977).

Van Gelder, Richard G., *Biology of Mammals* (New York: Scribner's, 1969).

Vanishing Species (New York: Time-Life Books, 1974).

Van Wormer, Joe, *The World of the American Elk* (Philadelphia: Lippincott, 1969).

Van Wormer, Joe, *The World of the Black Bear* (Philadelphia: Lippincott, 1966).

Van Wormer, Joe, *The World of the Coyote* (Philadelphia: Lippincott, 1964).

Van Wormer, Joe, *The World of the Moose* (Philadelphia: Lippincott, 1972).

Van Wormer, Joe, *The World of the Pronghorn* (Philadelphia: Lippincott, 1969).

von Frisch, Karl, *Animal Architecture* (New York: Harcourt Brace Jovanovich, 1974).

Wagner, Frederick H., *Wildlife of the Deserts* (New York: Harry N. Abrams, 1980).

Walker, Theodore, *Whale Primer* (Point Loma, CA: Cabrillo Historical Assoc., 1975).

Wallace, Robert A., *How They Do It* (New York: William Morrow, 1980).

Watson, Lyall, *Sea Guide to Whales of the World* (New York: E.P. Dutton, 1981).

Wendt, Herbert, *The Sex Life of the Animals* (New York: Simon and Schuster, 1965).

Whitaker, John O., Jr., *The Audubon Society Field Guide to North American Mammals* (New York: Alfred A. Knopf, 1980).

Whitfield, Philip (editor), *Macmillan Illustrated Animal Encyclopedia* (New York: Macmillan, 1984).

Whitfield, Philip, *The Hunters* (New York: Simon and Schuster, 1978).

Wood, Gerald L., *Animal Facts and Feats* (3rd edition) (Middlesex, England: Guinness Superlatives Ltd., 1982).

Wolfenden, John, *The California Sea Otter: Saved or Doomed?* (Pacific Grove, CA: Boxwood Press, 1979).

Yalden, D.W., and P.A. Morris, *The Lives of Bats* (New York: Quadrangle, 1975).

Zim, Herbert, and Donald Hoffmeister, *Mammals* (New York: Western Publishing Co., 1955).

Index of Common Names

313

DATE DUE

NOV 25 1993	
APR 1 7 1995	
OCT 2 0 1997	
OCT 2 3 1997	
OCT 2 5 1999	

DEMCO, INC. 38-2931